first 5.1

Library of Congress Control Number: 2015952299

Edition 5.1

Published by the
Flexographic Technical Association, Inc.
Printed in the United States of America

Inquires should be addressed to:
FTA
3920 Veterans Memorial Hwy Ste 9
Bohemia NY 11716-1074
www.flexography.org
International Standard Book Number ISBN-13: 978-0-9894374-6-2

Content Notes:

1. This reference guide is designed and formatted to facilitate ease of use. As such, pertinent information (including text, charts, and graphics) are repeated in the Communication and Implementation, Design, Prepress and Print sections.
2. Registered trademark products are identified for information purposes only. All products mentioned in this book are trademarks of their respective owner. The author and publisher assume no responsibility for the efficacy or performance. While every attempt has been made to ensure the details described in this book are accurate, the author and publisher assume no responsibility for any errors that may exist, or for any loss of data which may occur as a result of such errors.

The Mission of *FIRST*

FIRST seeks to understand customers' graphic requirements for reproduction and translate those aesthetic requirements into specifications for each phase of the flexographic printing process including: customers, designers, prepress providers, raw material & equipment suppliers, and printers.

The intention of *FIRST* is to provide all participants in the flexographic reproduction process with a common set of guidelines, tutorials, and data that can be used as communication and production tools.

FIRST Objectives

FIRST is a set of specifications, not standards. When followed, these specifications facilitate producing a predictable, consistent result. It is the responsibility of the customer to determine where, when, and how these specifications are implemented. This does not imply that a printer's capabilities cannot exceed *FIRST* specifications, or that the printer is limited to these specifications as a maximum quality level. The process and specifications supported in *FIRST* intend:

- To outline key flexographic procedures and guidelines to be used from the beginning of the process to the end, including the implementation, design, prepress, and print processes.
- To improve quality and consistency through improved communication and measurement procedures.
- To reduce cycle time and minimize rework through improved process control methodology.
- To control production costs through streamlined raw materials and process improvement methodology.
- To enable consumer product companies to obtain optimal flexographic print quality, which equals or exceeds offset lithography and gravure printing.
- To grow the overall flexographic printing industry through increased market share of an expanding market.

Historical Perspective of *FIRST*

Prior to *FIRST*, many consumer product companies were creating individual package reproduction specifications. The generation of too many individualized specifications can become overwhelming to an industry – resulting in manufacturing inefficiencies and confusion. In pursuit of a more universal approach, the FTA membership partnered with leading consumer product companies to create a universal set of flexographic specifications.

The resulting premier edition of *FIRST* (debuting in 1997) and subsequent editions consisted of specifications and tolerances representing the realistic capabilities of 70% of the industry. Data was derived from three years of industry input, three industry-wide surveys, and statistically controlled designed experiments. This edition of *FIRST* includes technical updates to maintain relevancy with the ever-evolving technology, as well as significant subject expansion designed to more fully encompass the entire flexographic process and various industry segments. With hundreds of industry experts, from around the world, contributing to the technical content over the past decade, *FIRST* has become the technical resource for the flexographic industry.

FIRST 5.1 CONTRIBUTORS

The Flexographic Technical Association would like to recognize the contributions and dedicated efforts of those involved in the development, editing, and evaluation of *FIRST* 5.1. These individuals exhibited tireless enthusiasm in spearheading the continuous advancement of the flexographic printing process.

FIRST 5.1 LEADERSHIP COMMITTEE

Mark R. Mazur	*FIRST* Chairperson	Retired
Robb Frimming	*FIRST* Vice-Chairperson Implementation Co-Chairperson	SCHAWK!
Lon Robinson III	FTA Board Representitive	Tension Envelope Corporation
Chad Fulwiler	Design Co-Chairperson	Leibold
Michael McGinnis	Design Co-Chairperson	RR Donnelley
Eric Ferguson	Prepress Co-Chairperson	Havi Global Solutions
Jennye Scott	Prepress Co-Chairperson	Berry Plastics Corporation
John Gleich	Prepress Co-Chairperson	Group 360
PJ Fronczkiewicz	Print Co-Chairperson	Flint Group
Bill Malm	Print Co-Chairperson	Flexowash
Jean Jackson	Print Co-Chairperson	Graymills
Joe Tuccitto	Director of Education Implementation Co-Chairperson	FTA Technical Education Services Team
Shelley Rubin	Manager of Educational Services	FTA Technical Education Services Team

ADDITIONAL *FIRST* 5.1 SECTION CONTRIBUTORS

John Babl	Paper Coverting Machine Company
Brian Berhdt	Lauterbach Group
Richard Black	All Printing Resources
Roy Bohnen	Epson
Steve Carter	Phototype
Larry Goldberg	Beta Industries
Andy Knapp	Flint Group
Bjorn Knutson	FTA Technical Education Services Team
Tim Loehrke	SCHAWK!
Rory Marsoun	Esko
Rose McKernon	FTA Technical Education Services Team
Adina Oree	Lohmann
Ted Pech	Group 360
Steve Smiley	Smiley Color & Associates
Garrett Taylor	A&V
Brad Taylor	DuPont Cyrel Packaging Graphics
Scott Thompson	Southern Graphic Systems
Katie Tuckwiller	DuPont Cyrel Packaging Graphics
David Watson	Tesa Tape
Torben Wendland	Flint Group

PREVIOUS *FIRST* CONTRIBUTORS

FIRST 5.1 is the continuation of the work done by all of those involved in previous editions of *FIRST*. The Flexographic Technical Association would like to recognize the contributions and dedicated efforts of those involved in the development, editing, and evaluation of all those previous editions. No attempt has been made to update company names or company affiliations, which change over time.

Richard Ahlborn	National Envelope Corporation
Larry Ahleman	Western Michigan University
Joe Aker	Hood Packaging, Inc.
Jeff Albaugh	Mastergraphics
Dr. John Anderson	FCA Associates
Frank Anthony	Chattanooga Times/Free Press
Roberto Arosemena	Grupasa
Steve Balschi	Printpack, Inc.
Jason Barrier	Printpack, Inc.
Chuck Bell	Multi Color Corporation
Maynard Benjamin	Envelope Manufacturers Association
Dr. Penny Bennet	California Polytechnic State University
Brian Berhdt	Lauterbach Group
Michelle Beuscher	Independent Consultant
Richard Black	All Printing Resources, Inc.
Denise Bloy	OEC Graphics, Inc.
Dr. Mark Bohan	Printing Industries of America
Roy Bohnen	Epson
Bill Bowers	Flint Group Flexographic Products
Alfred Bowers	RR Donnelley
Al Brancaccio	Overnight Labels Inc
Tom Cassano	MacDermid, Inc.
Carl Cecil	Color Resolution, Inc.
Steve Carter	Phototype
Mark Causey	Beck Compression
Ray Cheydleur	X-Rite, Inc.
David Chinnis	Flint Group Flexographic Products
Kevin Chop	Diageo
Brian Chwierut	Sun Chemical
Tom Cluck	Tyson Foods
Kern Cox	Clemson University
Randy Crutchfield	Prairie State Group
Sherry Cunningham	DuPont Cyrel Packaging Graphics
Edward DeBano	RR Donnelley
Raymond Delricki	GS1US
Angela Denmon	Procter & Gamble
Chris Deye	Phototype
Patrick Dillon	CL&D Graphics
Tony Donato	Harper Corporation of America
Dr. Lorraine Donegan	California Polytechnic State University
Dr. Martin Dreher	DFTA - Germany
Rich Emmerling	Flint Group Flexographic Products
Larry Wm. Evans	Clemson University
Jon Fehrman	Scotts Company
Bill Ferguson	Cincinnati Precision Plate
Eric Ferguson	Esko-Graphics, Inc.
Julian Fernandez	Esko
Michael Ferrari	Proctor & Gamble Company
James Ford	Color Resolutions Intl.
Michelle Ford	Monarch Color Corp.
Robb Frimming	SCHAWK!
John Gaber	Flint Group
Eddie Ghea	Bemis Company, Inc.
Scott Gilbert	Smurfit-Stone
Sam Gilbert	Sun Chemical
John Gleich	Group 360
Larry Goldberg	Beta Industries
Steve Goldfarb	DuPont Cyrel Packaging Graphics
Jordan Gorski	Flint Group
Justin Green	Anderson & Vreeland, Inc.
Larry Haas	Alcan Packaging
Joe Hamilton	Stevenson
Tom Hammer	Flint Group Flexographic Products
Jessica Harrell	A&V
Neil Harrelson	Schawk, Inc
Lesley Hide	EFTA - United Kingdom
Gary Hillard	Hood Packaging, Inc.
Ian Hole	Esko-Graphics, Inc.
Mike Impastato	Flint Group Flexographic Products
Dr. Sam Ingram	Clemson University
Jean Jackson	Praxair, Inc.
Alexander James	Harper Corporation of America
Dr. Malcolm Keif	California Polytechnic State University
Patricia Kent	Pamarco Global Graphics
Rob Kidwell	National Envelope Corporation
Andy Knapp	Flint Group
Bjorn Knutson	FTA Technical Education Services Team
Dick Koslowski	OEC Graphics, Inc.
Tom Kowalski	Eastman Kodak Company
Jim Kulhanek	DuPont Cyrel Packaging Graphics
Allison Lakacha	Techkon, USA
Paul Lancelle	All Printing Resources
David Lanska	Stork Cellramic, Inc.
Colleen Larkin Twomey	California Polytechnic State University
Dr. Brian Lawler	California Polytechnic State University
Paul Lodewyck	LIG Technology
Tim Loehrke	SCHAWK!
Rory Marsoun	Esko
Dan Martinez	Matthews International Corp.
Darlene Masek	Nestle USA, Inc.
Dr. Mark R. Mazur	DuPont Cyrel Packaging Graphics
Joe McCarthy	Burrows Paper Company
Terri McConnell	Phototype
David McDowell	NPES/CGATS
Michael McGinnis	RR Donnelley
Rose McKernon	FTA Technical Education Services Team
David Merkley	American Color Graphics
Patrick Mollman	Siegwerk
Jason Nelson	OEC Graphics
Doug Nelson	Water Ink Technologies
Arleen Neustein	New Excelsior Packaging
Dr. Dieter Niederstadt	Asahi Photoproducts
Randy Nienas	Vertis, Inc.
Roberto Nunez	GIPSA - Mexico
David Nunez	International Paper
Robert O'Boyle	Sun Chemical
Dr. Liam O'Hara	Clemson University
Wayne Peachey	Keating North America
Cherie Pierce	Printpack, Inc.
Bill Pope	FTA/FFTA
F. Cordes Porcher	Packaging Corporation of America
Todd Pressly	Group 360

Lou Prestia	Prestia Consulting Inc.
Dr. Joseph Rach	Chemence, Inc.
Jeffery Randazzo	Controlled Displacement Technologies LLC
Dan Reilly	Plastic Packaging Inc.
John Richardson	All Printing Resources, Inc.
Greg Robinson	Integrity Engineering, Inc.
Lon Robinson III	Tension Envelope Corporation
Steven Rose	National Envelope Corporation
Mark Samworth	Esko
Pete Santkuyl	Kimberly Clark
Kevin Schilling	OEC Graphics, Inc.
Andrew Schipke	W & D Machinery Company, Inc.
Jon Schlosser	OEC Graphics, Inc.
Mike Shanley	National Envelope Corporation
Marek Skrzynski	CSW Graphic Services
Steve Slater	X-Rite, Inc.
Steve Smiley	Smiley Color & Associates
Bob Smith	Great Northern Corporation
Herman Spencer	The News & Observer
Jay Sperry	Clemson University
Seetharaman Srinivasan	Sonoco
David Straten	Advanced Packaging Corporation
John Sweeney	IQ Color
Michelle Talko	Prairie State Group
Tom Thackeray	Weyerhaeuser
Shawn Thiessen	Fastik Label & Supply
Scott Thompson	Southern Graphic Systems
Kevin Trischett	National Envelope Corporation
Katie Tuckwiller	DuPont Cyrel Packaging Graphics
Kelly VandenBosch	ISRA Surface Vision
Rebecca Van Handel	RR Donnelley
Shridhar Varde	Agfa Corporation
Ryan Vest	MacDermid
Joan Wallace	Label Technologies Southeast
Melanie Ward	Wikoff Color Corporation
Bill Warner	Allsion Systems
Brian Watkins	Burrows Paper Company
Phil Wedding	Sonoco
Jarrett Westman	Tetra Pak Inc.
Frank Wheeldon	Schawk, Inc
Catherine Whitaker	Anderson & Vreeland, Inc.
Dr. Nona Woolbright	Clemson University

ADDITIONAL CONTENT AND GRAPHICS CONTRIBUTORS:

360 Imaging, 3M, AGFA, Asahi Kasei America, Beta Industries, BOBST, BST Pro Mark, CGS Publishing Technologies International, Color Resolutions International, Clemson University, C-P Flexible Packaging, DFTA, Dunwoody College of Technology, DuPont Cyrel Packaging Graphics, Eastman Kodak Company, EFI, E.I du Pont de Nemours & Co., Epson, Esko, Eudes Scarpeta, F. Cordes Porcher, Fischer & Krecke, Flint Group, FTA/FFTA, FUGIFILM Graphics Systems USA, Gallus, Graymills, Harper Corporation of America, Hood Packaging Corporation, IDEAlliance, Integrity Engineering, Inc., Interflex Laser Engravers, INX International, Just Normlicht, MacDermid Printing Solutions, Mark Andy, Inc., Michelle Beuscher, OEC Graphics, Opaltone, Inc., Paper Converting Machine Company, Phototype, Praxair Surface Technologies, Printing Industries of America, Right Angle Concepts, RIT, Smurfit-Stone Container Corporation, Spectrum Label Corporation, Sun Chemical Corporation, Smyth Companies, Sun Chemical Corporation, Techkon, VT Graphics, William Fox Munroe, Windmoeller & Hoelscher Corporation, X-Rite

THE FTA TECHNICAL EDUCATION SERVICES TEAM IS:

Joe Tuccitto, Rose McKernon, Bjorn Knutson, Duane Woolbright, Sharon Cox, Shelley Rubin and Mark Cisternino

DOWNLOAD *FIRST* 5.1 Extras
Referenced in the document at:
http://www.flexography.org/FIRST_extras

COMMENT

***FIRST* is a document created for the industry by the industry.**
Your feedback and contributions are extremely important.
If you have comments, suggestions, or changes for future editions please visit:
http://www.flexography.org/FIRST_comments

DOWNLOAD *FIRST* 5.1 Extras Referenced in this Section at: http://www.flexography.org/FIRST_extras

1.0 Implementation: *FIRST Benefits all segments of the flexographic printing industry.*

1.0 IMPLEMENTING *FIRST*: THE FOUNDATION

Communication is the foundation of *FIRST*. Using agreed upon terminology and a defined workflow, *FIRST* seeks to minimize miscommunication and promote the manufacturing of consistent, conforming flexographic materials. The role of the customer is crucial to the communication and implementation of specifications. The customer determines the requirements and establishes the objectives for each job. Printers and suppliers help the customer achieve their quality and cost containment goals within the given time frame. Each member of the supply chain must utilize process control methodologies if the final product is going to conform to the quality and cost goals established by the customer.

The customer and suppliers must have a detailed plan and well-defined workflow to efficiently produce the printed material. The *FIRST* Packaging Development Workflow diagram (page 5) provides an overview and clarifies the role of each party within the production workflow. The diagram facilitates an understanding of the workflow, for what and when each party is responsible, and a resolution path if the product fails to meet quality standards at any stage of the workflow. Each supplier must implement specifications, including process control of variables, to ensure the final product is compliant with the predetermined job specifications.

CGATS TR 011-2002 (Graphic Technology – Package Development Workflow – Design Concept Through Approved Production File) describes a model for the package development process beginning with the identification of a project through preparation of an approved production file. It is intended for use as a reference in the creation of workflow procedures for specific organizations or products. CGATS TR 012-2003 (Graphic Technology – Color Reproduction and Process Control for Package Printing) outlines the steps necessary to understand and objectively define the color and tone reproduction capabilities of a specific printing process. The steps identified in CGATS TR-012 provide information required in the package development workflow (CGATS TR-011). The Committee for Graphic Arts Technologies Standards (CGATS) technical reports 011 and 012 can be obtained from NPES (The Association for Suppliers of Printing, Publishing and Converting Technologies). Refer to the Appendix B for contact information. *FIRST* summarizes CGATS TR 011-2002 in Section 1.4.1.1. *FIRST* summarizes CGATS TR 012-2003 and provides a detailed explanation of its application to flexographic printing in Section 1.3.

1.1 The Value of *FIRST* Specifications

The mission of *FIRST* is to create a common set of specifications and communication protocols for the flexographic industry. *FIRST* seeks to identify the responsibility of each party in the package/product development process and facilitate communication between parties in order to quickly, efficiently and cost effectively produce attractive flexographic packaging. Effective communication during design development, final art and printing can result in packaging produced with:

- Fewer errors, thus decreasing cost and time to market
- Increased production efficiencies resulting in improved yields
- More consistent and predictable printed results, yielding improved quality and reduced downtime
- Reduced management effort due to roles, responsibilities and specifications being clearly established

To realize the maximum benefits of *FIRST*, all parties must understand *FIRST* concepts and participate fully in the process. These specifications are a quality tool for the entire product development supply chain.

1.2 Adopting and Implementing *FIRST* Specifications

Specifications are created, or adopted, when a problem exists or improved results are desired. *FIRST* specifications cannot be fully implemented without understanding, cooperation and adherence by all involved parties.

If the consumer product company requires justification, the printer and material suppliers, as experts in print production, can provide verification and data indicating how *FIRST* specifications bring value and predictability to the customer's end product.

Selling *FIRST* externally (within a design firm, prepress house or printing company) may prove to be a bigger challenge for many companies than customer buy-in. A firm, consistent commitment from senior management is required to dedicate resources and emphasize the role of *FIRST* in improving quality and solidifying a competitive advantage. Sometimes a commitment from senior management does not occur without the insistence of the customer. The same data supplied to the customer, to demonstrate how *FIRST* specifications bring value

1.1 *FIRST* 5.1

1.2 Consistent Packagaing: *One of the benefits to the Consumer Products Company (CPC) of implementing the FIRST Methodology is consistent graphics produced on a variety of substrates.*

and predictability to the finished product, can also be supplied to senior management to emphasize quantitative value.

1.3 *FIRST* Methodology

FIRST supports the methodical approach to process control outlined in CGATS TR012-2003 (Graphic Technology – Color Reproduction and Process Control for Packaging Printing). While it is true that each organization employs its own workflow based upon its individual needs, there are many common elements and certain fundamentals to flexographic printing that are universal.

As part of the graphics and product design process, the printer must be able to communicate the ability of the particular printing process to satisfactorily reproduce the graphic design. The printer must:

- Determine if the printing process can achieve the requirements of the job (color, tone reproduction, register, etc.)
- Identify design elements that are potentially troublesome.
- Provide guidance and suggestions to ensure the final design can be printed satisfactorily
- Maintain the printing process in a stable, consistent, repeatable condition as the job is printed time and time again

The need for testing usually becomes evident as a design progresses toward completion and production requirements are identified. The scope of testing required depends upon the printer's experience with certain design variables such as:

- The specified ink and substrate combination on a specific press
- The color or combination of colors specified on a specific press
- Design elements that may produce unexpected results (reverses, special effects, vignettes, etc.)

The scope of testing must balance acceptable predictability of results with budgetary considerations. Available historical data may preclude the necessity of some aspects of testing.

The *FIRST* Methodology is based upon CGATS TR 012-2003, which outlines the steps of optimization, fingerprinting, process control, characterization and process improvement. These steps are independent of organizational structure or specific design criteria.

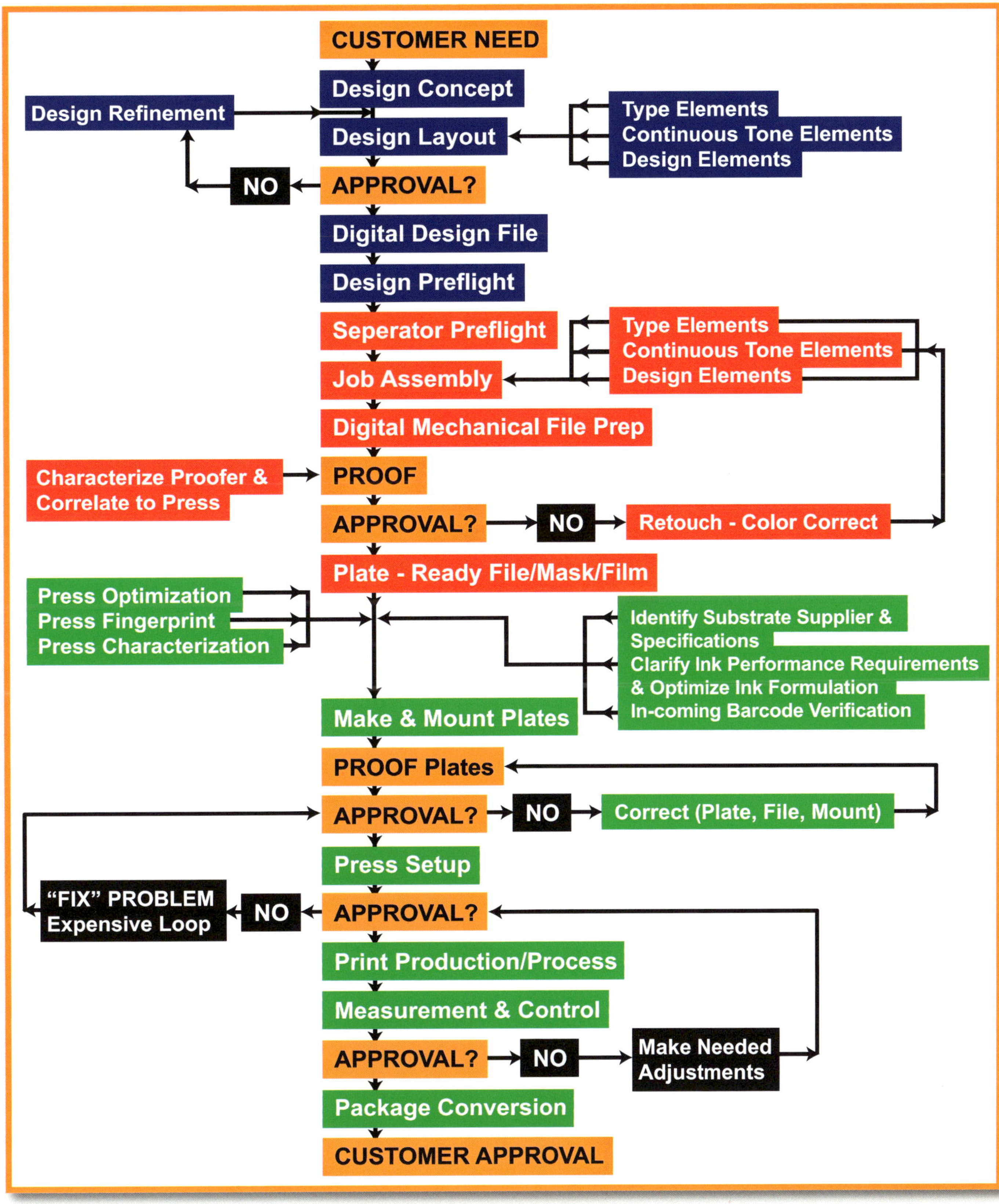

1.3 FIRST Packaging Workflow Diagram

Optimization: The first step requires the printer to analyze the printing process to determine preferred operating parameters. Preferred operating parameters strike a balance between print quality, print stability (ability to maintain throughout the run and across runs), and efficiency of operations (minimizing downtime and maximizing speed). The optimization process may or may not require special print trials. Section 1.3.1 details the optimization process.

Fingerprinting: This involves benchmarking the performance of the press under pre-established conditions, defined during optimization. Fingerprinting utilizes the test elements outlined in Section 1.3.2 to establish standard operating conditions (aim point specifications and tolerances) for a given set of materials and printing conditions. Fingerprinting provides the baseline against which future production runs and process improvements are measured. Statistical analysis is the primary tool utilized to summarize meaningful results and establish target aim point values & tolerances. The fingerprinting process requires print trials for each unique combination of materials and print conditions. Section 1.3.2 details the fingerprinting process.

Process Control: This step uses statistical analysis to establish and maintain specific operating conditions. Measuring various test elements during the fingerprinting process and quantifying the results to establish aim points and tolerances establish operating conditions. Operating conditions are maintained by measuring control targets during each job set-up and run. Information obtained by measuring the control targets is used to adjust the printing process to remain within the established control limits throughout the pressrun and across runs. Section 1.3.3 details process control methodology.

Characterization: The characterization trial utilizes the IT8.7/4 test target for color management calibration within the prepress system. It is used to calibrate monitors, scanners, imagesetters/platesetters, and proofers, to the printing press. The IT8.7/4 target is for process color image reproduction only. The characterization press trial must adhere to the aim points and tolerances established during the fingerprinting process. These same aim points and tolerances must be maintained during future production runs. Section 1.3.4 details the characterization process.

Process Improvement: Statistical analysis of optimization, fingerprinting, characterization and ongoing run data should be used to improve quality, efficiency, and productivity. Process improvement depends upon constant monitoring of the printing process every day, on every job. If this data is recorded, it can be maintained in a database, which allows the printer to:

- Evaluate the printing process by key process variables
- Identify improvement opportunities
- Measure the improvements once implemented
- Quantify the results

1.3.1 Press Optimization

The goal of the optimization process is to identify the best combination of print variables to achieve the design requirements. The test conditions must represent normal production behavior and quality. The optimization process occurs prior to fingerprinting. It involves identifying the print variables that will produce the desired results with the substrate and ink colors specified by the customer. The optimization step must be completed for the intended graphics of each print deck (process/line/combo/solid). It is usually not necessary to perform an optimization test for every print variable. If certain variables are standardized, and if sufficient experience or historical data is available, re-testing is not necessary. Typically, a printer will standardize print variables through optimization print trials over time for each type of graphic (process/line/combo/solid) on common substrates.

It is important to appoint a qualified person to design the optimization print trials, organize all of the materials, work with applicable suppliers, order test components, specify print variables that will not be evaluated, ensure all print variables are identified and available, work with production to schedule the press run, label the samples as they are printed, accurately measure the print samples, and interpret the results. Without thoughtful test design, sufficient control of the press run, and meaningful evaluation of the data generated, the press run is of little value. Often if new equipment and/or consumables are added, a re-optimization may be necessary. The scope of the test can quickly become unmanageable if too many variables are evaluated at one time. Begin the print optimization process by identifying primary print variables that influence print quality. The following items have a measurable influence on print results and image quality.

Job-Specific & Press Component Print Variables:

Print variables can be grouped into two general categories: Job Specific Print Variables and Press Component Print Variables. While these are reviewed in detail in Sections 20.0 and 21.0, the print variables are discussed in this section as they relate to the optimization process.

- **Press component** print variables are mechanical parts of the printing press and include: dryers, registration control, tension control and press mechanics (gears, bearings, cylinders, rollers, and drive systems). Optimization of press components determines if the press is mechanically holding impression settings and register, and evenly delivering and drying ink. Conduct print trials to test the various press settings and mechanical components verifying the press is able to hold its settings. The press should be optimized when it is installed and whenever the press is altered (new gears, rollers, dryers, etc). Changes from wear occur slowly and may go unnoticed unless measurements are routinely taken and compared to original performance specifications. This is a general press optimization and not design dependent; it identifies the basic mechanical ability of the press. Refer to Section 21.0 for a detailed discussion of the primary press component print variables.
- **Job specific** print variables include: substrate, inks, coatings, plate type, mounting tape, anilox rolls and sleeves. These are the variables that are typically optimized for a specific design. After testing to determine which combination is optimum for the graphics intended for each print deck (process/line/combo/solid), it is important to standardize each combination and consistently use the same set of materials. Changing the components will change the print outcome and make the existing data, used by the designer and prepress provider, incorrect. This may result in the inability to match the contract proof and achieve acceptable print quality on press.

A few examples of optimization print trial objectives include:

- **Anilox** – Use a banded anilox roll to determine optimum anilox volume, cell count, and engraving angle for each type of graphics (process/line/combo/solid) on a given substrate with a specified ink formulation. The engravings on the banded anilox will vary based upon the type of graphics being printed (process/line/combo/solid). Refer to Section 20.4.2 for more information.
- **Mounting Tape** – Determine optimum density and cell structure for each type of graphics (process/line/combo/solid) that will provide desired print quality and maximum longevity. Refer to Section 20.5.2 for more information.
- **Ink** – Determine ideal pigment load (strength) to achieve desired color and/or density and minimize dot gain/TVI while maintaining ink stability throughout the press run. Refer to Section 20.2 for more information.

- **Coating** – Determine minimum coating weight to achieve desired "Coefficient of Friction (CoF)" and gloss values while minimizing dry rate and cost.
- **Plate Type** – Determine optimum plate material for each type of graphics (process/line/combo/solid) that will deliver desired print characteristics while maximizing plate life and therefore, minimizing cost. Refer to Section 17.0 for more information.
- **Substrate** – Determine minimum surface tension level of film that will achieve acceptable ink adhesion. Refer to Section 20.1 for more information.
- **Process Color Gamut** – Evaluate ink strength, pigment selection, and ink film thickness to achieve the largest gamut possible while minimizing dot gain/TVI. Refer to Section 20.2.4 for more information.

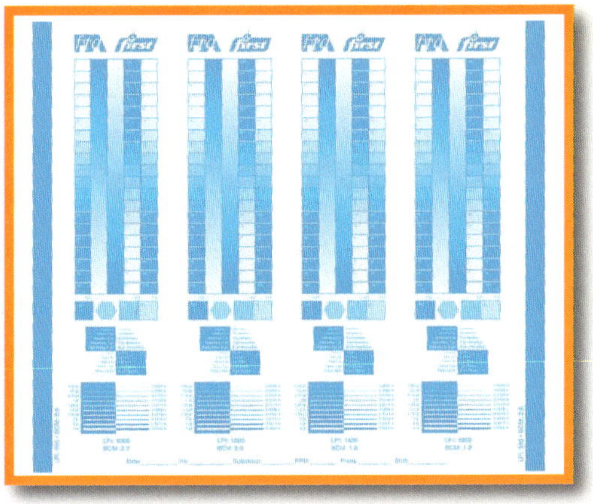

1.3.1.1 Press Optimization Test Form

1.3.1.1 Optimization Test Plan

The first step in the optimization process is preparation. The printer must review the preliminary design, layout, and printing requirements of the new product. Analyzing the job requirements and comparing them to existing jobs will determine if optimization press trials are necessary.

If optimization press trials are needed, a test plan must be created. The test plan identifies the variables to evaluate, the method to evaluate, and the test elements to use to determine which combination is optimal. The type of graphics printed on each print deck will determine the test elements needed to properly evaluate the print variables included in the test plan.

For Example: A print deck responsible only for process color will be primarily concerned with the size and quality of the dots printed. The goal will be to minimize dot gain/TVI, dot bridging, and hard edges while maximizing density and therefore, print contrast. The test target may require only tone scales and a solid patch. Two large vignettes (one with a holding line and one without) may also be included to evaluate the ability to print a smooth vignette without banding or hard edges. The test design may include 2-3 line screens to determine the optimum line screen for the print variables being tested. If so, both the tone scales and the vignettes should be in each line screen.

If a deck will be used as a combination (combo) deck, other test elements must be included, in addition to the elements listed above. A process/line combo deck should include both positive and reverse type elements in Serif and Sans Serif fonts to determine minimum font as well as positive and reverse lines to identify minimum rule widths. Applicable bar codes at varying magnifications and BWR's (bar width reductions) may be added to determine print capability of bar codes. A large solid should also be included to evaluate pinholing, mottle and ghosting.

For a deck printing only solids, the test design might include a much larger solid spanning the entire cylinder width to enable a better evaluation of smoothness and impression across the web. Larger type and line weights should be included, in both positive and reverse, to determine minimum capability on a solid deck. Also, the bar code test element varying magnifications & bar width reduction's (BWR's) might also be included to determine the ability to print bar codes on a solid deck, which will almost certainly be different than if the bar code prints on a combo or line deck due to ink film thickness.

Typically, optimization press trials are single color. The test plan must identify the press and print deck to be used as well as the run length of the press test, press speed, and sampling frequency. In addition, the test plan should define the settings for variables not being tested. It should quantify job set-up and run conditions that must be achieved before sampling begins.

While it is possible, and sometimes desirable, to create a more complex optimization test plan (ie: evaluating anilox engraving and ink strength simultaneously to identify the optimum combination), great care should be given to keep the test manageable. The test design should be controlled to include a maximum of three print variables (ie: anilox, ink strength & line screen). Three print variables will result in a very complex analysis requiring advanced statistical training to meaningfully evaluate.

1.3.1.2 Optimization Print Trials

Print trials provide physical samples to evaluate the performance of the printing system under various conditions. They must be thoughtfully planned, carefully controlled, and meticulously documented to be of value. It is generally desirable to have the optimization press trial built into the press production schedule so that adequate time and resources are allocated. Ideally, the shift supervisor and press crew are briefed on the press trial objectives and on the procedures to follow. The test coordinator should be present throughout the optimization print trial to verify proper set-up, document run conditions, and flag/pull print samples.

It is important to verify the performance of the chosen materials in everyday production. Each raw material should have documentation (ie. Certificate of Analysis) verifying conformance to established parameters. Ideally, have other trained press crews, not previously involved in the press trial, run the test and achieve similar results. This will confirm that the results are repeatable in the production environment.

1.3.1.3 Optimization Data Collection and Documentation

The outcome of the optimization process is a document that describes the optimum settings for the key print variables. It should include: job specific print variables for each print deck (ie. ink, plate type, mounting tape, and anilox roll) as well as settings for the more general press component print variables (ie. tension settings - unwind/rewind, dryer settings - temperatures & flow rates, and doctor blade settings, etc.).

Documenting the results of the print optimization trails is needed to complete the fingerprinting process. The optimization print trials determine how the press will be set-up (specifying each print variable) when the fingerprint press test is run. The optimization process establishes optimum settings for each of the print variables.

Careful consideration should be given to how the print characteristics will be evaluated, measured, and recorded. For the variables not considered and investigated during the optimization process, simply record and repeat them during the fingerprint trial, characterization trial, and future production runs. For the print variables evaluated during the optimization process, pull samples throughout the press run and record the measurements. Once the press samples have been measured and the data recorded into a spreadsheet, or database, statistical analysis can begin. The mean/median/mode of each element is a simple method to summarize the "average" of the data. A more informative analysis uses a control chart to summarize the range of measurements, which describes the variability of the process.

Appropriate software and statistical analysis training is required to properly evaluate print trial data. The sophistication of software and level of training required varies with the complexity of the test design and the sophistication of the information desired. Simple print optimization trials can be successfully evaluated with widely available software, such as Microsoft Excel, and moderate statistical analysis training.

Caution! The number of data points to be summarized multiplies exponentially with each variable. For example, a press optimization trial that uses a banded anilox with four bands and

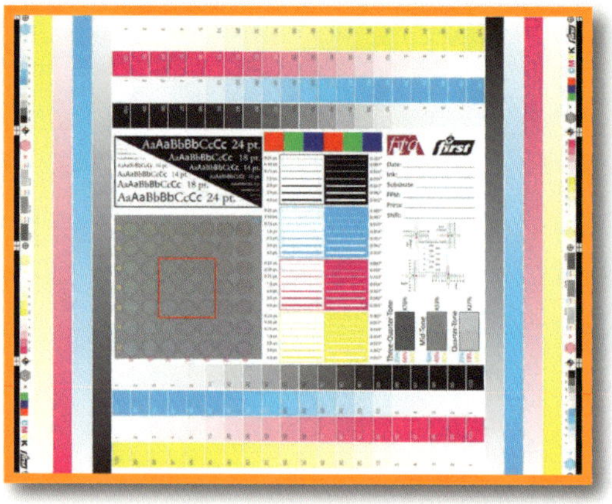

1.3.2 Press Fingerprint Test Form: *Should include solid color, gray balance and overprint patches as well as tone scales, vignettes, line and type elements, bar code, registration and impression targets.*

two ink strengths produces eight possible combinations. If this test is run on a process only deck with the following measures: 2%, 30%, 50% and 70% dot, solid, vignette with and without a holding line (measuring dot bridging – yes/no; and hard edges – yes/no), the result is 36 measurements per sheet. If there are 100 samples pulled for each ink strength, the result is 7,200 raw data points that must be analyzed. If three line screens are evaluated as well, the number balloons to 21,600 raw data points! As this example demonstrates, the scope of the test can quickly become unmanageable which is why prior planning is so critical. As the number of possible combinations expands, the statistical complexity increases as well.

1.3.2 Press Fingerprint

The primary objective of the fingerprint trial is to measure and record the print characteristics of a particular press, operating with specific settings and materials. The specific settings and materials are determined, in part, by the customer's design requirements (substrate, ink colors, design elements), and by printer experience or optimization trials (best anilox for each deck, optimized ink formulation, best mounting tape, etc.). The fingerprint trial must include all process and spot colors to be used on the job because the fingerprint trial results in data, which is used to define the printing process specifications (including aim point values and tolerances) and establishes the benchmark upon which all future runs, including the characterization trial, will be compared.

The goal of the fingerprint trial is to document the press conditions and the resulting print quality across the sheet/web, under stable and repeatable conditions. A job set-up sheet detailing how the press is set-up and the aim points and tolerances for key quality measures (density, dot gain/TVI, bar code scan rate, Delta E of custom colors) should be developed from the fingerprint trial results.

1.3.2.1 Fingerprint Test Plan

While the test elements included on the fingerprint test form will be more comprehensive than on "live" production jobs, it should also contain the same process control target that will be used for process control on production jobs. This is important for consistency and ease-of-use. When the same control target is included on the optimization, fingerprint and characterization trials as on "live" jobs, the press crews are clear about which test elements they should measure, what the measurement result should be and when to adjust the press, thereby improving consistency.

When designing the fingerprint test form, the test elements typically included can be divided into two general categories:

1. **Process Control Parameters/Elements**
2. **Mechanical Control Parameters/Elements**

1. Process Control Parameters
Process control parameters indicate the print capability of a particular press, operating with specific settings and materials as required by the graphic design. These test elements measure key quality attributes of the graphic design. Typical process control parameters on a fingerprint test form include:

Solid Color Patches:

- **Process Colors:** With process printing, the solid test patches represent important "anchor points" for the color reproduction process, so it is important to quantify them. The process color solids can be measured using both densitometry (solid ink density) and spectrophotometry (L*a*b*C*h°). Refer to Section 19.4.4 for additional information.
- **Custom/Spot Colors:** Using the best press set-up and ink formulation (as determined during the optimization process), quantify using spectrophotometry (L*a*b*C*h° DE) and visual evaluation, how well the printing process is able to match the specified color. A sample of the fingerprint test form, printed with the same substrate, press, ink formulation, etc. as future production, should be sent to the customer for color approval. Once approved, the solid custom color printed on the fingerprint test form should be used to create the "standard" in the color measurement software to which all "live" jobs will be compared. Refer to Section 19.3 for more information.

Gray Balance Patches: This test element should include patches representing highlights, ¼-tones, mid-tones, ¾-tones and shadows. Gray balance is influenced by the color and density of the process inks, dot gain/TVI, and ink trapping. It is the most sensitive parameter to monitor and the easiest to evaluate visually. While gray balance will not identify which variable has deviated, it is the first place where a change in the printing process is usually detected. Refer to Sections 12.9.1 and 19.4.2 for more information.

Overprint Trap Patches: These patches are used to assess how well inks trap, or print over one another. In addition to the process color solids, process color overprints (red, green, blue) represent additional "anchor points" in the color reproduction process. They can be quantified either

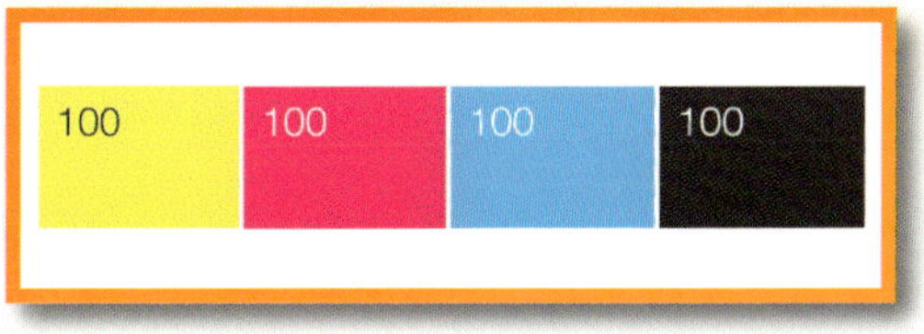

1.3.2.1a Solid Color Patches

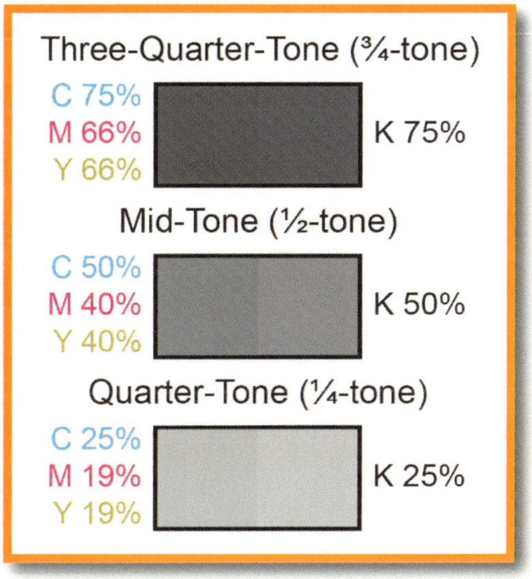

1.3.2.1b Gray Balance Patches

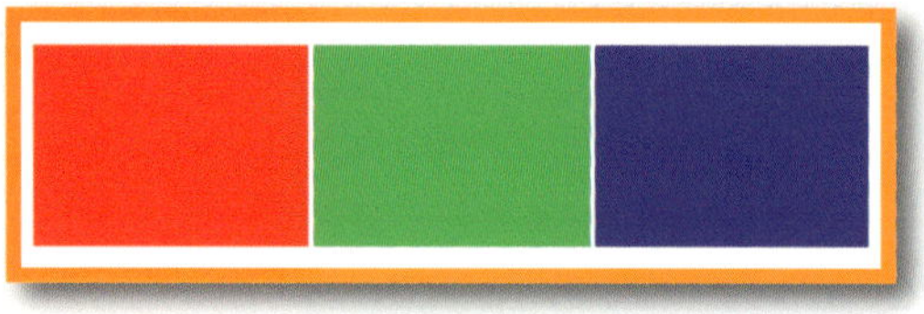

1.3.2.1c Overprint Trap Patches

1.3.2.1d Tone Scales

1.3.2.1e Vignettes

1.3.2.1f Bar Codes

densitometrically (trap equations) or colorimetrically (L*a*b*C*h°). Also, if there are any non-process overprints (ie. spot with process, spot with spot), they should be quantified as well. Traps should appear smooth and free of pinholes. Refer to Section 19.4.6 for additional information.

Tone Scales: Tone scales are flat tints typically ranging from 1%, as measured in the film/file, to as much as 98%. They can be used to measure dot area, dot gain/TVI, print contrast, and tint density. Regardless of the measurement equations used, it is important to understand how halftone dots are reproduced on the press and to quantify the results (ie. establish aim points and tolerances) for future runs. In addition to evaluating tone compression, tone scales are useful for identifying the best screen ruling and screening technology (conventional/stochastic/ hybrid) for the design requirements. Refer to Section 19.4.3 for additional information.

Vignettes: A vignette is a continuously graduated scale that smoothly transitions from a solid to the minimum dot held on the plate. Vignettes may be included in several screen rulings to visually determine the optimum ruling to use on "live" jobs given the press set-up and design parameters. Refer to Section 19.3.3 for additional information.

Bar Codes: For the appropriate symbology, include several different combinations of magnification (size) and bar width reduction (BWR). Orient the symbols in both the web and cross directions to determine the optimum combination to use on the "live" jobs given the press set-up and design parameters. Refer to Section 19.3.4 for additional information.

Line & Type Elements: These test elements are used to determine the minimum type size and rule width to use on "live" jobs given the press set-up and design parameters. These elements should be printed on every deck since the minimum printable size will vary with deck set-up conditions. The test elements should include both positive and reverse lines/rules and Serif & Sans Serif fonts. Include the point size of each line of text for easy identification. Other common design elements to consider include: ™, ©, kosher symbol, and www. http://. Refer to Section 19.3.1 for additional information.

2. Mechanical Control Parameters

Mechanical control parameters are used to ensure the press is mechanically operating as it should be. These elements should be included in several places across the fingerprint test form (gear, center and operator sides) to ensure the press is balanced across the sheet/web. Typical mechanical control parameters on a fingerprint test form include:

Impression & Slur Targets: These elements evaluate plate-to-substrate and anilox-to-plate impression, ink volume, ink dry rate, and fluting (in corrugated applications). Refer to Section 19.4.8 for additional information on the types of impression & slur targets and when and how to use them.

Plate Exposure Guides: These test elements confirm optimum exposure and processing of the printing plates. Refer to Section 17.9 for additional information.

Registration Targets: There are several registration targets used to determine the accuracy and consistency of color-to-color registration on press. These test elements also monitor the position of the image (center marks) and confirm the image is square to the lead edge (in corrugated and envelope blank applications). Refer to Section 19.4.7 for additional information on various types of registration targets and when and how to use them.

Solid Ink Color Patches: In addition to the process control parameters that are evaluated using these test elements, they can also be used to evaluate mechanical control parameters such as anilox-to-plate and plate-to-substrate impression as well as paper mottle. Refer to Section 19.3 and 19.4 for additional information.

Vignettes: Vignettes can also be used to identify defects that result in light and dark areas, or bands that stretch across the entire web/sheet width. Common names include barring, gear marking, or gear chatter. This defect is typically attributed to gear problems, form layout, or doctor blade chatter. There can also be localized banding problems as a result of imaging problems in platemaking. Higher line screens are more likely to show barring. Refer to Section 19.3.3 for additional information.

Flat Tint: This test element is very useful to quickly identify press problems during the fingerprint trial associated with: ink volume & dry rate, anilox volume, bad bearings, plate impression settings, and fluting (in corrugated applications). The printer should select the best tint for their operation; a 30%-35% tint is typical.

1.3.2.2 Fingerprint Print Trial

The primary goal of the fingerprint press trial is to provide samples printed under "normal operating conditions" for the particular press, using specific settings and materials. The fingerprint trial samples are measured and analyzed to define

1.3.2.1g Line & Type Elements

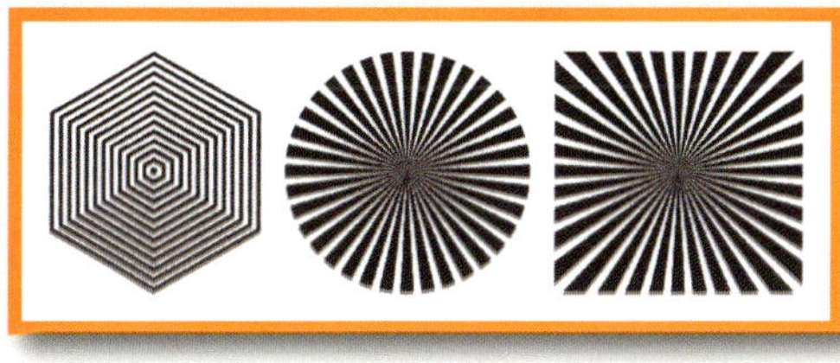

1.3.2.1h Impression & Slur Targets

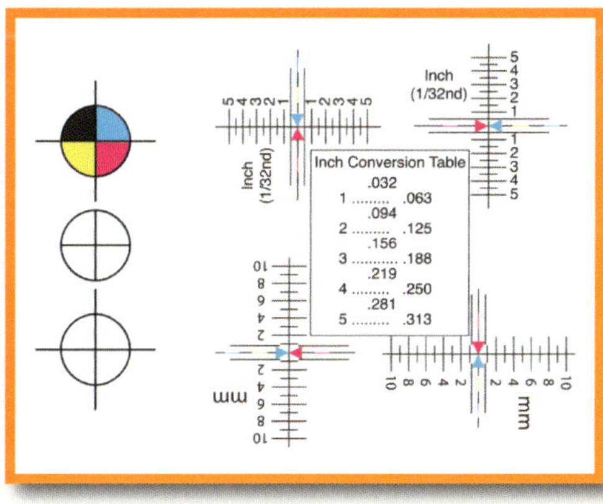

1.3.2.1i Registration Targets

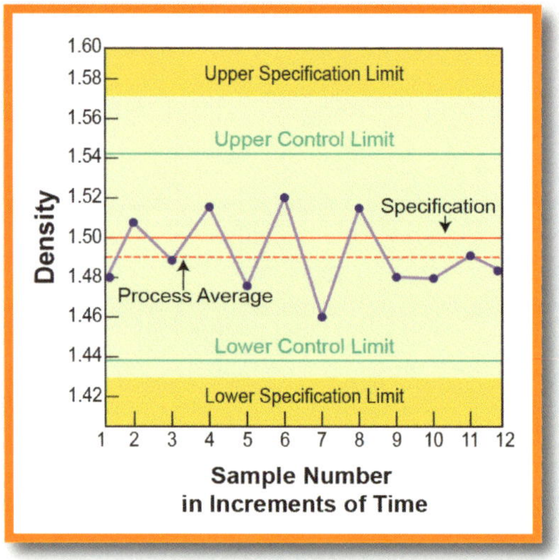

1.3.2.3a Control Charts: *Control charts are the easiest and most efficient way to quickly understand and react to changes in the printing process.*

FTA *FIRST* 5.1 Press Operating Data Sheet

PREPRESS RECORD

PRESS RECORD

MEASUREMENT DATA

1.3.2.3b Press Operating Data Sheet: *Should be customized to reflect printer-specific conditions. It is used during optimization, fingerprint and characterization trials as well as on live production jobs.*

capability settings of key design elements (ie. type size, bar code size & compensation, screen ruling) as well as establish aim points and tolerances for key quality variables (ie. density, dot gain/TVI, print contrast, custom color matches, overprint traps, etc.).

The press must be run under "normal operating conditions" if the information is to be predictive and useful during "live" production runs. "Normal operating conditions" include controlling variables such as: press speed, correct ink formulation for the job, optimum ink pH & viscosity, proper blade settings (ie. angle, pressure, balance), appropriate anilox roll in each deck for the "live" job design, appropriate tension settings for the specified substrate (same as "live" job), appropriate mounting tape/sleeve/plate combination for each deck, etc.

During the fingerprint press trial, the press crew should work to achieve a "steady-state" condition in which the press is stable and the process control parameters are printing consistently without drifting. Once a "steady-state" condition is achieved, the press run should continue for at least 30 minutes, pulling a minimum of 25 random print samples. In addition to the 25 random samples, another 100 consecutive impressions should be pulled in order to measure and quantify short-term "natural" variation.

The fingerprint press trial should be performed as a separate dedicated press run. In some situations, the fingerprint trial is incorporated into the characterization press run, but should only be done by qualified color experts. With the appropriate control targets and similar job set-up, it is also possible to obtain the necessary data from multiple production runs. The appropriateness of combining the fingerprint trial with other press tests or "live" production runs largely depend on how much data and knowledge exists about the press using these specific settings and materials.

1.3.2.3 Fingerprint Data Collection and Documentation

Measure at least 25 random print samples and 100 consecutive impressions to quantify both long-term and short-term variation inherent within the specific printing process. A control chart (for this type of sampling protocol it is usually an X-bar/R control chart) identifies the process average (mean) and control limits for each process control parameter. The control limits (UCL/LCL) should be specified as +/- 3 sigma (ie. +/- 3 standard deviations from the average). The control chart for each process control parameter identifies the range in which the printing press, given the specific settings and materials required by the design, is going to print 99.6% of the time, as long as the printing conditions are consistent. By comparing the printing process average and upper & lower control limits with the aim point and tolerance (defining what is acceptable quality) – the process capability (Cp/Cpk) can be statistically determined. Process capability explains what percentage of the time the printing process will produce "out-of-specification" material. Ideally, the tolerance (upper & lower specifications – USL/LSL) should be larger than the process control limits (UCL/LCL) – indicating the printing process is capable of operating within the specifications 100% of the time, as long as the printing process is operating in control. When the tolerance is tighter than the process control limits, the printing process is not capable of producing acceptable product without generating considerable waste. Therefore, the specification or the printing process conditions, or both, must be evaluated and altered.

Two key items should be documented from the press fingerprint. First, a "Press Operating Data Sheet", begun during the optimization phase, should be finalized during the press fingerprint. This is the job set-up form identifying all press, substrate, and ink variables that must be replicated on future production runs. This information is required in order to achieve the process control parameters (density, etc.) documented from the fingerprint trial. Generally, when press variables are altered, the printed result deviates as well. In addition, the process control parameter data can be summarized and included on the "Press Operating Data Sheet" (ie. density, dot gain/TVI, print contrast, % trap, bar code scan grade, etc.).

The second document derived from the fingerprint press trial is a "Press Fingerprint Report". The "Press Fingerprint Report" is comprised of three main parts:

1. **Summary of data collection methods** – Explains how the press was operating when samples were pulled (press speed, ink viscosities, etc.), the duration of the press run (actual run time when samples were pulled), the number of samples pulled, and the frequency with which they were obtained.
2. **Process capability summary for each process control parameter** – Identifying the average (mean) and upper & lower control limits for each process control parameter, as well as the recommended aim point and tolerance, and an explanation of process capability.
3. **Summary of the results for additional parameters evaluated** – Provide analysis of more subjective process control parameters such as the print quality of vignettes and positive & reverse type elements.

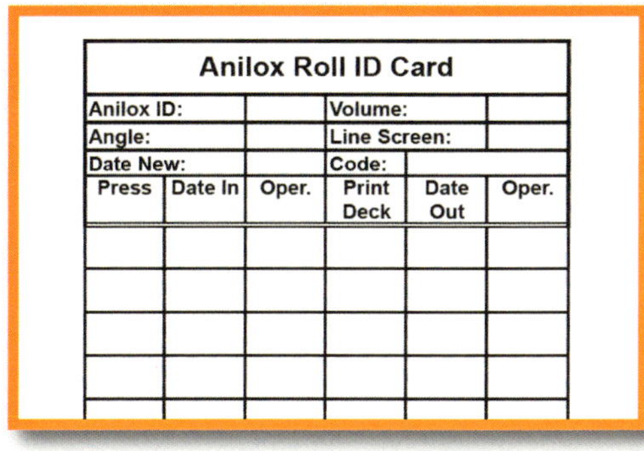

Anilox Roll ID Card					
Anilox ID:		Volume:			
Angle:		Line Screen:			
Date New:		Code:			
Press	Date In	Oper.	Print Deck	Date Out	Oper.

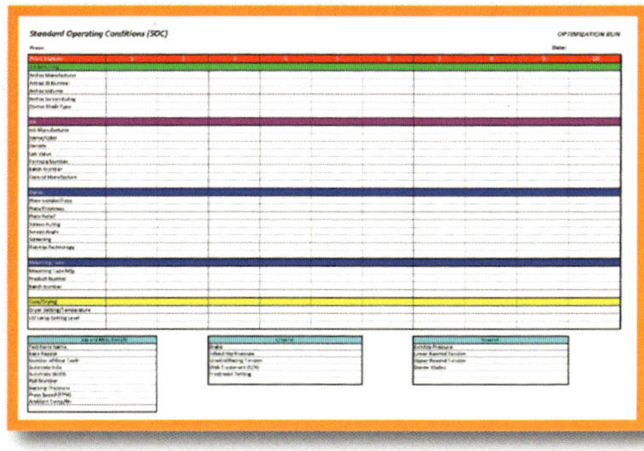

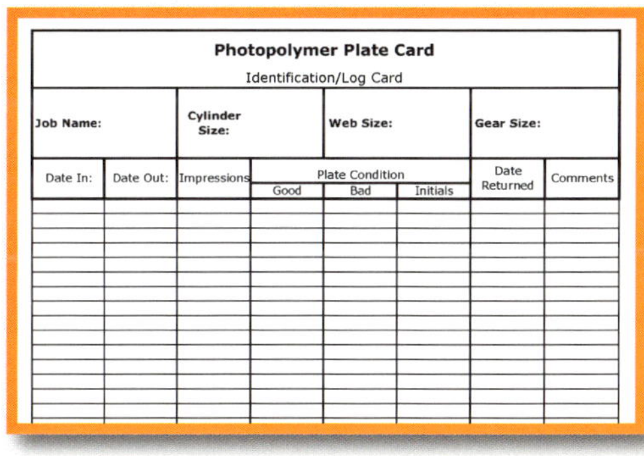

Photopolymer Plate Card

Identification/Log Card

Job Name:		Cylinder Size:	Web Size:			Gear Size:	
Date In:	Date Out:	Impressions	Plate Condition			Date Returned	Comments
			Good	Bad	Initials		

1.3.2.3c Documentation: *These forms are examples of typical documentation used to control the printing process for repeatable results.*

The information contained in the "Press Operating Data Sheet" and the "Press Fingerprint Report" will be used on "live" production runs utilizing the same press settings and materials. The digital files of future jobs will be adjusted based on the data obtained during the fingerprint trial and included in the "Press Fingerprint Report".

1.3.3 Process Control

If the customer desires a consistent, repeatable product on the store shelf, space must be allocated (either on the design or in the waste matrix of the job) to accommodate an appropriate control target. The same control target (ie. run target) included on all optimization, fingerprint and characterization trials must also be included on "live" jobs. It is impossible to control what is not measured. Measuring the process control parameters (ie. density, dot gain/TVI, print contrast, % trap, gray balance, spot color DE, bar code scan rate, etc.) and key mechanical control parameters (ie. ink viscosity, pH, film surface tension, etc.) then comparing the results "in real-time" to the average and control limits established from the fingerprint data, makes it possible to control the printing process for repeatable results. Real-time control charting, with appropriate training, makes it easy for press crews to determine when press adjustments are necessary and, as importantly, when they are not necessary to maintain print quality and consistency. Over-adjusting a press can actually increase the amount of variation in print quality and generate more scrap and downtime. Basic statistical training clarifies for press crews what variation is inherent in the process (should not attempt to correct) and what variation is due to a change in the process or special cause (should be corrected).

In addition to using real-time measurement and control charting to evaluate the process and mechanical control parameters, there are other tools a printer can use to improve the repeatability and consistency of the process. These other tools can be divided into two general categories:

1. **Press Maintenance** – A properly maintained press is capable of achieving similar print results run after run. However, a press in need of maintenance will have a difficult time duplicating previous results. Press maintenance is an example of "special-cause variation" which must be corrected in order to print "in control". Any major press maintenance needs to be done prior to starting the optimization step of implementing *FIRST* into your workflow. An ongoing press maintenance schedule is key to making sure additional variables are not being introduced into your everyday "live jobs".

2. **Process Documentation** – Strong documentation systems make it easy to monitor and control the components coming to press and to set-up and run the press the same way every time a job is run. Some examples of process documentation that help to reduce variability include:
   - **Press Operating Data Sheet** – Ensures the press is set-up and run the same way every time. It is described in Section 19.2
   - **Anilox Roll Identification Code & Card** – By labeling each individual anilox roll with a unique code, the printer can ensure the same roll is put in each deck (not just the same cell count, volume, and angle). The card that accompanies the anilox roll includes manufacturer, date new, date last cleaned, cell count, screen angle, and actual cell volume measurement. The anilox roll card makes it easy to determine the last time a roll was cleaned, how old it is, and the latest volume measurement.
   - **Printing Plate Labeling & Card** – Each printing plate should have clearly imaged, in a non-print zone, the appropriate CMYK, Spot 1, Spot 2, etc. label. This provides a quick and easy check throughout the manufacturing process to confirm each plate is printing in the correct deck. Also, the plate card tracks date new, plate type, each date it is used, and the number of impressions on the plate. This information helps to determine when a plate should be replaced and the plate material to use when making a replacement plate.

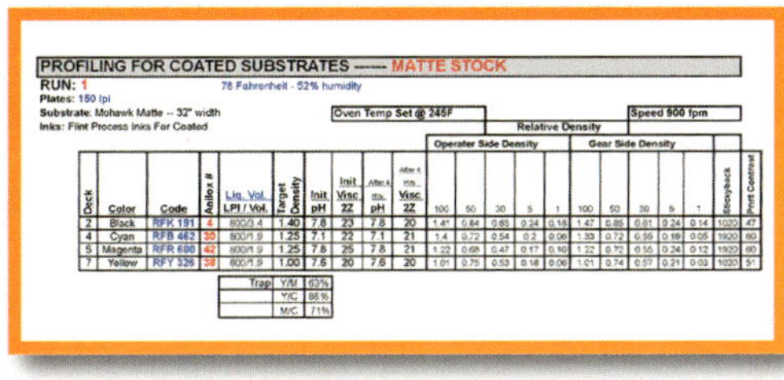

PROFILING FOR COATED SUBSTRATES ------ MATTE STOCK

RUN: 1 — 78 Fahrenheit - 52% humidity
Plates: 150 lpi
Substrate: Mohawk Matte -- 32" width
Inks: Flint Process Inks For Coated

Oven Temp Set @ 245F — Speed 900 fpm

										Relative Density											
										Operater Side Density					Gear Side Density						
Deck	Color	Code	Anilox #	Liq. Vol. LPI / Vol.	Target Density	Init. pH	Init Visc. 2Z	After 4 Hrs. pH	After 4 Hrs. Visc. 2Z	100	50	30	5	1	100	50	30	5	1	Strayback	Print Contrast
2	Black	RFK 191	4	800/3.4	1.40	7.8	23	7.8	20	1.41	0.84	0.63	0.24	0.18	1.47	0.85	0.61	0.24	0.14	1020	47
4	Cyan	RFB 462	30	800/1.9	1.25	7.1	22	7.1	21	1.4	0.72	0.54	0.2	0.06	1.33	0.72	0.56	0.18	0.05	1920	60
5	Magenta	RFR 600	42	800/1.9	1.25	7.8	25	7.8	21	1.22	0.68	0.47	0.17	0.10	1.22	0.72	0.55	0.24	0.12	1920	60
7	Yellow	RFY 328	38	800/1.9	1.00	7.6	20	7.6	20	1.01	0.75	0.53	0.18	0.06	1.01	0.74	0.57	0.21	0.03	1020	51

Trap		
	Y/M	63%
	Y/C	88 %
	M/C	71%

1.3.3.1 Process Documentation: *Documenting key run variables makes it easier to achieve similar print results run-to-run.*

1.3.3.1 Process Control Data Collection and Documentation

Throughout the press run, the press crew should measure and record, on a standardized form, spreadsheet, or database, the process control parameters and appropriate mechanical control parameters. This information is used "real-time" to control the printing process. There are two additional uses of this data:

Internal Communication – Collect data on key process and mechanical control parameters in a spreadsheet. In addition to the actual quality measurements, include columns for important identifying information such as: design code, order number, date, substrate, press, ink color, roll number, etc. These additional columns allow the data to be sorted into meaningful subgroups, facilitating a better understanding of the process capability. This database provides the foundation of knowledge required to pursue process improvement, which is detailed in Section 1.3.5

1.3.4 Press Characterization Test Form: *The IT8.7/4 Target is the key test element required for a press characterization trial.*

External Communication – A database of run data enables the printer to provide the customer job-specific "Certificate of Analysis" data, detailing the quality achieved on any given production run. Additionally, the printer is able to provide the customer a quarterly analysis by design (run-to-run quality analysis) or design-family quality analysis. It also makes it easier to locate and isolate "out of spec" material from the production floor or from inventory. A comprehensive database, and employees who understand how to sort and statistically analyze the data, can be a powerful sales tool. Summarizing process control data can clearly communicate to existing customers, or potential customers, a company's competitiveness in the marketplace (ie. what the company is capable of, and why it is better than the competition) in quantitative terms. It provides an additional "yard stick" to compete on instead of simply price and delivery.

1.3.4 Press Characterization

The press characterization trial utilizes the IT8.7/4 characterization target and applies only to process-color image reproduction. The IT8.7/4 target, printed during the press characterization trial, is used for color management calibration within the prepress system. The data derived from the printed IT8.7/4 target is used to create an ICC profile, which is then used to calibrate monitors, proofers and other color output devices to simulate the printing condition. It defines the relationship between input data and the printed result.

Characterization data is of little value unless the printing condition being characterized is known and can be repeated. In order for the various devices upstream to accurately simulate the appearance of the printed image, the characterization trial must adhere to the aim points and tolerances established during the press fingerprint trial. These aim points and tolerances must also be maintained during future production runs.

The fingerprint trial establishes the aim points and tolerances the printing conditions able to achieve under normal production conditions. The press operator is responsible for achieving the aim points, within established tolerances, on the characterization trial and all future production runs. Without this repeatability, the characterization trial is of little value. Aim points and tolerances should be established during the fingerprint trial for key process color quality attributes such as: solid ink density, dot gain/TVI, print contrast, gray balance and ink trap. Refer to Section 1.3.2 for additional information.

The result of a successful press characterization is optimized color separations and proofs that accurately predict the printed result. This leads to less make-ready time, reduced production waste, and improved press speeds while satisfying the customer's quality expectation.

1.3.4.1 Characterization Test Target IT8.7/4

FIRST supports using the industry standard, IT8.7/4 characterization target. This target is the result of collaborative efforts of several standards committees, and represents the best technical approach to a standardized target. The target can be formatted in a variety of ways to be read by various color measurement devices and sized to fit a variety of press sizes. Typically, the IT8.7/4 target is evaluated using an automated scanning spectrophotometer instead of manual measurements with a hand-held spectrophotometer. This not only saves time but also allows for the efficient measurement of multiple samples. Therefore, it is the responsibility of the application creating the target to provide the mapping between the physical location and the ID number of each patch as well as to confirm that each patch is larger than the aperture of the measurement device. *FIRST* recommends the printer confirm the target format and measurement device compatibility prior to printing.

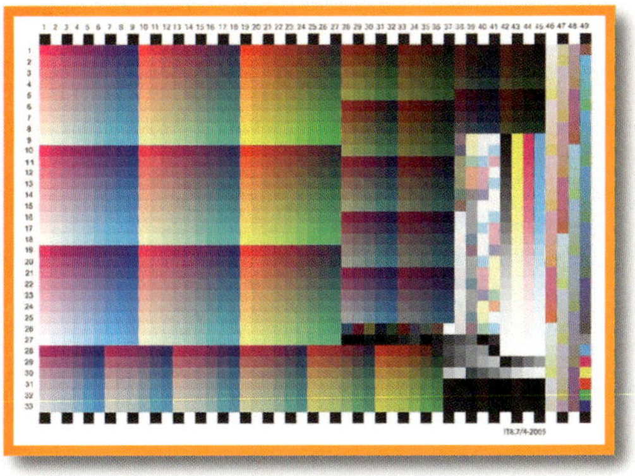

1.3.4.1a IT8.7/4 Target: *Ordered Layout*

1.3.4.1b IT8.7/4 Target: *Random Layout*

There are two layouts of the IT8.7/4 target, the "visual-layout" (also referred to as the "ordered format") and the "randomized-layout". Both contain the same combinations of CMYK patches, just in a different pattern. The random-layout target should be used for print characterization analysis. By randomizing the locations of the various CMYK combinations, ink takeout across the target area is more uniform. Errors introduced by press anomalies will be randomized with respect to CIELAB color space. The ordered, or visual-layout may be useful for those applications where manual measurement is required. *FIRST* recommends utilizing the full print width of the web to incorporate multiple random-layout IT8.7/4 targets. Ideally, the target should be oriented in two directions with a 90-degree rotation.

The IT8.7/4 target, both ordered and random layouts, contain the following elements:

1. **Solid Ink Patches:** Used to measure the colorimetric properties of yellow, magenta, cyan and black.
2. **Tint Patches:** Contain known tint values of the four single colors (CMYK). Twenty tint patches are included for each process color.
3. **Trap or Overprint of Solids:** Two-color overprints of

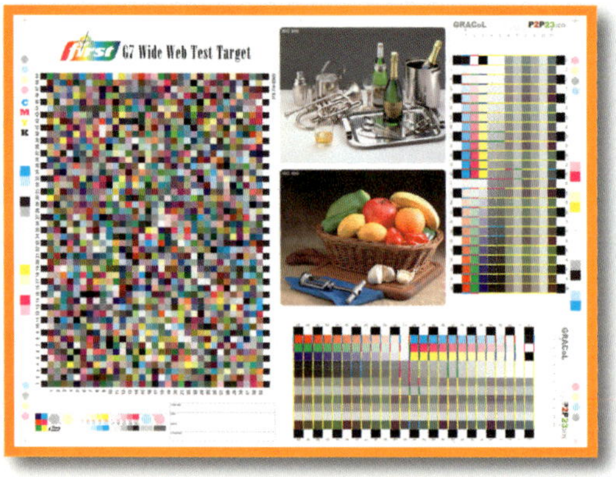

1.3.4.2 Press Characterization Test Form: *This is another example of a characterization test form containing the IT8.7/4 target.*

red, green and blue, are used to measure the colorimetric properties of the overprints.

4. **Gray Balance/Near Neutral Calibration:** Highlight, ¼-tone, midtone, ¾-tone and shadow gray balance patches are included on the IT8.7/4 target.
5. **CMYK Combinations:** Various two, three and four-color combinations at varying dot percentages of the process colors (CMYK) are included to colorimetrically quantify the full color space and determine the achievable color gamut.

In total, the IT8.7/4 target contains 1,617 elements or patches. The L*a*b* values of each element or patch are derived from measurements using a spectrophotometer. The data is used to build the press ICC profile that defines the color gamut that can be printed. The press ICC profile is used to calibrate the digital proofing system to match the printed gamut.

1.3.4.2 Characterization Print Trial

The accuracy of the characterization print trial is paramount to the successful implementation of color management because the data derived from the IT8.7/4 target is used to create the ICC profile, which will be applied to actual contract proof. In order for the ICC profile created from the characterization press run to be useful, it must be predictive of future press runs. Prior to the press characterization run, the printer must establish aim points and tolerances for press set-up and run conditions by completing the optimization and fingerprint steps detailed in Sections 1.3.1 and 1.3.2. A control target, as described in Section 19.2, must be included on all characterization plates. The printer should use the control target to measure and control solid ink density, dot gain/TVI, gray balance, print contrast and overprint traps during the characterization trial. Samples should not be pulled until the press is operating within the tolerances established during the press fingerprint trial.

Process documentation is key to success. The printing process must be standardized and the set-up repeated for each press trial: optimization, fingerprint and characterization as well as on future production runs. Key process variables that must be standardized, documented and repeated include: plate imaging, plate material, mounting tape, anilox count/volume/angle, ink deck, color, formulation, viscosity, and pH as well as substrate, tension settings, dryer and chill roll temperatures, winding and run speed. A Press Operating Data Sheet should incorporate all necessary information to ensure the press is set-up the same way each time. The conditions identified on the Press Operating Data Sheet are determined from the results of optimization trials and/or previous production experience. The quality measures, aim points and tolerances are determined during the fingerprint trial. Averaging the measurements from multiple targets taken from the same characterization press run, results in a more accurate profile. CGATS recommends evaluating at least six samples that are within the aim points of the printing condition. It is important to print at production speed for an extended period prior to pulling samples. Generally, place a flag in the rewind roll every 200 - 300 feet to cut down to and pull the samples after the print run is complete.

1.3.4.3 Characterization Data Collection and Documentation

Multiple press sheets (at least 6) pulled from the characterization press run, that are within the aim points of the printing condition, should be measured using an automated scanning spectrophotometer (if possible) and averaged together. This approach leads to more accurate results because it minimizes the "noise" that is typical of the printing process, caused by small variations in the printed target.

A report should accompany the characterization data and include a detailed description of the printing specification aim points and tolerances, materials used and actual run conditions. It should also summarize the data collection and analysis steps followed. Refer to CGATS "Developing a Color Characterization Data Set – Analyzing & Reporting". Refer to http://www.npes.org/standards/workroom.html website for availability. This report will address the issues of data analysis for the development of characterization data in detail.

A proof is not required for the characterization press run. The operator's job is to "run to the numbers" and achieve the print condition aim points, within the established tolerances. After a press profile has been created, it is a good idea to make a proof representing the press profile. The ordered-layout IT8.7/4 target makes it easy to visually identify areas that are inconsistent with the printed result.

1.3.5 Process Improvement

Process improvement uses statistical analysis of printing process data to improve quality, efficiency, and production yield. The database created as part of the process control step, is used to analyze the printing process to identify potential areas of improvement. Customer complaint data is another very useful tool when pursuing process improvement. Refer to Section 1.3.3 for additional information on process control.

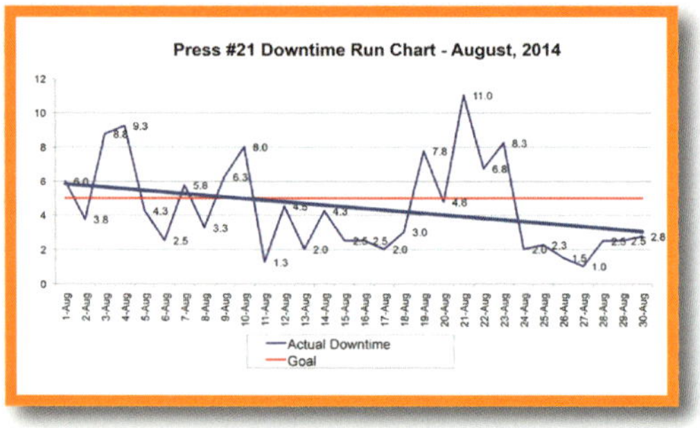

1.3.5 Process Improvement: *By measuring and controlling the printing process, process improvements, such as reduced downtime, become easier to achieve.*

Work with suppliers to identify potential methods to improve the process. Designed experiments can be created to determine the best method to improve the process.

By continually monitoring and measuring the printing process, and capturing the data in a database, it is easy to determine if the corrective action implemented in process improvement actually results in the desired process improvement over time and across jobs. It enables the printer to quantify the quality improvement of an investment in the printing process, which can be converted to dollars and estimated payback time.

The "Process Control Database" facilitates an understanding of the observed process trends, and cause-and-effect relationships that can be used to build a process knowledge that allows future jobs to be more efficiently specified and/or higher quality products to be produced with existing equipment. Pursuing process improvement without quantitative benchmarks of current performance, that can be used to quantify the level of improvement and cost-benefit analysis is highly inefficient and open to debate.

1.4 Communication and Implementation

Communication is not only the foundation of *FIRST*, but it also plays a critical role in implementing *FIRST* throughout the flexographic workflow. Through using standardized terminology in each defined workflow, *FIRST* seeks to minimize miscommunication to help promote the manufacturing of consistent, conforming flexographic print materials.

FIRST identifies the Package/Product Development Project Team as a key communication element for successful implementation. Communication among the members of this entire team plays a vital role in keeping the customer, designer, prepress provider and printer involved in the package development process.

1.4.1 External Communication

In order to provide you with the tools you need to foster communication throughout the entire team, we will break the Package/Product Development Project team into two categories: the external team and the internal team. By breaking the team into two categories, we can provide you with the terminology and tools needed to successfully convey the project goals of the external team (package/product development team) to the internal team (prepress artist, platemaking, plate mounting, press operators and quality control).

1.4.1.1 The Package/Product Development Process

The package development process, outlined in this section, defines the roles and responsibilities of each organization, as well as detailing the objectives of key meetings during the design development stage of the process. These meetings are recommended to aid in the collaboration process necessary for large product introductions. Introducing *FIRST* early in the production cycle helps to ensure specifications are followed throughout the workflow.

1.4.1.1.1 The Package/Product Development Timeline

The project schedule is directed by the customer in cooperation with all suppliers and usually coincides with product introduction dates, factory changes, or changes in government regulations. The project team should work together to refine the steps to accommodate the available timeframe. The *FIRST* Extras Download contains the customizable timeline outlined on pages 26-27.

The *FIRST* Package Development Timeline provides a general task list and fill-in-the-blank timeline for packaging development. The timeframe for each step is blank because it will change based on the individual company requirements and project scope. The tasks listed in the Package Development Timeline should be used as a starting point.

1.4.1.1.2 The External Package/Product Development Project Team

The primary role of each organization in the product development process can be summarized as:

- **Consumer Product Company** is the customer.
- **Graphic Designer** creates the original design file.
- **Prepress Provider** evaluates and applies the color characterization data to art file, provides contract proofs and makes image carriers (plates, cylinders, etc.).
- **Printer** mass-produces the printed product.

Superior print production is achieved by utilizing the skills of many professionals throughout the product development workflow. The *FIRST* product development project team may be comprised of, but is not limited to, the following organizations:

Consumer Product Company

Direct Responsibility: Packaging Director/Manager
Indirect Responsibility: Design Manager, Purchasing
Key Responsibilities:

- Design Brief
- Package Structure (primary and secondary)

Package/Product Development Process: Timeline

STEP	*FIRST* Section	Leader	Support Role	Duration	Start Date	End Date
Assemble Team: Designer, Prepress Provider, and Printer		CPC				
Competitive Store Audit	1.4.1.1.4	CPC	Designer			
Design Brief	1.4.1.1.3	CPC	Designer			
Identify Printer Capabilities (ink and coating requirements, substrate requirements, graphic capabilities)		CPC	Printer			
Select Substrates (General Guidelines)	3.2.2; 20.1.1	CPC	Designer Printer			
Press Optimization (as required by the design and press operating conditions)	1.3.1	Printer	Prepress Suppliers			
Template Layout (General Guidelines)	3.2.1; 12.5	CPC	Designer Printer			
Develop Concept: Type and Design Elements	4.0	Designer				
Develop Concept: Photography	5.0	Designer				
Template Layout (Detailed)	3.2.1; 12.5	Designer	Prepress Printer			
Design Refinement	1.4.1.1.5	Designer	CPC			
Regulatory and Legal Review of Content		CPC				
Pre-Production Meeting	1.4.1.1.6	CPC	Designer Prepress Printer			
Identify Substrate Specifications and Supplier (where applicable)	20.1	Printer	CPC Substrate Supplier			
Clarify Ink Performance Requirements and Optimize Ink Formulations	20.2; 20.3	Printer	CPC Ink Supplier			
Digital Design File	1.3.5	Designer				
Pre-Production Meeting (Mechanical)	1.4.1.1.7	CPC	Designer Prepress Printer			
Final Design Approval		CPC				
Bar Code Evaluation and Guidance (BWR, color combination, size, orientation)	22.0	Printer	Prepress			

Package/Product Development Process: Timeline *(continued)*

STEP	*FIRST* Section	Leader	Support Role	Duration	Start Date	End Date
Press Fingerprint and Characterization Trials	1.3.2; 1.3.4	Printer	Prepress			
Proofer and Press Correlation (as required)	16.0	Prepress	Printer			
Job Assembly (image trap, text elements, vignettes, bar codes, control target, template)	12.0	Prepress	Printer			
Color Separations	13.0; 14.0	Prepress	Printer			
Digital Mechanical File Preparation	1.4.1.1.8; 11.0	Prepress				
Contract Proof (proof-to-press characterization correlation, proof production, customer approval, CoA)	16.0	Prepress	CPC Printer			
Final Films/Mask/Files (dot shape, image stagger, registration marks, distortion, etc)	15.0	Prepress	Printer			
Printing Plates (production, evaluation, mounting)	17.0	Printer	Prepress			
Print Production: Process Measurement and Control (control targets)	1.3.3	Printer	CPC			
Package Converting	21.0	Printer	Converter			
Ship Finished Package to CPC Locations		Printer	Converter			

- Consumer Research Strategy
- Promotion Strategy
- Legal Copy
- Product Development
- Driving the process utilizing the *FIRST* packaging workflow

Design Firm

Direct Responsibility: Account Manager
Indirect Responsibility: Designer, Production Staff
Key Responsibilities:

- Design Strategy
- Final digital production art created in adherence to specifications
- QC all deliverables, providing process control run targets and verification documentation within the design

Prepress Provider

Direct Responsibility: Account Manager
Indirect Responsibility: Customer Service, Technical Representative, Production Staff
Key Responsibilities:

- Color separations created to specifications
- Film/file assembly created to specifications
- Contract/color proofing created to specifications
- QC all deliverables, providing process control run targets and verification documentation
- Platemaking (when applicable) conforming to specifications

Printer

Direct Responsibility: Account Manager
Indirect Responsibility: Design Manager/Art Director, Print Production Staff, Platemaker, Ink Company Representative and other suppliers as needed
Key Responsibilities:

- Press optimization, fingerprint, characterization, process control & improvement, as defined in Section 1.3 and in CGATS TR-012
- Platemaking (when applicable) conforming to specifications
- Producing the product to specifications
- QC finished product, providing process control & verification documentation

1.4.1.1.2 Package Development Responsibilities: *In short, the designer creates the image, the prepress provider manipulates the image, and the printer mass produces the image.*

It is important to bring the project team together as early in the design process as possible. *FIRST* suggests meeting after the initial comps are complete. Although the designs are not final, the group can begin to understand the overall project objectives and provide initial feedback on the designs. The communication in the group must focus on how the designs can be reproduced, instead of how the designs are impossible to complete within the specifications. Often alterations to design elements are required; this can be done while maintaining the intended graphic look and feel. *FIRST* specifications are not intended to limit production capabilities but rather to outline a controlled process and a common set of technical specifications to achieve customer satisfaction.

1.4.1.1.3 Design Brief Meeting

1.4.1.1.3 Design Brief & Project Initiation

Participants: Consumer Products Company, Design Firm

Purpose: The purpose of this meeting is to review all aspects of the product marketing and design objectives. Preliminary budget requirements and timing should also be discussed. This meeting should provide the information necessary for the design firm to develop a range of design options that meet the marketing and design objectives and comply with the production information provided.

Responsibilities: The Consumer Product Company supplies the design firm with all relevant information, such as:

- Project Description
- Product Name and Description
- Background (needed especially in the case of redesigns or sub-brands)
- Product Positioning and Promotional Strategy
- Product Design Objectives and Strategies
- Target Consumer
- Competition
- Product Messages
- Project Timing
- Communication Priorities
- Legal Copy
- Print Specifications (number of colors, die line, print process, printing substrate)
- Supplier Names and Contacts (prepress provider, printer(s), etc.)
- Utilization of *FIRST*

1.4.1.1.4 Competitive Store Audit: *Used to evaluate the competitive material and environment in order to optimize product positioning.*

The design firm creates design concepts that meet objectives while managing the client's expectations of color representation and reproducibility based on an understanding of production realities.

Deliverables: The design firm presents product design concepts as face panel designs on color concept proofs or agreed upon displayable electronic files (such as Adobe Acrobat display of PDF files).

Outcome: The Consumer Products Company selects one or two preferred design(s) that will advance to design refinement. Decisions are made regarding the timetable, project and design expectations.

Recommendations: To manage project costs and timeline, the number of concepts selected should be limited to one or two. Minutes of the meeting(s) should document the decisions reached.

1.4.1.1.4 Competitive Store Audit

Participants: Consumer Products Company, Design Firm

Purpose: The purpose of the competitive store audit is to evaluate the competitive material and environment in order to understand:

- Product category
- Packaging forms
- Color and category positioning norms
- Product merchandising

Outcome: The outcome of the audit provides the designers with a frame of reference and guides them to concentrate on design solutions that enable the product to visually compete within the specific store environment.

1.4.1.1.5 Design Refinement

Participants: Consumer Products Company, Design Firm, Prepress Provider, and Printer

Purpose: The purpose of this step is to refine the concept(s) selected by the customer for further development. Additionally, the design should address all panels and show the graphic placement of all copy that is required on the product. Additional refinement phases may be necessary to address issues such as side-panel design or promotional efforts.

Responsibilities:
The Design Firm:
- Produces a refined concept that achieves the client's design objectives
- Provides a 100% scaled proof of design concept for all product variations
- Confirms receipt of all information and material from client that is required for the development of production artwork
- Provides color targets to the prepress provider and printer if the design contains custom colors

The Consumer Products Company:
- Provides approval of refined design from primary decision makers within the CPC
- Coordinates the delivery of the final design concept proof to the prepress provider and printer, along with basic production information explaining the designer's intentions

The Prepress Provider & Printer:
- Analyzes the concept proof and provides suggestions for color reproduction and design integrity
- Recommends the contract proofing method

Deliverables: The approved concept proof is sent to the prepress provider and printer for further analysis. The project team must also agree and document the approved method for all contract proofs (including how to represent custom colors).

Outcome: The outcome of the Design Refinement Step is to select one concept, approved by the project team, for advancement to production art.

Recommendations: The minutes of the meeting, including all agreements reached, should be distributed.

1.4.1.1.6a Pre-production Meeting

1.4.1.1.6 Pre-production Meeting

Participants: The complexity of the project determines who should attend the meeting: Consumer Product Company, Design Firm, Prepress Provider, Printer and, as required, Package Converter, Ink & Substrate Suppliers (other suppliers as needed)

Purpose: The purpose of the Pre-production meeting is to critically review the final design. All project team members are involved to ensure a smooth transition from design to production. The design intentions and expectations are clearly defined in order for the team to determine the optimum

1.4.1.1.6b Pre-production Meeting: *The project team agrees on the production methods that will be used to create the mechanical art for the job.*

approach to creating mechanical artwork that achieves the marketing and design objectives. The meeting agenda includes these items for discussion:

- Product design objectives
- Design review
- Design specifications
- Product sizes (similarities and differences)
- Discussion of best production techniques for the design
- Number of colors needed
- Electronic assembly
- Contract proofing requirements
- Method to review, revise and approve color
- Timetable
- On-press approvals
- Image carrier/plates

Responsibilities:

The Consumer Products Company:

- Initiates the meeting
- Approves the approach chosen for mechanical art creation
- Ensures design and production procedures achieve marketing and design objectives

The Design Firm:

- Provides final design file and annotated concept proof
- Provides color targets for artwork

The Prepress Provider & Printer (Ink & Substrate Suppliers & Package Converter if applicable):

- Reviews concept proof and provide expert advice as to the best approach to producing final mechanical files
- Recommends contract proofing method that the client can approve and the printer is able to match

NOTE: Many of the topics reviewed at this meeting come from the printers' press characterization data. It is important that the printer fully understands and controls the press characterization process so the data is reliable and repeatable during production. Section 1.3 (and CGATS TR-012) outlines the steps necessary to understand and objectively define the color and tone reproduction capabilities of a printing process. These steps provide information required in the product development workflow summarized in this section (and CGATS TR-011).

Deliverables: The meeting minutes, which outline agreed upon mechanical art production methods, is distributed to all members of the project team.

Outcome: The project team agrees on the production methods to be used to create the mechanical art.

Recommendations: This is one of the most critical steps in the Product Development Workflow and should not be omitted. Production problems and delays occur when this step is skipped or the appropriate people are not involved.

The decisions reached during this meeting must be documented and distributed to all members of the project team.

1.4.1.1.7 Pre-production Meeting – Mechanical

Participants: Consumer Products Company, Design Firm, Prepress Provider, and Printer

Purpose: The purpose of this meeting is to review all information relating to the digital art production of the product prior to the execution of any mechanical production.

Decisions needed at this time include the number of colors on the job, the method of printing, and the designation of the printing substrate. Topics for discussion at this meeting should include:

- Separation (printer concerns)
- Specific trapping/output assembly issues
- Process control targets (placement of run and control targets)
- Press characterization requirements and layout (if necessary)
- Finishing requirements, review template: glue zones, cuts, scores, non-print areas, eyemark location, die lines, etc.
- Timetable
- Photo retouching (if necessary, who will execute, who will direct)

At this meeting, a final approved layout of the product and photography needs to be available for evaluation. The number of printers involved with the print production must be determined.

Outcome: The design file is released to the prepress provider for mechanical production. Meeting minutes should document all agreed upon decisions and be distributed to all project team members.

1.4.1.1.8 Digital Mechanical File Preparation

Participants: Design Firm, Prepress Provider, Consumer Products Company

Purpose: The purpose of this step is to create the digital files required for smooth and effective print production.

Responsibilities:
The Design Firm:

- Creates digital design files, in consultation with the customer, that represent the visual "look and feel" of a particular product, but are not production ready
- Delivers, to the prepress provider, the digital design files of the final product design, including the overall layout, text, low-resolution images, illustrations and identification of mechanical, assembly and die cutting requirements for the final product
- Constructs the digital design files in compliance with the prepress provider's electronic requirements; *FIRST* facilitates the process by providing common specifications among various providers

The Prepress Provider:

- Creates digital mechanical files from the digital design files in accordance with *FIRST* specifications
- Produces production-ready files that have been corrected and compensated according to the press characterization data

Outcome: The outcome of this step is a digital file ready for platemaking.

1.4.1.1.9 Contract Proof

Participants: Prepress Provider, Consumer Products Company, Printer

Purpose: The purpose of this step is to create a customer approved contract proof.

Responsibilities:
The Prepress Provider:

- Correlates the proofer to the press using the press characterization (IT8.7/4) and proofer profiles
- Creates the contract proof
- Generates "Certificate of Result" to accompany the contract proof.

The Consumer Products Company:

- Approves the contract proof (confirm accuracy of copy, image placement and color)

The Printer:

- Approves the contract proof (confirm the appropriate control target, die lines, etc. are applied to the job)

Outcome: The outcome of this step is an approved contract proof which functions as a communication document between all parties.

1.4.1.1.11 Corrugated Press

1.4.1.1.10 Final Films/Digital Mask/Plates

Participants: Prepress Provider and/or Printer

Purpose: To create the compensated image on the printing plate.

Responsibilities:
The Prepress Provider:

- Applies appropriate dot shape, image stagger, register marks, distortion, etc. to the job
- Makes digital or analog plates (if applicable)
- Constructs plate/die package if it is a corrugated job

The Printer:

- Makes digital or analog plates (if applicable)
- Inspects and mounts plates

Outcome: The outcome of this step is a finished plate package mounted and ready for press.

1.4.1.1.11 Print Production

Participants: Printer, Consumer Products Company, Prepress Provider (if acting as a representative for the CPC)

Purpose: The purpose of this step is to produce the printed material to the agreed upon specifications, as defined by the contract proof.

Responsibilities:
The Printer:

- Sets up the press consistent with the conditions during the fingerprint and characterization runs (determined with control targets on both the fingerprint and characterization runs and the production run)
- Maintains consistency throughout the production run (verified by process control data collection)
- Converts, either in-line or downstream (where applicable)

1.4.2.1 Press Set Up

The Consumer Products Company:

- Approves the job (either CPC or the prepress provider as CPC representative), if desired
- Releases the order and provides delivery requirements

Outcome: The outcome of this step is an award winning flexographically printed product/package!

1.4.2 Internal Communication

No one person can be responsible for a solid implementation of *FIRST*. Instead, everyone involved in the workflow must be accountable for the internal communication required for successful implementation.

A key element of effective communication is team building within and among the various teams involved in each project. Each team has many responsibilities and communication requirements in the development of the workflow. For example, printers are responsible for the communication related to critical elements of the workflow such as: calibration, standard operating procedures and day-to-day equipment maintenance.

1.4.2.1 Internal Communication and Packaging Workflow

In Section 1.4.1.1.2, we discussed the roles of the package product development team. In this section, we will break down the roles and responsibilities even further and identify what each team member needs to bring back to his/her element of the workflow. For example: we will focus on the printer. As *FIRST* explains, the responsibilities of the printer are as follows:

- Set up a press consistent with the conditions during the fingerprint and characterization runs
- Maintain consistency throughout the production run
- Convert, either in-line or downstream (where applicable)

Setting up the press consistent with the conditions during the fingerprint and characterization is critical. But what happens when we only fingerprinted the press on the first shift with one of our best press operators? If we are a 24-hour day, seven day a week production facility, will the other three press operators know how to setup the press consistent with the conditions during the fingerprint and characterization? When complying with *FIRST*, it is important to not assume that everyone is doing everything exactly the same way all the time. Rather, we must establish guidelines and procedures for everyone to follow consistently.

Perhaps just as important as the fingerprint and characterization process outlined in *FIRST*, optimization is the critical step that needs to happen just prior. The optimization process occurs prior to fingerprinting. The goal of the optimization process is to identify the best combination of print variables to achieve the design requirements under test conditions that represent normal production behavior and quality. It involves identifying the print variables that will produce the desired results with the substrate and ink colors specified by the customer. The optimization step must be completed for the intended graphics of each print deck (process/line/combo/solid). It is usually not necessary to perform an optimization test for every print variable. If certain variables are standardized, and if sufficient experience or historical data is available, re-testing is not necessary. Typically, a printer will standardize print variables through optimization print trials over time for each type of graphic (process/line/combo/solid) on common substrates.

Attempting optimization without having all of the members of the internal press team present is not recommended. The parameters that are included in the optimization print trial include anilox roll configuration, mounting tapes to be used, ink types, plate types and substrate types. In addition, it is during this optimization step that press housekeeping procedures as well as print station settings and press configurations are established.

1.4.2.2 Defining Internal Teams

Many factors are forcing flexographic printers to look for new ways to meet market demands quickly and efficiently. These factors include: the need for speed, the need to respond to vast technological change, the trend toward globalization, improved quality and increased market pressures.

One of the most efficient ways to implement significant change within a workflow is by utilizing process improvement teams. Adhering to the *FIRST* methodology is the same as implementing continuous process improvement.

To successfully implement *FIRST*, the knowledge, skills, experience and perspectives of the entire production environment must be brought together. Everyone involved in the production workflow must be involved in the implementation. This includes press operators, prepress operators, plate makers, plate mounters, ink technicians and floor supervisors; teams are a key element for implementing *FIRST*. Teams create

1.4.2.2 Production Meeting

environments in which participants can keep up with needed changes, gain skills in collaboration and install ownership into the new processes being implemented.

As a general example, teams outperform individuals when:

- The task is complex
- Creativity is needed
- The path forward is unclear
- More efficient use of resources is required
- Fast learning is necessary
- High commitment is desirable
- The implementation of a plan requires the cooperation of others
- The task or process is cross functional

When selecting the right members for each team, it is important to remember that each member will contribute something important and unique. Defining the primary objective(s) for each team at the beginning of the project/process will aid in selecting the team members. For example: One of the first steps in complying with *FIRST* is to establish the optimization parameters for the anilox rolls, ink densities, mounting tapes and all of the other variables that we will be using throughout our workflow. Since we know that the objective of the team involved is to set the optimization parameters, we know that we need to select team members who can address all of the variables involved in the optimization process.

Although the focus of this section is on internal communication and team building, it should be noted that *FIRST* recommends including a member of the external team (i.e. package product development team) to participate in internal team meetings whenever possible. The purpose of including an external team member is to continuously maintain communication among and between all involved team members.

In order to keep external teams, internal teams and individuals focused on the same objectives, it may be helpful to incorporate other continuous quality improvement tools and strategies like Six Sigma, Kaizen, SMED and LEAN (as well as others) in conjunction with *FIRST*. The purpose of incorporating such tools is to supply all involved members with common terminology, frames of reference and evaluation strategies that help each team and member understand other team's and member's points of view. Once all of the teams and members are speaking the same language and using the same set of tools, it is much easier to maintain open channels of communication.

Further, using progress reporting and data collection strategies outlined in various continuous quality improvement methodologies fosters communication among and between teams and members as the strategies require all teams and members to become active participants in the process.

Finally, a key element in fostering and maintaining communication among and between teams and members is using effective meeting planning and facilitation strategies. Knowing when to hold meetings, who to invite to the meetings, what the meeting objectives are and who is responsible for the action items at the end of the meeting is critical for maintaining communication and ultimately successfully implementing *FIRST*.

1.4.2.3 Internal Team Roles and Responsibilities to *FIRST*

Each internal team plays a vital role in a successful implementation of a *FIRST* compliant workflow. For example, when assembling teams in a pressroom, it is often essential to include press operators from each shift working on the same press, plate mounters, ink technicians and possibly a graphics person or two. The purpose of assembling teams with these types of members is to provide them with the skills and knowledge that relates to each workstation's standard operating procedure. The primary function of assembling teams with these types of members is to tear down the walls that divide departments/work areas.

In addition to breaking down the barriers that divide departments, this structure also provides some consistency in the way each individual is performing his/her job functions. This eliminates the finger pointing, by giving the internal teams the tools to track down problems by using data collected.

1.5 *FIRST* 5.1: *Approaching the printing process as a science, rather then an art, results in a more predictable, consistent and repeatable product.*

1.5 Applying Specifications to Art Through Print

Beginning the journey to better printing involves adopting and implementing *FIRST* specifications. Customers and suppliers will enjoy the benefits of an improved relationship when all components of the product development process are clearly specified, accurately measured and controlled. Moving printing from an art to a science results in a more predictable, consistent and repeatable product.

Establishing a process to verify conformance to specifications by all parties (including the customer, designer, prepress provider, and printer) is critical. At each stage of the product development process, the receiving party should check all in-

1.6 *FIRST* Logo

coming materials (art, plates, proofs, etc.) to verify conformance to specifications. The receiving party may choose to require a CoA from the supplier, confirming conformance to the agreed upon specifications. In order to achieve the customer's quality and cost objectives, the deadline may need to be adjusted when non-conforming materials are passed down the supply chain. The specific reason for a rejection must be documented and communicated to improve the process and avoid future rejections.

1.6. *FIRST* Certification

Quality is integrated into all aspects of the *FIRST*. The *FIRST* methodology involves the following five-step process adopted from CGATS TR 012-2003 (Graphic Technology – Color Reproduction and Process Control for Packaging Printing): optimization, press fingerprint, process control, press characterization, process improvement. These five steps work together to build outstanding, repeatable results. Each step builds upon the previous step to play a vital role in quality. Step three, process control, is particularly central to quality. The vast majority of focus on quality stems around the notion that if we can control a process, we can control its outputs. If we can control the outputs, we can repeat the process, over and over. Process control is truly foundational to producing a quality product.

1.6.1. *FIRST* Operator Certification

The *FIRST* Operator Certification program, administered online via the TEST Virtual Campus, offers three, 3-level certification programs. Each program leverages proven e-learning technology to empower working professionals to hone their expertise and give their plant a decided advantage over the competition.

Choose to become *FIRST*:

- Press Operator Certified
- Prepress Operator Certified
- Implementation Specialist Certified

Regardless of which designation you choose, you will:

- Maximize efficiency
- Boost productivity
- Achieve consistent, repeatable and color-accurate results
- Gain a decided advantage over the competition
- Increase and sustain customer satisfaction
- Win repeat orders
- Build enthusiasm, initiative and innovation among employees

FIRST Press Operator Certification

This certification is designed for press operators wanting to gain an expert proficiency of the flexographic printing process. After successfully completing this 3-level program, students will have a solid understanding of *FIRST* and how it applies within their daily routine of press operations, specifically how to:

- Achieve optimal results for consistency and color/tone accuracy
- Measure and control print variables
- Effectively manage inks, substrates, plates and other press components
- Identify methods to reduce waste
- Utilize process control techniques throughout the workflow
- Evaluate press conditions and perform proper maintenance
- Streamline communication with internal/external customers and suppliers
- Enhance personal productivity during press make-ready
- Ensure each element within the workflow complies with *FIRST* methodology

FIRST Prepress Operator Certification

This certification is designed for prepress personnel wanting to gain an expert proficiency of flexographic prepress operations. After successfully completing this 3-level program, students will have a solid understanding of *FIRST* and how it applies within their daily routine of prepress operations, specifically how to:

- Achieve optimal results for consistency and color/tone accuracy
- Measure and control print variables
- Effectively manage digital assets throughout the workflow
- Perform calibration procedures as they relate to the graphic artist
- Utilize process control techniques throughout the workflow
- Streamline communication with internal/external customers and suppliers
- Enhance personal productivity as it contributes to the workflow
- Ensure each element within the workflow complies with *FIRST* methodology

FIRST Implementation Specialist Certification

This certification is designed for non-production personnel, including supervisors and managers, as well as suppliers of

1.6.1a *FIRST* Press Operator Certification Logo

1.6.1b *FIRST* Prepress Operator Certification Logo

1.6.1c *FIRST* Implementation Specialist Certification Logo

COMPANY CERTIFICATION

1.6.2 *FIRST* Company Certification Logo

flexographic products and services. After successfully completing this 3-level program, students will have a solid understanding of *FIRST*, and how to communicate both internally and externally to assure ongoing workflow compliance. Specifically, students will learn how to:

- Achieve optimal results for consistency and color/tone accuracy
- Cultivate a culture of communication among internal/external team members
- Develop internal workflow teams to establish ownership of the *FIRST* process
- Orchestrate the functions of the package/product development team
- Enhance and optimize workflow productivity
- Perform print evaluations and compose documentation procedures
- Ensure each element within the workflow complies with *FIRST* methodology

Renewing *FIRST* Operator Certification

To maintain an active certification you must renew you *FIRST* (Press, Prepress or Implementation Specialist) Certification every three (3) years. If your certification is currently up for renewal, you can take the online *FIRST* Recertification Exam. This is an open-book exam, based on the methodology supported by the most current edition of *FIRST*. The 50-question exam contains a variety of questions that will assess your understanding of the flexographic print process. Subject matter includes: optimization, fingerprinting, process control, characterization, continuous improvement, expanded gamut, measurement devices, best practices and more.

1.6.2. *FIRST* Company Certification

The purpose of the *FIRST* Company Certification Program is to recognize flexographic printing companies that are applying *FIRST* methodology and have attained compliance with the specifications and tolerances related to Communication and Implementation, Design, Prepress and Print as detailed in the most current *FIRST* document.

FIRST Company Certification is valid for a three-year period provided the company:

- Complies with required on-site audit criteria
- Maintains the required number of certified staff
- Complies with the *FIRST* workflow methodology through adhering to the ongoing data submission requirements

Eligibility
In order for a company to be eligible for *FIRST* Company Certification, the following criteria must be met:

- 40% of pressroom staff must be *FIRST* Press Operator Certified
- 40% of prepress staff must be *FIRST* Prepress Operator Certified
- 40% of management/production staff must be *FIRST* Implementation Specialist Certified

Additional criteria which is outlined in full in the *FIRST* Company Certification Policies document include:

- *FIRST* Workflow Compliance
- Workflow mapping
- Device/equipment calibration schedule/Manufacturer's Recommended Operating Parameters (MROP) documentation
- Workflow tagging
- Tonal value aim points

Maintaining *FIRST* Company Certification
To sustain a certification for the three-year duration, a company must verify its ongoing commitment to workflow compliance through the weekly uploading of workflow data to an FTA designated portal. The purpose of this is to ensure the business is maintaining its *FIRST* recommended workflow compliance throughout its tenure as a *FIRST* certified entity.

Renewing *FIRST* Company Certification
FTA will notify a company six months in advance of its *FIRST* Company Certification expiration date. It is the responsibility of the company seeking recertification to submit the renewal application form in enough time to allow for FTA to process and plan for an on-site audit prior to the expiration date. The process for renewing the *FIRST* Company Certification is identical to the initial certification process.

For detailed information regarding *FIRST* Certification visit the FTA website at www.flexography.org

DOWNLOAD FIRST 5.1 Extras Referenced in this Section at:
http://www.flexography.org/FIRST_extras

2.0 DESIGN INTRODUCTION

2.1 Overview

FIRST was created to facilitate communication among all participants involved in the design, preparation and printing of flexographic materials. The designer is responsible for creating a graphic design that achieves the marketing objectives of the Consumer Product Company (CPC) and that can be successfully reproduced on press. The Design Section is intended to assist the designer in understanding the flexographic print considerations necessary to create reproducible designs. The better the designer understands the flexographic process, the easier it will be to create aesthetically pleasing designs while optimizing production efficiency and reducing the time-to-market. A primary objective of the Design Section is to provide guidance on how to create electronic files that will enhance quality and speed of manufacturing while minimizing cost. This must be accomplished while allowing the designer to maintain creative control of the project. This can be best accomplished when everyone in the supply chain has a clear understanding of the requirements of flexography and when these requirements are addressed during the design phase of development. Because designers and production artists often have overlapping responsibilities, the information in this section applies to both parties.

Depending on the methods and practices of the companies involved and the complexity and frequency of the work among them, *FIRST* recommends establishing ground rules and procedures for designing products before actual production begins. This is a necessary step when providing services to the flexographic industry because of the complexity of the graphics, print issues and converting equipment considerations. A dialogue regarding design and production considerations should be initiated among the production team (designer, consumer product company, prepress provider and printer). *FIRST* provides guidelines to facilitate the project flow through the design and manufacturing processes.

2.2 Responsibility

As packaging graphics continue to increase in complexity and production timelines continue to compress, clear assignment of responsibilities is necessary to ensure a quality printed product in a timely manner. The assignment of responsibilities requires planning and collaboration among all involved parties.

Consumer Product Company (CPC): Ultimately, the customer defines expectations and therefore must drive the collaboration

2.2 Product Development Responsibilities: *In short, the designer creates the image, the prepress provider manipulates the image, and the printer mass produces the image. All members of the supply chain must work together utilizing FIRST to achieve a desirable outcome.*

process. The customer determines the effort expended to reach satisfaction. The CPC must facilitate communication between the supply chain parties: designer, prepress provider and printer.

Designer/Production Design: The designer must work with both the prepress provider and the printer to understand the capability of the printing/converting process being utilized. Based upon the print capability, the designer must provide a design concept that will enable the printer to meet the expectations of the customer (CPC). The earlier in the design development process the prepress provider and printer are involved, the better the team is to determine specific capabilities that will ensure the final product meets the customer's design objectives. Additionally, the designer is responsible for:

- Establishing a color scheme and palette before final files are sent to production
- Checking all copy for spelling and kerning
- Treating common elements and logos consistently in the layout
- Building all copy and vector-based elements in accordance with the specifications of the print provider
- Image positioning

Prepress Provider: The prepress provider must work with the printer to understand the capability of the printing/converting process being utilized. The prepress provider supplies the designer with accurate and timely information regarding print capabilities at the beginning of the design phase to facilitate the creation of a printable design. Based upon the print capability, the prepress provider produces appropriate films/files/plates that will enable the printer to meet the expectations of the customer (CPC). They must document the controls that ensure the consistency and accuracy of the supplied media (films/files/plates). Additionally, the prepress provider produces a contract proof calibrated to accurately predict the printed result. The prepress provider must give the printer the ability to objectively confirm the accuracy of the prepress work and the printing process. This can be accomplished through the use of agreed-upon control targets.

Printer: The printer is responsible for consistently reproducing the graphic design to the satisfaction of the customer (CPC). They utilize and document the process controls necessary to ensure that accuracy and consistency are achieved. They work with the other parties and suppliers to define the capability of the printing process. The printer provides the designer with accurate and timely information regarding process capabilities at the beginning of the design phase to facilitate the creation of a printable design.

2.3 Assumptions

In order to keep the content focused and pertinent, the following assumptions were made when creating these guidelines:

- The audience consists of professionals who are using current versions of software and hardware (designers who expect their project to efficiently move through the production workflow should be using current versions of software and hardware proven compatible with downstream processes)
- Certain programs and manufacturers are mentioned (*FIRST* recognizes these are not the only solutions)
- The audience is familiar with electronic design terminology and workflow in a digital environment (if you are not familiar with electronic design terminology and/or digital workflows, visit www.flexography.org for more information)
- Technology continues to change rapidly (to help address this issue, additional training and support documentation will be updated and available at www.flexography.org)

3.0 GETTING STARTED

3.0.1 General Outline/Definition of a Creative Brief and Style Guide

Creative Briefs and Style Guides play an important role in the development and ongoing continuity of packaging designs. In the simplest terms, a Style Guide is a set of standards for a brand (the do's and don'ts of the brand identity), while a Creative Brief provides specific details concerning the execution of a single package design (ie. a set of objectives for a package).

Not all projects will be accompanied by either of these documents, but it is common for larger brands, or the agencies that represent them, to have both. Regardless of the size of the brand or project, access to the information contained in documents such as these makes things easier for everyone in the supply chain. The following describes typical categories of information that make up the contents of both Creative Briefs and Style Guides respectively.

Creative Brief

Project Title	
Client:	(Main contact/decision maker)
Key approvers:	(Legal)
Project Manager:	(Coordinator/account manager)
Objective:	(Purpose of materials and desired outcome of project)
Messaging:	(Supplied copy, priorities/hierarchy, translations/languages)
Supporting Info:	(Target audience, market/trend research, competition, brand/product info)
Budget:	(Estimated costs with approval)
Project Scope:	(Services, quantities, exclusions)
Deliverables:	(PDFs, comps/mockups, presentation boards, prototypes)
Timeline:	(Concept deadlines, production/delivery schedule)

Style Guide

Content	
Overview:	(Standards)
Building our Brand:	(Alternate heading)
Mission/Vision:	(Meaning/purpose)
Logo/Identity/Brandmark:	(Consistency)
Tagline:	(With and without)
Lockup:	(Spacing/positioning)
Primary Branding:	(Logo/lockup)
Secondary Branding:	(Icons/alternate colors)
Full Color Logo:	(Versions)
One Color Logo:	(Versions)
Logo Variations:	(Special usage)
Proper Usage:	(Size, clearance)
Improper Usage:	(Modified logo)
Color Palettes:	(Pantone/CMYK)
Secondary Colors:	(Pantone/CMYK)
Specifications:	(Print vs. web)
Font Usage:	(Specified)
Brandmark Placement:	(Newsletter, blog, website)
Stationery:	(Collateral)
Signage/Collateral:	(Logo/color palette/icons)
Apparel:	(Alternate style)
Proposals/Presentations:	(Format/medium)
Case Studies:	(Format/results)
Acquiring/Sending Assets:	(Login/download)

3.1 Recognizing Attributes of the Flexographic Printing Process

The use of spot colors, specialty inks and a wide variety of substrates are just a few choices available with flexography. Designers must be informed about the advantages of the flexographic printing process in order to make use of them during the design process. The designer must communicate with the print provider to understand their capabilities and how they can jointly optimize the quality and effectiveness of the final product.

3.1 Flexographic Market Segments: *The flexographic printing industry offers designers broad choices of packaging types, substrates, inks and in-line converting capabilities.*

3.2 Materials and Information Needed to Begin

- Template or Die line: A die template or drawing (supplied by the customer, prepress provider or printer) must include bleeds, glue or heat seal areas, live areas and dimensions. There may also be other pertinent information on the template (ie. die number, size, count number, etc.) that the designer should reference in the digital file.
- Production information gathered by the design team such as substrate, number of ink colors and whether the specified color is a spot or process color build should be documented in the digital file
- Customer specifications
- Design brief
- Brand style guide and corporate art guidelines
- Legal and government regulations

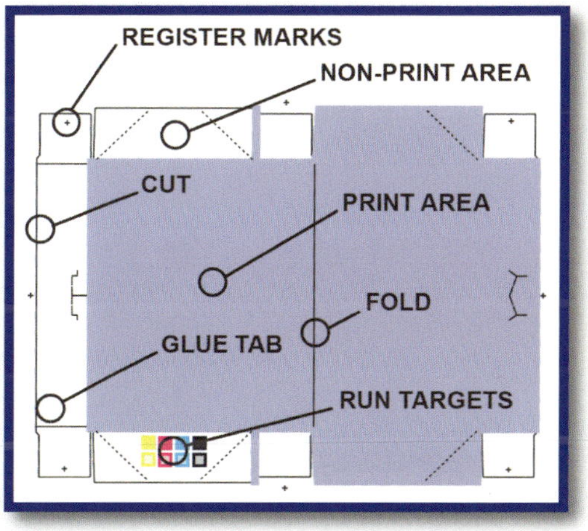

3.2 Materials & Information Needed to Begin: *Template layouts along with general production information and customer specifications are critical for successful design development.*

3.2.1 Template Layout/Die-Cut Specifications

Die line/Electronic File

A final die line or electronic file must be provided with the art, prior to final assembly, for all die-cut jobs. All supplied die lines must indicate cuts, folds and scores as well as non-print areas. The designer, in conjunction with the packaging buyer, should indicate the area in which the print control target may be placed. Refer to Sections 1.3.3, 3.7 and 12.7 for print process measurement and control.

Using the Template Layout

A template layout is also referred to as a keyline, die line or full-scale drawing. It is the responsibility of the printer and the customer (CPC) to provide the design firm with the appropriate electronic template file, including layout dimensions, prior to the conceptual design phase. The template should include non-image area, non-print area, print direction, varnish area, seal area and "inside view" identification. It is the responsibility of the design firm to consider the non-print areas during the design process.

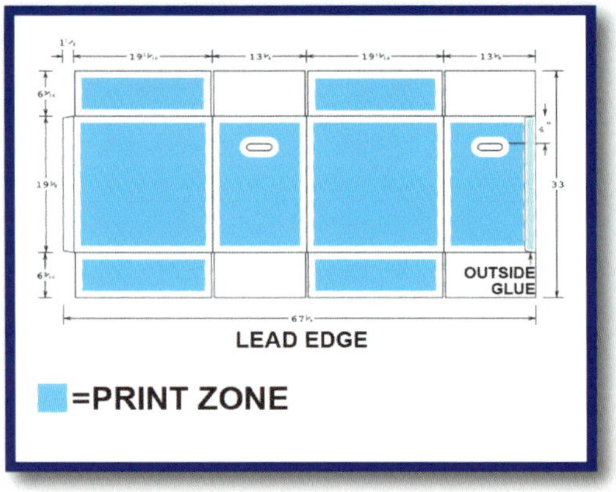

3.2.1 Template Layout: *It is the responsibility of the design firm to consider the non-print areas during the design process.*

The designer forwards the final template to the prepress facility where all job elements are verified and correctly positioned for product assembly. Refer to Section 12.5 for additional information.

Die Origin

Dies are designed using a graphics program or CAD system. Files generated from these systems can be translated into a format recognizable by design and prepress software. Incorporation of dies, bleeds, or pressmarks (internal and external) should be determined on a case-by-case basis. Early communication about who will build a die line and how it will be used is essential.

Printing Form Layout Considerations

The printing form layout communicates how individual die-cut units are arranged on a sheet or web. This may influence control target placement and create additional design considerations. If certain knives are common, or shared, between individual units, the design may be affected at the perimeter of the unit. This information can only be obtained through contact with the printer. Designers must work with the customer (CPC) and the printer to receive this vital information.

Print-to-print and print-to-cut production tolerances should also be verified with the printer or the customer (CPC). These tolerances may vary depending on several factors including press width and press type (ie. central impression, stack, in-line). Important elements should be placed away from cuts and scores. Die position tolerance is typically smaller for thin board stock and larger for thicker stock. Consult the printer for job specific print-to-print and print-to-cut production tolerances.

Electronic Format

It is important for the designer to work with an accurate physical representation of the unit's form to avoid downstream adjustments to the design. Sometimes the die is modified to match graphic elements (windows, cutouts, or coupons). Most translation programs provide a link from the more common package design programs to CAD formats (ie. DXF, DDES2, IGES, PDF). The structural designer should indicate what formats can be produced.

Measurement of Die Drawings

Indicate measurements on the electronic die line file including the dimensions and marks for the live print area.

3.2.2 Print Substrate

A sample of the substrate should accompany the project as soon as it is available. The whiteness, color and texture of the substrate should be considered. Printing on foil or colored paper, or printing white behind the graphics, will influence the printed color gamut. Often, the colors on the printed product will deviate from the approved contract proof if the proof is not made to reflect the substrate and/or printed white ink. White ink can appear darker (dirtier) and typically less opaque than white paper or film. In addition, various packaging substrates exhibit different color properties when printed. For example, some paper substrates will inconsistently absorb ink producing a 'muddier' image.

3.3 File Naming Conventions

Alternate versions of an electronic file should have separate and distinct names from the original version. File naming conventions for live, high-resolution images should be in accordance with the criteria of the collaborating parties. For example, workflow may dictate file names, SKUs, job numbers, or UPC references. When naming a file, special characters such as "!", "@", "#", "$", "%", "/", "\" and "*" should never be used. Suffixes identify and distinguish formats and variations of working files. Examples of this are as follows:

asparagus.tif/asparagus.eps/asparagus.psd
or
abcdefgh.raw/abcdefgh.rgb/abcdefgh.cmyk

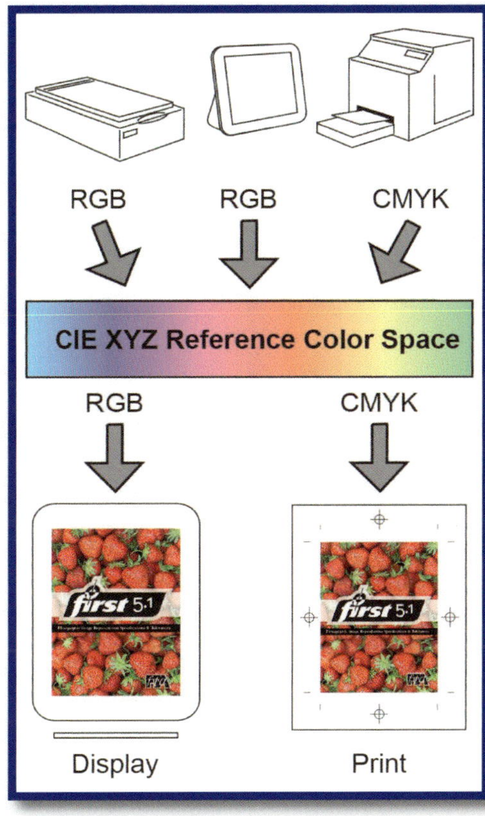

3.4 Color Management: *Color Management Systems (CMS) are a collection of software tools that quantify and reconcile the color differences among monitors, scanners, image setters, proofers, and printing presses to ensure consistent color throughout the reproduction process.*

3.4 Understanding Color Management

The number of colors the average human eye can perceive is much larger than the number of colors that can be reproduced on workstation monitors, proofing devices and printing presses. An important key to understanding color management is to have a familiarity with the concept of color space. Digital cameras and scanners record images in the RGB color space, while proofing devices and film/plate setters output images in other color spaces such as: CMYK, or expanded gamut (ie. CMYKOGV).

Color Management Systems (CMS) are a collection of software tools that quantify and reconcile the color differences among monitors, scanners, imaging devices, proofers and printing presses to ensure consistent color throughout the reproduction process. Typically, the available color gamut diminishes as a job progresses through the production cycle. A CMS will map colors from a larger gamut and indicate what colors are achievable in a device with a smaller gamut, such as a printing press. This process allows for realistic expectations to be set during the proofing process.

3.5 Standard Viewing Conditions:
Standardizing viewing conditions between the customer, designer, prepress provider and printer will minimize color discrepancies.

Although digital tools can make the process seem as simple as a click of a mouse, converting from one color space to another is the first place where color fidelity and contrast can be significantly compromised. Once information is lost in the conversion process it cannot be restored. Even when sending an RGB image to a digital proofing device, there is an automatic conversion. The proof is actually a CMYK rendering that was run through default color management settings unless a more specific profile has been generated and applied.

Each color output method has limitations based on the type and number of colorants, the imaging engine, colorant delivery technology and the substrates being used. The more a designer understands these limitations, the better the design concept is managed. In the event that a known output source (a specific printing press) is identified prior to the creative stage, the photographer/designer may contact the prepress provider and request a color profile, referred to as an "ICC (International Color Consortium) profile," for that print condition. An ICC profile of a standardized color space such as GRACoL 2006 can be utilized, allowing the prepress to synchronize press conditions back to this standard. With this profile, the designer can control the conversion process more effectively. Section 14.4 outlines a more comprehensive explanation of color management.

3.5 Viewing Artwork, Proofs & Printed Material

Application: A color-viewing booth is used to view printed images, proofs, or transparencies under a controlled and standard light source. Accurate and consistent visual perception of color requires the image to be viewed in a standard, chromatically neutral, controlled environment. If the designer, printer, prepress provider and customer standardize viewing conditions color discrepancies can be minimized.

Industry Standard: *FIRST* supports the standards set for proper viewing conditions in ANSI 2.30 1989. However, *FIRST* recognizes the light source may not be optimal for all print segments. The designer should consult with the customer to identify the preferred color viewing conditions.

Instrument Agreement: The illumination used in the light booth should be the same as the equipment illuminant setting. For example, if the measurement equipment (spectrophotometer) is using D65 instead of D50, the light booth should use 6500° Kelvin bulbs (D65) instead of 5000° Kelvin bulbs (D50).

ANSI 2.30 - 1989: Color Viewing Conditions	
Parameter	Specification
Lighting	5000° Kelvin Bulbs
Light Source	Color Rendering Index >/= 90
Viewing Surface	Luminance = 204 +/- 44 Foot Candles
Viewing Area Color	Neutral Gray: Munsell N8 or E
Angles of Illumination	Minimize Glare

Table 3.5

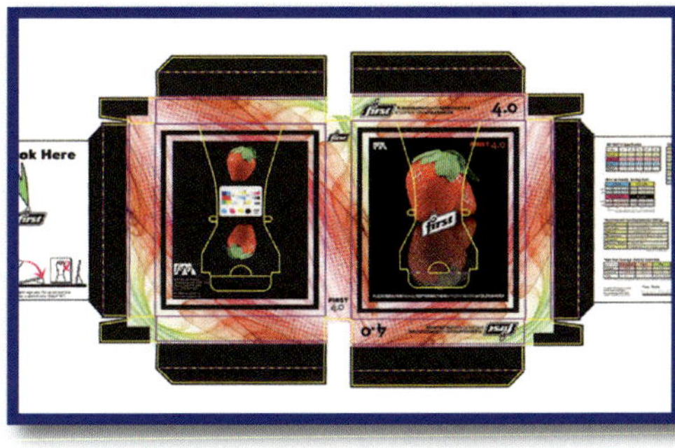

3.6a Profiled Contract Proof: *The contract proof must include a control target as well as template layout markings.*

Communication: Regardless of the settings used, it is important that the settings are communicated and agreed upon by all parties receiving data measurements.

3.6 Types of Proofs

All parties involved with a project must agree upon the process and terminology used to evaluate and communicate the design, including color. Specifically, every proof created throughout the workflow should be clearly labeled to communicate:

- The purpose of the proof
- The system or device on which it was created
- Whether the output device was profiled and which profile was used
- The proof's suitability for judging color

Types of Proofs

Concept Proof: The concept proof is common in the early creative stages of the project. It is used to capture input from all partners in the supply chain during initial design development and is also referred to as a "collaborative proof". This proof is typically not color profiled, therefore not used for matching color.

Color Target Proof: The color target proof is often the selected "concept proof". It represents the ideal color intent of the designer and client, independent of the print process or the ability of an individual press to achieve that color. Some of the color in this proof may not be achievable in the final print. To avoid rework costs and unachievable expectations downstream, it is helpful when possible, to produce this proof based upon the known or expected capabilities and color gamut of the anticipated printing process(es).

3.6b Type of Proofs: *Before a contract proof can be accurately used, the entire reproduction system must be characterized so that the proofing system is calibrated to match the printed result.*

Comprehensive Proof (Comp)/Mock Up: The comp is formed to the shape of the final product and should indicate whether or not it is color accurate.

Profiled Contract Proof: This represents what the customer is expecting to receive off press. The profiled contract proof represents the customer's complete content and color expectations for the final printed product and is the basis for negotiations on project performance. It illustrates how the printed image is expected to look when reproduced on press and is an important quality control tool and communication device. It is profiled using a color management system (CMS) and is prepared using a profile provided by the specific printer or prepress provider and produced according to *FIRST* specifications. The contract proof does not have to be a dot-for-dot reproduction, but it must be an overall visual simulation of the expected print results. Therefore, it must simulate the dot gain/TVI, color attributes, detail and contrast of the printed image. It must also contain a control target that is processed and imaged as part of the proof. The control target is used to verify accuracy and consistency throughout the design, proofing and printing process. It must contain specific screen values, which should be determined with the printer, for all colors printing dots (including vignettes). Although most digital proofing devices may not reproduce a conventional dot pattern, the tonal scales should be measured using a densitometer (or spectrodensitometer) in the dot area function. Each one of the tonal scales must equal the weight (dot area) identified by the press profile. Before a contract proof can be accurately used, the entire reproduction system must be characterized so that the proofing system is calibrated to match the printed result. Afterward, both press and proofing systems must be maintained for consistency and repeatability. Refer to Section 14.0 Process Color Calibration, for additional information on profiling.

A "Proof Compliance Cover Sheet" or label must accompany the contract proof submitted for color match at press once approved by the customer. It should identify the proofing product or system used and the company supplying the proof (contact name, telephone and fax numbers) as well as operator, date, job number and customer. The cover sheet must also contain information needed to verify the proof's compliance to the technical attributes required for that proofing type. Refer to Section 16.5 for more information. It is a best practice approach for all proofs to include a "Certificate of Result". It should include all pertinent measurements: density, dot area, Delta E (@ 100% and 50%), trap, print contrast, bar code scan analysis, etc. Proof densities should be within the printer's on-press density specifications. The Proof Compliance Cover Sheet and Certificate of Result can be combined into one document. Refer to Section 19.4.4 for *FIRST* guidelines on solid ink density by print segment.

Soft Proof: The soft proof consists of viewing a job on a color-calibrated monitor. It is used at any point in the product development process from a concept proof to a contract proof, depending on how well the system is calibrated. Components include a color consistent monitor and a color management system (CMS).

3.7 Process Control Test Elements

Application: If consistency and repeatability are important to the customer, then space must be allocated on the sheet, web, or package for appropriate process control test elements. Measuring at set-up and throughout the run enables the printer to produce repeatable, consistent and accurate results on every job. The test elements used to measure the print characteristics outlined in Sections 12.8 line work and 12.9 process color work, can be used for print optimization and fingerprint trials as well as on every "live" jobs to facilitate process control. The test elements included will vary based on the print characteristics that are pertinent to the job being printed and space constraints. Using similar test elements on the fingerprint trial as on live production jobs enables the printer to verify current print conditions and flag any changes since the press was last fingerprinted. Refer to Section 1.3 for a detailed explanation of print optimization, fingerprint, and characterization trials.

Placement: In order for the printer to deliver the desired print results, the customer and design team must include key test elements in the product design. Some packaging lends itself to placing test elements under flaps, in a glue zone, or on the waste matrix; other packaging requires the test elements to remain visible on the finished package. Therefore, each print application should determine where to place the individual elements to be monitored throughout the production run. The designer should consult with the printer and CPC on the necessary test elements and properly place them on the package/sheet/web when creating the design.

Format: Sections 12.8 and 12.9 describe the key print characteristics for both line and process work, and the test

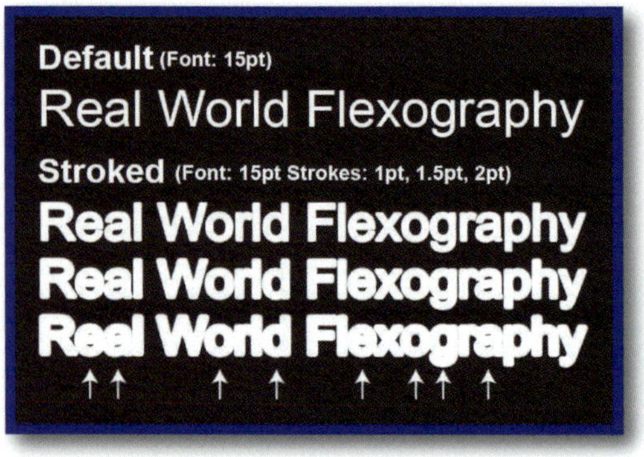

4.1a Typography: *If type is stroked, swelled, or framed to increase its thickness, the "counters" may fill in. Type can be stroked to increase its thickness, but the "counters" (holes in letters such as a, d, o, e and R) may fill in, so care must be used.*

element used to measure each characteristic. Previous editions of *FIRST* have supplied the *FIRST* control target. Beginning with this edition, all of the test elements discussed in Sections 12.8 and 12.9 will be supplied for construction into a suitable control target, optimization or fingerprint test design for each print application. The test elements are available to all members and nonmembers through the FTA as an electronic file and are included in the *FIRST* Extras Download folder. Sample run targets are also included for review but should not be considered more than working examples of what can be used.

Test Element Construction:

Size: The designer must be careful to allocate enough room for the necessary elements of the process control target. ANSI/CGATS.5 (2003 Graphic Technology – Spectral Measurement and Colorimetric Computation for Graphic Arts Images) provides the minimum and recommended apertures (and therefore minimum test element size) specified by line screen listed in the following table. While these guidelines are useful, the print application must also be considered. The minimum acceptable aperture may be larger for some print applications. The designer and prepress provider should confirm individual test element size with the printer. For direct-print corrugated, each test patch (solid or tint) should be 2-3 times the flute width to provide a stable measurement target.

CGATS.5 Densitometer & Spectrophotometer Aperture Size					
Screen Frequency		Round Sampling Aperture (mm)		Non-Round Sampling Aperture (mm)	
Lines per inch	Lines per cm	Minimum	Recommended	Minimum	Recommended
65.0	26.0	3.8	5.7	11.3	25.5
85.0	33.0	3.0	4.5	7.1	15.9
100.0	39.0	2.6	3.9	5.3	11.9
120.0	47.0	2.1	3.2	3.5	7.8
133.0	52.0	1.9	2.9	2.8	6.4
150.0	59.0	1.7	2.6	2.3	5.1
175.0	69.0	1.4	2.1	1.5	3.5
200.0	79.0	1.3	2.0	1.3	3.0

Table 3.7

Imaging: All test elements must be imaged at the same time and with the same care and accuracy as the live job. The test elements must be imaged at the same line screen, angle, dot shape, etc. as

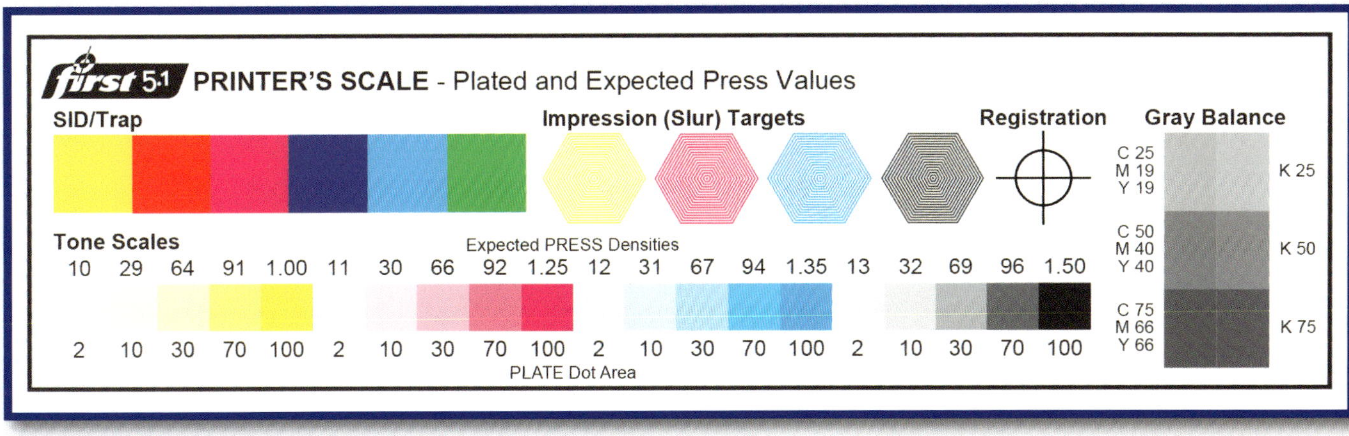

3.7 Process Control Test Elements

the actual image. Surprinting, plate slugs, or plate build up of the test elements is not an accurate representation of the live image area and are, therefore, not acceptable.

Special attention must be given to imaging tone scales. Refer to Section 12.9.2 for a detailed explanation of the type of tone scales required on press trials and production runs.

Process Control Test Elements: *FIRST* recognizes certain press configurations (narrow web) and product types (ie. poly bags, envelopes and newsprint) may not have large enough trim areas or glue zones to maintain all recommended process control elements throughout the production run. On these products, the test elements used to verify density and at least one dot area should be placed on the live area of the product to remain consistent throughout the press run. The more test elements included on production jobs, the better equipped the printer is to achieve the desired print result. Ideally, these five test elements should be on all process color jobs:

1. Registration: color-to-color and print-to-cut
2. SID/Trap
3. Tone scales
4. Impression: anilox-to-plate and plate-to-substrate
5. Gray balance

4.0 TYPE AND DESIGN ELEMENTS

4.1 Typography: Know the Print Process Capabilities

Due to the nature of the flexographic process, text that prints positive will tend to fatten while text that is reversed out will tend to fill in (lose fine lines and Serifs) and become plugged. Therefore, when selecting fonts, care and attention is critical.

4.1b Minimum Type Size: *Using type sizes below the printer's minimum recommended size can result in type filling and is not supported by FIRST.*

Minimum Type Size: General Guidelines
Minimum type size is print system dependent; determine minimum type size with press fingerprint (ref. 1.3.2)

Segment		Substrate	Positive		Reverse		Printer Specific			
							Positive		Reverse	
			Serif	Sans Serif	Serif	Sans Serif	Serif	Sans Serif	Serif	Sans Serif
Wide Web	Preprint Linerboard	All	8 pt	6 pt	10 pt	8 pt				
	Combined Corrugated	White Top	8 pt	6 pt	10 pt	8 pt				
		Coated Paper	6 pt	4 pt	8 pt	6 pt				
	Folding Carton	All	6 pt	4 pt	8 pt	6 pt				
	Multiwall Bag	Coated Paper	8 pt	6 pt	12 pt	10 pt				
		Uncoated Paper	10 pt	8 pt	18 pt	12 pt				
	Film Products	Polyester	8 pt	6 pt	12 pt	10 pt				
		Polypropylene, Polyethylene & Metallized	8 pt	6 pt	10 pt	8 pt				
	Newsprint	Uncoated Paper	10 pt	7 pt	11 pt	10 pt				
Narrow Web	Paper Products	All	6 pt	4 pt	8 pt	6 pt				
	Film Products	All	6 pt	4 pt	8 pt	6 pt				
	Envelope	All	6 pt	4 pt	8 pt	6 pt				

Table 4.1a

Tables 4.1a and 4.1b provide general guidelines by flexographic print segment. Because the minimum type size and rule width are print system dependent, the designer should confirm rule width and type style and size with the print provider.

When attempting to increase the weight of a Serif font, it is not always effective to use the bold, heavy, black, or ultra versions. When fonts are changed to a heavier version, verify the text did not reflow. Type can be stroked to increase its thickness, but the "counters" (holes in letters such as a, d, o, e and R) may fill in, so care must be used. Refer to Section 12.2 for additional information on text elements.

Type Size Considerations

Serif vs. Sans Serif: Sans Serif can be printed at a smaller type size than Serif print. Sans Serif type stays cleaner because it does not have the fancy details on the ends of the letters that tend to fill-in and run together at smaller sizes.

Positive vs. Reverse: Positive type can be printed clearly at a smaller type size than reverse type. Reverse type is more

Minimum Rule Width: General Guidelines *Minimum rule width is print system dependent; determine minimum rule width with press fingerprint (ref. 1.3.2)*						
Segment		**Substrate**	**Positive Rule**	**Reverse Rule**	**Printer Specific**	
					Positive Rule	**Reverse Rule**
Wide Web	Preprint Linerboard	All	0.010"	0.015"		
			0.254mm	0.38mm		
	Combined Corrugated	White Top	0.013"	0.020"		
			0.33mm	0.51mm		
		Coated Paper	0.007"	0.010"		
			0.18mm	0.254mm		
	Folding Carton	All	0.006"	0.008"		
			0.15mm	0.20mm		
	Multiwall Bag	Coated Paper	0.007"	0.010"		
			0.18mm	0.254mm		
		Uncoated Paper	0.013"	0.020"		
			0.33mm	0.51mm		
	Film Products	All	0.007"	0.013"		
			0.18mm	0.33mm		
	Newsprint	All	0.007"	0.013"		
			0.18mm	0.33mm		
Narrow Web	Paper Products	All	0.005"	0.010"		
			0.13mm	0.245mm		
	Film Products	All	0.004"	0.008"		
			0.10mm	0.20mm		
	Envelope	All	0.007"	0.010"		
			0.18mm	0.254mm		

Table 4.1b

vulnerable to ink volume and impression settings resulting in type filling in and becoming illegible.

Single-Color vs. Multi-Color: Single-color type can be printed clearly at a smaller type size than multi-color type. Multi-color type size is restricted by the press registration tolerances.

Design Variables: Other variables that influence minimum type size includes: ink coverage, substrate absorbency and compression, etc.

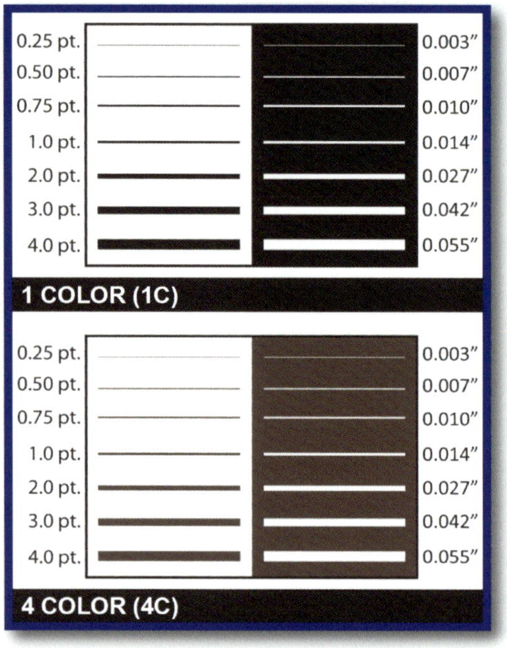

4.1c Line Weight: *The acceptable line thickness will vary depending upon whether the line is positive or reverse printing and whether it is a single color or multicolor line.*

4.1.1a Image Trap: *When trapping two colors, FIRST recommends "spreading" or enlarging the lighter color under the dominant color.*

4.1.1 Registration Tolerance

When one word is printed in one color and another word next to it is printed in a second color, register shifts can cause these two words to overlap or misalign. Due to this register shift, different color text should be more than twice the image trap dimension away from each other. Table 4.1.1 Total Trap Tolerance provides general trap guidelines by print segment. Confirm the trap tolerance with the print provider.

Total Trap Tolerance: General Guidelines
Trap tolerance is print system dependent; determine minimum trap with press & print optimization (ref. 1.3.1) and press fingerprint trials (ref 1.3.2).

Segment			Color-to-Color	Printer Specific
Wide Web	Preprint Linerboard	Total Trap	</=0.0156" (1/64")	
			</=0.3969mm	
	Combined Corrugated	Between Station	</=0.0625" (1/16")	
			</=1.5875mm	
		Through the Press	</=0.125" (1/8")	
			</=3.175mm	
	Folding Carton	Total Trap	</=0.0156" (1/64")	
			</=0.3969mm	
	Multiwall Bag	Total Trap	</=0.0313" (1/32")	
			</=0.7938mm	
	Film Products	Total Trap	</=0.0156" (1/64")	
			</=0.3969mm	
	Newsprint	Total Trap	</=0.0156" (1/64")	
			</=0.3969mm	
Narrow Web	Paper Products	Total Trap	</=0.0156" (1/64")	
			</=0.3969mm	
	Film Products	Total Trap	</=0.0156" (1/64")	
			</=0.3969mm	
	Envelope	Total Trap	</=0.008" (1/125")	
			</=0.2032mm	

Table 4.1.1

4.1.2 Process Color Type

When identifying colors for text copy, the designer should be aware which colors would be built from process and which will use dedicated spot colors. In general, text copy should be printed with a single color or built from two process colors. As text size increases, a third process color may be introduced. Using more than one color to create text should be discussed with both the prepress and print providers to determine capability.

4.1.3 Process Reverse/Knockout

A holding line should be used when type is reversed and comprised of more than one color. The holding line should be a single, dark color to hide any slight misregistration that is likely to occur during the printing process. The weight of the holding line should be twice the registration tolerance for the print segment as identified in Table 4.1.1 Total Trap Tolerance. Because the values provided are general guidelines, the designer should confirm the trap requirements with the prepress and print providers.

If a holding line is not used, the darkest or predominant color should be made full size and the remaining color must be choked back the width of one row of dots as determined by the screen ruling. If possible, the background color should be limited to one color.

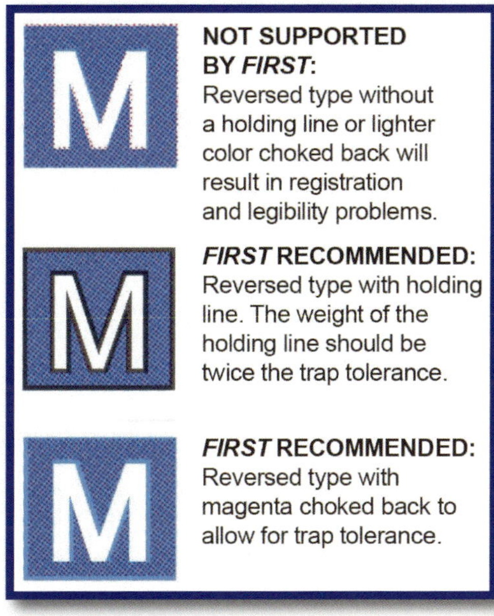

4.1.3 *FIRST* Process Reverse/ Knockout Recommendations

4.1.4 Line Reverse/Knockout

Reverse copy should be limited to one color. If copy is to be reversed from two or more colors, a holdback or choke must be created for register. Refer to Table 4.1.1 Total Image Trap Tolerance and the specific print segment. Because the values are general guidelines and print system dependent, the designer should confirm the trap requirement for reverse text with the prepress and print provider.

4.1.5 Drop Shadow

If a drop shadow is abutting another color, it will need to trap. Be sure to move the drop shadow by more than twice the specified image trap for the appropriate print segment. Refer to Section 4.1.1 for segment specific guidelines on total trap tolerance.

It is best to use drop shadows only for larger type, unless the color selected for the type is darker than the color it is abutting; remember, these abutting colors will be required to overprint each other to form the image trap.

Drop shadows that fade should be limited to a single color to allow special screening to support the light tones of the gradient. Refer to Section 7.5 for additional information on blends/ vignettes/gradations.

4.1.5 Drop Shadow: *If inappropriate image trap tolerances are applied, objectionable type will result.*

4.1.8a Font Utility Programs: *There are many font utility programs to help manage fonts effectively.*

4.1.6 Spaces and Tabs

Always use tabs rather than multiple spaces to position text. If a font change is required, the spaces will change size, while tabs will not change.

4.1.7 Text Wrap

Most programs will wrap text around imported images. If an image is replaced in production, text will reflow if automatic text wrapping features were used to define the text wrap area. Use the polygon tool or other shape to define the text wrap or run around instead of letting the text automatically wrap around the image. When the high-resolution image is placed into the file, the program may see its edges differently and rewrap the type. The prepress provider will have to rebuild the desired wrap to get the text to reflow the same way.

4.1.8 Fonts

It is possible for a font to have the same name but exist in different file formats. For example, two different companies that make the font (sometimes called foundries) may name the font the same. Substituting a different font file format may cause the text to reflow and change the original design.

Fonts may be selected and used from a variety of sources. It is possible for a font to vary in appearance or performance in downstream operations based upon its source. For that reason, it is recommended that, in addition to the original file, a copy of the file be supplied with type converted to outlines.

PostScript/Type 1

A PostScript font is a Type 1 font and is created from two components: a printer font and a screen font. The printer font contains the outlines that allow the output device to accurately render the font in any size. The screen font allows the font to be viewed on a computer screen (monitor). Type 1 fonts require both pieces to work properly. PostScript fonts are the de facto standard for professionals in the creative and print environments.

OpenType Fonts

There are several advantages to the OpenType format. First, as with TrueType, the entire font is housed in a single file. Second, this file is cross platform, the same file can be used on a Mac or Windows platform with consistent results. Third, an OpenType font can contain either PostScript or TrueType outline data. Lastly, OpenType can support Unicode information, which can contain thousands of characters including high quality ligatures, swash glyphs and other advanced typographical features. This is a significant benefit over PostScript Type 1, which is limited to 256 characters.

Manufacturers
Sometimes downstream companies (such as prepress providers and printers) working on a design file may not have easy access to fonts used. If so, the design firm (or whoever is creating the content) should convert these fonts to outlines or paths.

Styles of Fonts
In some applications, there is a style menu with type attributes such as bold, italic, outline, shadow, small caps and all caps. Do not use this feature. Use only the actual font, such as Times Bold, rather than Times with the bold attribute. When using attributes, results vary depending on the RIP, printer drivers and application being used. Selecting style attributes usually creates a pseudo version of the typeface, which is a degradation from the original font design. Many newer RIPs, printer drivers and applications ignore pseudo commands and simply use the plain printer font. For example, if the italic command from the style menu is selected for Humanist 541 Condensed Bold (which has a corresponding printer font), the font will display as condensed bold italic on screen but will typically not print in italics.

Outline Effect
To create an outline only, use a vector program and give the type a stroke in the desired color and a fill of 'none' or 'white'. To stroke only the outside, use a copy of the type with no stroke and a white fill exactly on top of the stroked copy. Be sure the stroke is at least twice the specified image trap for the applicable print segment. Refer to Table 4.1.1 for print specific total trap tolerance guidelines.

Proprietary Fonts
Fonts designed for a specific client or job are considered proprietary and should be included with the submitted files for the job.

Other Font Architectures
Multiple Master, TrueType GX and other font architecture should be avoided. If their use is unavoidable, confirm the prepress provider can work with the required font architecture.

4.1.8b Outline Effect: *To create an outline only, use a vector program, and give the type a stroke in the desired color, and a fill of 'none' or 'white'. Be sure the stroke is at least twice the specified image trap for the applicable print segment.*

4.1.8c Converting Type to Outlines: *Type converted to outlines minimizes font problems but cannot be edited.*

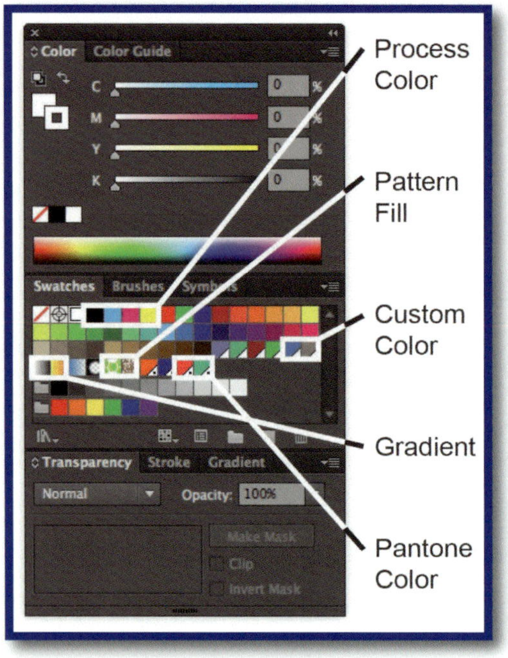

4.2a Custom Colors: *Most products are printed with colors other than CMYK. Correct usage of "custom colors" can expedite the production process.*

4.2b Color Proof vs. Production Files: *If a file includes custom colors that overlap to create a third color, produce two files: One file to produce a color comp proof, and a second file for production plates.*

Poorly Written Fonts

Poorly written fonts may be node heavy (built with too many points), have bad kerning pairs, or incomplete character sets. They should be avoided. If there is a typeface that absolutely must be used, test it first through an imaging device. If using a font that is not available from the output supplier, convert it to outlines. If the font is public domain, send it with the files.

Supplying Type Fonts

To avoid copyright infringements or unauthorized use of type fonts, the licensing responsibility resides with both the creator of the file and the company outputting the file. The creator must check with the supplier of the fonts to confirm that the license held allows the fonts to be used by both the creator and the output supplier.

Converting Type to Outlines

A common practice for handling type is to convert type to outlines in order to prevent font problems and lock content. However, this makes the text no longer editable and may alter its appearance. When converted to outlines, small type may appear heavier and should be reviewed prior to the final conversion.

- When a file with outlined type is supplied, it is advisable to also send a copy of the original file (including fonts) prior to outlining the type
- Electronic files (.ai, .eps, .psd) containing text that are to be placed in another document, should also have all text converted to outlines (fonts in placed images often are not reported as missing until the file is RIPed)
- Converting fonts to outlines helps identify poorly written or corrupt fonts

4.2 Custom and Special Colors

"Custom Colors" as defined in a file should represent only the actual inks, or tints of those inks, that will be printed.

A designer should specify or confirm the actual colors that will be used on press. Many products are printed with both spot colors and process colors. Correct identification of "custom colors" versus colors built from process inks, can expedite the production process. A file containing 15 or 20 custom (spot) colors is not printable; therefore, requires the prepress provider to attempt to interpret the intentions of the designer.

In some programs, the designer can specify whether a custom color is meant to be created using a CMYK (process color) mix, or single custom color ink. The designer must be sure the color specification is clearly indicated. On the annotation layer, it must be specified how each color is created. Using industry standard ink color designations such as Pantone®, TOYO®, etc., will assist with proper color communication and allow standard colorimetric data/values to confirm the final match.

CMYK equivalents of custom colors do not always match. If the custom color is to be built with process colors (CMYK blend), the prepress provider must know if they are expected to use exact percentages or if they are responsible for verifying that the necessary tints are used to match as close as possible to the custom color callouts.

It is not uncommon for special colors to be used in process illustration, either as an enhancement or as a replacement for one of the traditional process colors. In these cases, special separation and proofing techniques are required.

Differentiating White Ink from Unprinted Areas

If white is to be an ink, a custom color is created and used to specify which areas print white, as opposed to not printed. This color should be named "white ink" in the color palette. To further distinguish areas that are to be left unprinted, create an additional color named "Unprinted" or "Clear." Either the white ink or the unprinted area needs to be filled with a differentiating tint.

Custom Color Proofing: Color Proof Files vs. Production Files

If a file includes spot colors that overlap to intentionally create a third color, it is necessary to set the top color transparency to "multiply". This will display a created third color.

The best way to predict the third color result of overprinting two spot colors is to have the printer (or the ink supplier) create overlapping ink drawdowns of the two inks. If it is necessary to create a proof that accurately represents the overprint, it may be necessary to create a separate proofing file with the color of the overprinting area defined by CIELab data obtained from the overlapped portion of the ink drawdowns.

4.3 Bar Code Design Considerations

Formerly, the Uniform Code Council (UCC) was responsible for managing the bar code system in the USA. The UCC is now the GS1 US organization. GS1 US manages the GS1 system and assigns GS1 company prefixes to companies/organizations in the USA. The most common use of a GS1 assigned company prefix is the creation of UPCs (Universal Product Codes), which contain a 12-digit Global Trade Item Number (GTIN).

4.2c The Pantone Matching System: *The Pantone Matching System (PMS) is a common way to specify custom colors.*

9 01234 56789 8

UPC - A

9 012345 678906

EAN - 13

901234567890

GS1 - 128

901234567890

Code 39

4.3.2a Bar Code Type: *The type of bar code depends on many factors including where it will be scanned and how it will be printed.*

The GS1 US publishes the following electronic data interchange guidelines based on the ANSI ASC X12 standard:

- Industrial/Commercial EDI
- Uniform Communication Standard (UCS), used in the grocery industry
- VICS EDI, used in the general merchandise retail industry

The GS1 US is also the code manager for the United Nations Standard Products & Services Code (UNSPSC). The UNSPSC provides an open, global, multi-sector standard for classification of products and services. Identify applicable commodity codes on UNSPSC website (www.unspsc.org).

For more information on prepress and print considerations for bar codes, refer to Sections 12.8.4 and 19.3.4 respectively. The GS1 US & UNSPSC contact information is included in Appendix A.

4.3.1 Bar Code Specifications

Bar code print specifications are produced by combining three types of related specifications:

1. Application Standards are published by accredited standards organizations.

Bar codes are used in many different applications with different scanning conditions. For example, one application involves packaging for retail check out lanes while another application is for coding shipments for conveyor lane routing in distribution centers. The specifications for bar codes used in these two applications are different because the conditions for scanning the bar codes are different. Accredited standards organizations (refer to Appendix A) provide specifications in the form of guidelines and standards to assist in:

- Selecting the bar code type to be used
- Structuring the data inside the bar code
- Defining the printed human-readable information that is inside the bar code
- Selecting bar code size within the acceptable range
- Understanding where the bar code should be placed on the product
- Defining the minimum print quality requirements

2. *FIRST* Print Specifications prescribe a minimal level of capability for all compliant printers.

These specifications fall within the acceptable limits of the appropriate Application Standard for the bar code being printed and will assist in:

- Determining the minimum size for a bar code depending on the printing press and substrate
- Identifying the preferred bar code orientation given the direction the web or sheet will travel

3. Job Specifications should be published for film or plate output.

These specifications should assist in:

- Identifying optimum film/plate output resolution
- Determining bar width reduction (BWR) required by the specified print conditions.

4.3.2 Designer Responsibilities

The designer, prepress provider and printer all bear responsibility for producing quality bar code symbols. Designers play a critical role in assuring a bar code conforms to all applicable Application Standards and *FIRST* Print Specifications. When creating an FPO (for position only) symbol, the designer must determine and communicate the symbol type and size, the color(s) used to print it, as well as the location and orientation on the printed product. Refer to Section 12.4 for prepress bar code considerations and Section 19.3.4 for bar code print considerations.

Because designers are often involved in the substrate and color selection process, as well as the bar code placement, orientation, and size determination, they should be aware of the design parameters for bar code performance. The designer should consider if the current design specifications might create scanning problems. Common design revisions requested because of the selected substrate or color include: a larger symbol, a different symbol orientation, an extra layer of background ink, or a dedicated bar code print station

1. Selecting the Appropriate Symbology

The type of bar code selected depends on many factors including the Application Standard, where it will be scanned, and how it will be printed. The designer must defer to the customer to identify which bar code type to use. Some of the common bar code types printed flexographically include:

- UPC--Version A and Version E (including add-on and composite component)
- GS1-128 (formerly known as UCC/EAN-128)
- EAN 8 (including composite component)

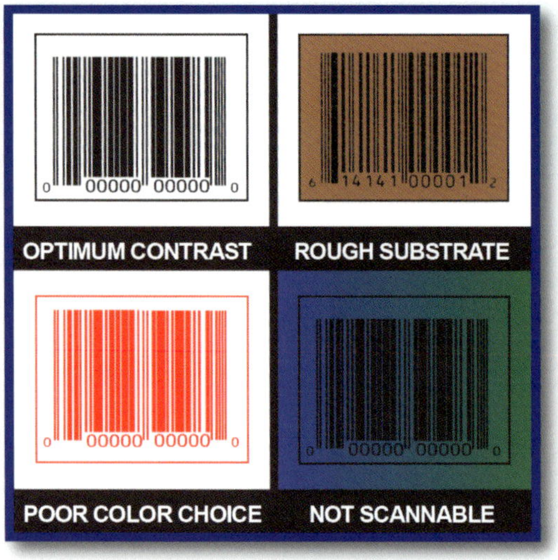

4.3.2b Color Considerations: *The optimum bar code color combination is opaque black ink for the bars and opaque white substrate or ink for the background.*

- EAN-13 (including add-on and composite component)
- ITF-14 (Interleaved 2-of-5 also referred to as Code 25)
- Code 128 (full ASCII character set supported)
- Code 93 (full ASCII character set supported)
- Code 39 (supported with and without check code)
- MSI (including option to display data)
- JAN 13 (variation of EAN 13 used in Japan)
- JAN 8 (variation of EAN 8 used in Japan)
- Plessey (hexadecimal character set)
- Telepen (including compressed numeric mode)
- 2D Codes
- Codabar (both USS and Traditional format supported)
- USPS 4CB (United States Postal Service Intelligent Mail Barcode)

2. Designing for Printability and Symbol Contrast

Substrate Considerations

Texture & Porosity: Bars and spaces are most accurately produced on smooth substrates with high ink holdout. The rougher, more textured and more porous a substrate, the greater the potential for printing bars with voids and/or printing specks in the spaces, either of which can reduce scanning rates. Textured and more porous stocks also tend to increase bar edge roughness, bar growth and bleeding. Any of these substrate characteristics can negatively influence scanning rates.

Color & Transparency: Bar codes scan most successfully with an opaque white background that provides white spaces and quiet zones with the maximum reflectance possible. When printing on a transparent or colored substrate, a solid light-colored (white is optimum) background, with maximum opacity, is recommended in the area where the bar code is to be located. Special consideration for the background ink formulation and press setup (anilox, double bumps of background color, mounting material selection, etc.) may be necessary in order to achieve maximum opacity.

Color Considerations

The optimum bar code color combination is opaque black ink for the bars and opaque white substrate or ink for the background. Bars printed in opaque black, dark blue, or dark green and backgrounds (spaces and quiet zones) printed on an opaque white material or on a white, red, orange, pink, peach, or yellow ink generally scan successfully. It is important to remember that colors with acceptable ANSI/ISO Symbol Contrast on an opaque substrate may not be acceptable on an opaque substrate of another color or on a translucent or transparent substrate. When printing on a transparent substrate or colored substrate, a solid light-colored (white is optimum) background with maximum reflectance is recommended in the area where the bar code is located. It is recommended that the bar code symbol not be placed on a printing plate used to print a large solid ink coverage. Printing plates that print large solid areas typically have requirements for extra impression and higher ink volume, which are not conducive to printing bar codes. Ink color specifications should be evaluated individually for different substrates.

Bar codes require bars with sharp edges in order for the scanner to perform successfully. Because scanning accuracy is reduced when variation in register occurs, the bars comprising a bar code must be printed in one color, using a solid line image on a single print station. Refer to Sections 12.4 and 19.1.3 for more detailed information on bar code color considerations.

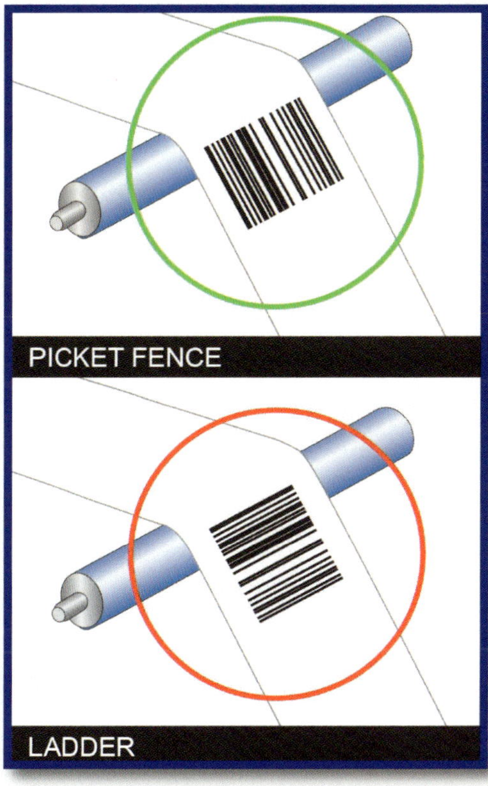

4.3.2c Bar Code Orientation: *Bar code orientation is critical. The left figure illustrates the bars on the UPC symbol traveling in the machine direction. The right figure illustrates the bars running across the press direction.*

3. Determining Optimal Size and Location

Location Considerations

Bar codes are placed in different locations based on the shape of the product and where the product will be scanned. The designer should check with the product manufacturer for placement specifications based on these factors. The designer should also consult with the package engineer to ensure the symbol will not be creased, scored, sealed, or folded. Placement of the codes in these areas may cause the ink to crack, producing voids in the bars or spots in the symbol background. Correct placement of the bar code is crucial to meet regulations and for accurate scanning.

Orientation Considerations

It is strongly recommended that the bars in a bar code be printed parallel to the direction the web is moving through the press to avoid slurring. In certain situations, the bars in a bar code must be placed in the transverse (across the web) direction. In these cases, the printer should be consulted. It may be necessary to use a larger symbol to meet the minimum print quality requirements specified by the appropriate application standard. If print slur occurs with the symbol printing in the machine direction, the bars grow in length only and are still scannable; however, if the symbol is printed in the transverse direction, the bars will grow in width, likely causing the code on the printed product to fail to meet specifications. Printing bar codes in the transverse direction is not supported by *FIRST*. Refer to Section 12.4 for additional information.

Minimum Bar Code Magnification: General Guidelines *Bar code magnification is print system dependent; determine optimum magnification with press fingerprint (ref 1.3.2)*			
Segment		**Magnification** *(Machine Direction)*	**Printer Specific Magnification** (*Machine Direction)*
Wide Web	Preprint Linerboard	100%	
	Combined Corrugated *(flute dependent)*	UPC: 110% - 200% ITF -14: 100%	
	Folding Carton	100%	
	Multiwall Bag	115%	
	Film Products	100%	
Narrow Web	Paper Products	80%	
	Film Products	100%	

Table 4.3.2

4.3.2d Quiet Zones: *Quiet zones allow scanners to detect when a bar code starts and stops. Minimum quiet zone specifications depend on the symbol specified and its magnification.*

Size Considerations

The area reserved for a bar code depends on several interrelated specifications. First, it is important to know what symbol type is specified based on where the product will be scanned. For example, if the product will be scanned at the retail POS (point of sale), an EAN/UPC symbol is typically specified. After the symbol type is known, it is important to know the allowable range of dimensions (height and width) for the symbol, including the human-readable text associated with it. It is important to note that certain symbols have a fixed relationship between their height and width, while others have minimum heights specified.

Bar code truncation is a reduction of a symbol's height below the application standard or symbol specification and is not supported by *FIRST.*

All compliant printers will be able to meet the minimum bar code sizes (outlined in the table 4.3.2). However, the smaller the symbol's size, the tighter the tolerance on bar width growth; therefore, larger symbols are better. Printing a bar code below the minimum size specified by the bar code application standards is not acceptable. Refer to Sections 12.4 and 22.2 for more detailed information on bar code size considerations.

Quiet Zone Considerations

The quiet zone is the area, free of printing, that precedes the left bar and follows the right bar in a bar code symbol. The quiet zones allow scanners to detect when a bar code starts and stops. Quiet zones are based on multiples of the symbol's narrowest element width (X-dimension). Minimum quiet zone specifications depend on the symbol specified. For example, the UPC-A symbol requires a quiet zone of 9 times the "X" dimension on each side, while a ITF -14 symbol requires a quiet zone of 10 times the "X" dimension on each side.

4.3.3 USPS Intelligent Mail Bar Code

The Intelligent Mail Bar Code (CB4), used by the United States Postal Service (USPS), is a 4-state bar code that consists of 65 bars. The information in this section was obtained from the United States Postal Service Intelligent Mail Bar Code specification USPS- B-3200C. For additional information, reference the USPS-B-3200C specification from the US Postal Service. Contact information is included in Appendix A. Refer to Sections 12.4 and 22.3 for additional information.

Dimensional Parameters

Horizontal Dimensions: The overall bar code width must be within 20-24 bars per inch

Vertical Dimensions: The overall bar code height must be within 0.134" (3.4mm) and 0.23" (5.84mm)

Quiet Zone:

- Minimum 0.040" (1.02mm) above and below bar code
- Minimum 0.125" (3.18mm) on either side of bar code

Specifications for Human-Readable Information

Horizontal Position: The human-readable information, when required, shall be printed so that the left edge of the leftmost digit aligns with the leftmost bar of the Intelligent Mail Bar Code.

Vertical Position: When-human-readable information is required, it shall be printed immediately above or below the bar code but outside of the quiet zone. The human-readable information shall be at least 0.04" (1.02mm) above or below the bar code but not more than 0.50" (12.7mm) above or below the bar code. No other printing is allowed between the bar code and the human-readable information.

Content: When human-readable information is required, it shall consist of the 20-digit tracking code and the 5-, 9-, or 11-digit routing code, if present. The tracking code shall include a space between each data field. When the bar code contains a routing code, the 5-digit ZIP code, the 4-digit add-on and the remaining 2 digits shall be separated with a space between data fields.

Font Specification: The human-readable information, when required, shall be printed using a Sans Serif font and a minimum 10 to 12 point type size.

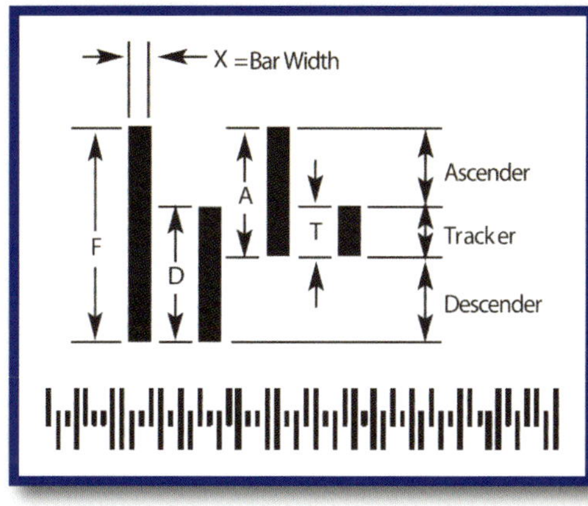

4.3.3 USPS CB4 Bar Code: *The Intelligent Mail Bar Code (CB4) is a 4-state bar code that consists of 65 bars.*

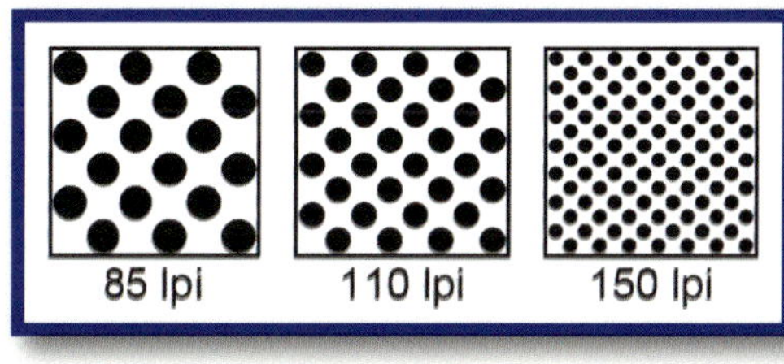

4.4 Screen Ruling: *The higher the line screen ruling, the more dots per square inch and the smaller the diameter of each dot. Generally, dot gain increases with higher line screens.*

Line Screen (lpi & lpcm): General Guidelines *Line screen is print system dependent; determine optimum line screen with press fingerprint (ref 1.3.2)*					
Segment		**Substrate**	**Sheet Photopolymer**	**Liquid Photopolymer**	**Laser Engraved Rubber/ Cured Polymer**
Wide Web	Preprint Linerboard	SBS Board	110 - 133 lpi (43 - 52 lpcm)	85 - 110 lpi (33 - 43 lpcm)	110 - 133 lpi (43 - 52 lpcm)
		Uncoated	100 - 133 lpi (39 - 52 lpcm)	85 - 110 lpi (33 - 43 lpcm)	100 - 120 lpi (39 - 47 lpcm)
	Combined Corrugated	All	55 - 110 lpi (22 - 43 lpcm)	55 - 150 lpi (22 - 59 lpcm)	55 - 110 lpi (22 - 43 lpcm)
	Folding Carton	SBS Board	120 - 150 lpi (47 - 59 lpcm)	85 - 110 lpi (33 - 43 lpcm)	110 - 133 lpi (43 - 52 lpcm)
		CRB Board	110 - 133 lpi (43 - 52 lpcm)	100 - 120 lpi (39 - 47 lpcm)	110 - 120 lpi (43 - 47 lpcm)
	Multiwall Bag	Coated Paper	75 - 120 lpi (30 - 47 lpcm)	65 - 110 lpi (26 - 43 lpcm)	75 - 110 lpi (30 - 43 lpcm)
		Uncoated Paper	65 - 85 lpi (26 - 33 lpcm)	65 - 100 lpi (26 - 39 lpcm)	65 - 100 lpi (26 - 39 lpcm)
	Newsprint	All	85 - 100 lpi (33 - 39 lpcm)	85 - 110 lpi (33 - 43 lpcm)	N/A
	Film Products	All	110 - 133 lpi (43 - 52 lpcm)	85 - 120 lpi (33 - 47 lpcm)	85 - 133 lpi (33 - 52 lpcm)
Narrow Web	Film Products	All	110 - 133 lpi (43 - 52 lpcm)	N/A	85 - 133 lpi (33 - 52 lpcm)
	Paper Products	Coated Paper	133 - 175 lpi (52 - 69 lpcm)	N/A	110 - 133 lpi (43 - 52 lpcm)
		Uncoated Paper	110 - 133 lpi (43 - 52 lpcm)	110 - 133 lpi (43 - 52 lpcm)	100 - 120 lpi (39 - 47 lpcm)
	Envelope	Coated Paper	133 - 175 lpi (52 - 69 lpcm)	N/A	N/A
		Uncoated Paper	85 - 133 lpi (33 - 52 lpcm)	N/A	N/A

Table 4.4

4.4 Screen Ruling

Screen rulings vary based on imaging method, plate material and print conditions (such as press width, anilox configuration and substrate). The range for both conventionally and digitally imaged plates is determined by print and substrate constraints. The graphics and process images to be used should be selected carefully because some print conditions require lower screen rulings. The screen ruling should be specified by the printer and considered by the designer. Table 4.4 provides general line screen guidelines by market segment and substrate category. The designer should consult the prepress and print provider to determine the optimum line screen for a specific design.

4.5 Tints

When tints are used, the values are adjusted during output using a print curve to compensate for the dot gain/TVI experienced in the printing process. A 2% minimum dot typically prints between 8% to 15%, while a tint value of 75% may print as 100%. Consult the print or prepress supplier for more information about profile specific dot gain/TVI considerations. The prepress provider applying the compensation curves can provide guidance on dot gain/TVI compensation.

4.6 Ink Colors

A designer should collaborate with the printer and consumer product company to determine how many colors are available for a product line. Many products are printed with additional colors other than CMYK. Transparent and/or opaque inks may be used and must be identified and listed in the color palette. The characteristics and print sequence of the inks used may require special considerations during the prepress phase.

In an effort to improve color matching across the product line, twelve ink pigments have been identified by color index (CI) name and number and recommended by *FIRST*. These twelve pigments are combined to create custom line colors (ie. PMS 186 or "Al's Soda" Red). These pigments are recommended because they provide the largest color gamut with reasonable fade resistance required by most packaging applications. Standardizing ink pigments improves the consistency of the color match between press runs and between printers while minimizing metamerism. This results in a more cohesive product appearance on the store shelf. When these twelve pigments are plotted to create a color gamut, colors within the gamut can be reasonably matched. When a designer or consumer product company selects a color that falls outside of the gamut, the printer will not be able to achieve an accurate color match using *FIRST* pigments. In such cases, the printer may opt to include additional pigments that expand the color gamut in order to achieve the desired color. However, due to limitations in the pigments available for a given ink chemistry or application requirement, it is not always possible to match a color precisely. Any combination of ink pigments, proofing/printing methods and substrates result in color matching limitations. The designer must consider the potential color match limitations of the inks, printing method and substrate specified for the project. Refer to Sections 20.2.2 and 20.2.3 for additional information on *FIRST* recommended pigments.

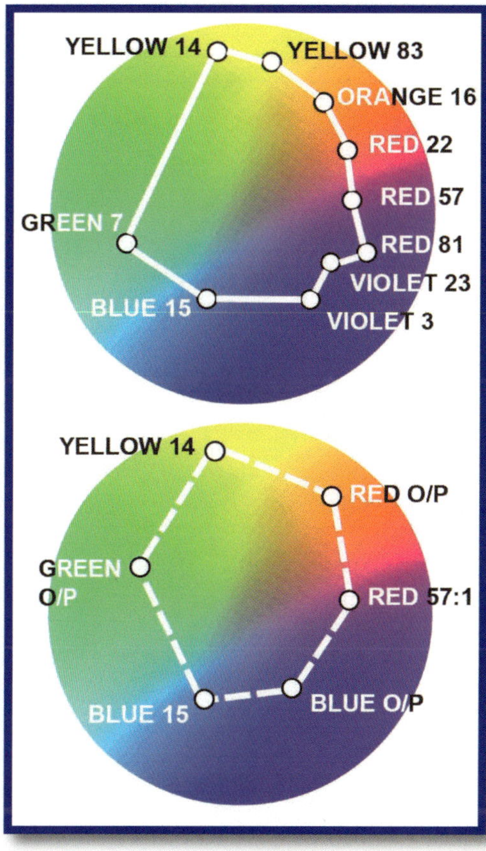

4.6 *FIRST* Ink Pigments: *The top graph illustrates the gamut created using FIRST recommended line pigments. The bottom graph illustrates the color gamut using FIRST process inks.*

5.1 Digital vs Conventional: *An RGB image must be converted to CMYK in order to provide a color proof.*

In Image 4.6, the *FIRST* recommended pigments for line inks have been proofed and plotted to create a color gamut (top graph). The bottom graph depicts the printable gamut using *FIRST* recommended process inks. Printers should proof *FIRST* pigments on substrates typically printed and, using a spectrophotometer, plot the color gamut that will best predict their ability to match color on press. All colors are dependent on the substrate to be printed. The designer and consumer product company should see drawdowns of the specified color match on the intended substrate before any job is approved for prepress. Substrate substitution in this approval process is not recommended.

5.0 DIGITAL PHOTOGRAPHY

In this section, workflows and points of measurement are identified to ensure that the aesthetic integrity of the photographer's digitally captured image is maintained. In addition, the responsibilities for handling, processing and repurposing must be clearly identified and communicated, regardless of who works with the digital file.

5.1 Digital vs. Conventional

An RGB image must be converted to CMYK in order to provide a color proof. The detail and vibrancy of an RGB captured image is greater than a converted CMYK image due to differences in the respective color gamuts. The photographer will generally review a digitally captured image on a computer display in RGB; however, color proofing is accomplished in CMYK which has a much smaller color gamut. Variability is introduced during the RGB to CMYK conversion and could be significantly different when performed by two different people using two different look-up tables or color profiles.

Refer to Section 5.3 for camera setup recommendations and Section 5.7 for image capture and communication of digital photos provided in RGB or CMYK color space. The camera setup recommendations are intended to capture the full range of the item being shot and do not consider special photographic effects or stylized techniques that may be desirable and intended, but cannot be achieved with strict adherence to the highlight, shadow setting and grayscale aim point. In this instance, special comments should be added to the file stating that a creative license has purposely been taken. Section 3.6 describes the accompanying color proof(s) to be identified according to *FIRST* recommendations.

5.2 Digital Proofs for Digital Photography

The digital proof generated from the digital photograph is often the color target proof. This proof represents the ideal layout and color intent of the designer and client, independent of the print process or the ability of the individual press to achieve it. Some of the color in this proof (and photograph) may not be attainable in the final print. To avoid rework costs downstream, it is helpful, when possible, to produce this proof based upon the known or expected capabilities and color gamut of the anticipated print process. In order to better predict the printed result, the designer or production designer should consider variables such as:

- Line screen
- Substrate
- Ink densities
- Ink hue
- Color rotation
- Special color simulation
- Dot structure
- Screen angles

To define the variables listed, the designer should contact the printer and/or prepress provider to obtain these and any other job-specific requirements including the press profile. Refer to Section 3.5 for additional information on proofing requirements.

It is always helpful to include a print control target which has test elements such as: color patches of the minimum dot %, 10%, 30%, 70% and solid ink density for all inks to be printed. A highlight and shadow gray should also be incorporated into the control target to assist in the evaluation of color balance. Refer to Sections 4.9, 19.2, 19.3 and 19.4 for additional information on *FIRST* recommended process control test elements.

5.3 Camera Setup Recommendations

Photographer's Recommended Computer (shooting) Settings

- Photoshop Working Space (RGB): Adobe RGB (1998)
- Photoshop Color Management Policies: preserve embedded profiles
- Recommended (Calibrated) Display Settings: Gamma 2.2, White Point 6500K

Photographer's Recommended Camera Settings

- Recommended Color Space: Adobe RGB (1998); many cameras default space is sRGB
- Recommended Capture Settings: raw or raw + largest TIFF available

5.3 Grayscale Aim Point: *The aim point of the shot should be the 40% neutral gray swatch or the number 3 or 4 block on the photographer's scale.*

Black Settings
In the RGB color space, a highlight setting that can still produce a dot structure should be used. The highlight setting should be between 236 and 240, which typically translates to a maximum dot of approximately 94% on the resulting halftone.

White Settings
In the RGB color space, a shadow setting that will still hold the detail without filling in should be used. The shadow setting should be between 18 and 22.

Grayscale in Photo Shot
It is imperative to use a standard photographer's grayscale for setting up any digital shot. The grayscale should be in all shots and positioned to best capture the scale within the outline of the shot. If there are several dropout shots and the scale cannot be placed in the shot, then start with shooting the scale in test shots to obtain correct grayscale settings.

When creating mood images or images where the light is filtered for an effect, photograph the grayscale with and without the filter on the light. Then supply both shots to the prepress provider noting the difference between the two for color reproduction.

There will be occasions when, for aesthetic reasons, visually pleasing color may be more desirable than technically accurate color. In these instances, it may not always be practical to also provide a completely color neutral reference image. However a second image with accurate color reference for any color critical subjects within the shot, should be provided along with clear direction as to how that reference image should be utilized for color correction.

The X-Rite ColorChecker product series and Kodak Q-14/Q-60 are examples of special color and grayscales that should be used as grayscale targets for digital photography to measure density and color. Place this grayscale in the main light source of the image. If a full grayscale cannot be used, use patches of white, black and a midtone neutral gray for studio photography.

Grayscale Aim Point
The aim point of the shot should be the 40% neutral gray swatch or the number 3 or 4 block on the photographer's scale.

Camera capture color should be neutralized (when neutral color is desired) utilizing either an industry standard Kodak Gray Card (18% reflection) or X-Rite ColorChecker (24 Patch #22,

Neutral 5) or comparable product. Gray reference cards should be replaced at least every two years for consistent color fidelity.

5.4 Photographic File Format

All shots sent to the prepress provider should be uncompressed, 8 bit or greater RGB TIFF files. 16 bit color is recommended for optimal color reproduction. CMYK conversions require using the printer profile and should be done from the original (or retouched) RGB file by the prepress provider. Photographers should NOT supply CMYK conversions, but can use soft proofing to emulate CMYK appearances on screen, when necessary. When photographers must supply CMYK conversions, the full gamut RGB files must also be provided in addition to the CMYK files. Providing the RGB files allows for subsequent adjustments and corrections as required by the printing application.

5.5 Unsharp Mask: *Unsharp masking produces the appearance of sharpness and detail within an image.*

RGB Conversion

Though many off-the-shelf programs are capable of converting from RGB to CMYK color space, there are many factors to consider including: ink pigments, printing substrate, screen ruling, etc. It is critical to identify which party is best equipped and responsible for making color conversions and for documenting the color status of any digital files and accompanying proofs. Traditionally, this is a core responsibility of the prepress provider who has advanced knowledge of the many variables involved.

5.5 Unsharp Masking

Unsharp masking is a technique that produces the appearance of sharpness and detail within an image, by accentuating edges where different densities and contrasting colors meet. The amount of sharpening applied is determined by image content and other factors. The prepress provider usually has the needed information to make the necessary unsharp masking determination. The correct amount of sharpening should be determined by a technician who is knowledgeable about the printing process and the effect of sharpening on images destined for flexographic printing. Too much sharpening can make an image look bad, a result of too much digital noise being added to the image by the sharpening process.

5.6 Resolution

The number of pixels (picture elements) in a given area determines the resolution of an image (typically specified as number of pixels per linear inch). 300 pixels per inch (ppi) is the typical resolution for color images at 100% for 133-150 line screen. The formula for calculating the optimum resolution is

5.6 Image Resolution: *Image resolution determines the printed image quality. Generally, the optimum resolution is 2 times the output screen ruling. 300ppi is the typical resolution for images printed at 100% using 133 - 150 line screen.*

two times the output screen ruling. Although this is the "rule of thumb," the amount of captured resolution is related to the final image quality. The enlargement of the image, the screen ruling and the image content (particularly detailed content) must be taken into consideration. For example: Original resolution (1,240ppi) divided by enlargement (350%) equals (354) lines of resolution at the reproduction size (pixels per inch) divided by screen ruling (175lpi) = (2.02). There should be no noticeable loss in detail as long as the answer is approximately 2.0.

Screen Ruling/Resolution		
Screen Ruling (lpi/lpcm)	Resolution (ppi/ppcm)	Key
55 / 22	110 / 43	lpi = Lines per Inch
65 / 26	130 / 51	lpcm = Lines per Centimeter
85 / 33	170 / 67	ppi = Pixels per Inch
100 / 39	200 / 79	ppcm = Pixels per Centimeter
110 / 43	220 / 87	
120 / 47	240 / 94	
133 / 52	266 / 105	
150 / 59	300 / 118	
175 / 69	350 / 138	
200 / 79	400 / 157	

Table 5.6

To convert from English measurement (lpi or ppi) to metric measurement (lpcm or ppcm), divide the number of lines/pixels per inch by 2.54.

5.7 File Transfer Recommendations

The receiver of any digital file should be contacted to determine the preferred transfer media. File Transfer Protocol (FTP) is a common method of file transfer, which may be available on the prepress or print providers' web site. Removable media such as a DVD may also be used to transfer files. Note: there are different security levels based on the selected transfer system used.

A hard copy proof must accompany every digital file, even if the hard copy proof is delivered the following day. Regardless of the file transfer method, all jobs processed should be accompanied by:

- A list of file names relevant to the job
- Files organized by directories/folders
- All high-resolution images embedded or linked in the job folder
- All supporting profile files (source and destination)
- A hard copy reflecting all files included
- Screen and printer fonts (when applicable)

6.0 Raster Images: *These files have a fixed resolution when created or scanned and cannot be enlarged without losing detail.*

6.0 PROGRAM APPLICATIONS

Operating System: PC or Mac

The packaging industry commonly uses the Macintosh platform for graphics production, though there are PC/Windows versions of many popular applications available.

Program Applications

Applications used in package design are divided into three categories:

- Drawing Programs: Adobe Illustrator (ie. which create vector files)
- Photo Editing Programs: Adobe Photoshop (ie. which create raster files)
- Page Layout Programs: Adobe InDesign or QuarkXPress

1. Drawing Programs: Drawing programs create files that contain objects and, are referred to as "vector" (mathematical coordinate) files. A line is created by identifying two points and providing the instructions to connect the points with a line of particular weight and color. Shapes have more points and indicate a fill color. There is no resolution to these graphics, thereby allowing an element to be scaled up or down with no loss of detail. Furthermore, they are inherently accurate and are best for graphics with a fixed set of colors (line copy). Most drawing programs also include the ability to create gradients, vignettes and blends.

When composing a job in a drawing program, always include the die drawing or template information on a separate layer or use a unique spot color such as "die line" so it can be isolated at output. *FIRST* recommends die-cut jobs (labels, cartons, corrugated) be produced entirely within a drawing program.

2. Photo Editing Programs: Photographic images or art created in photo editing programs may contain thousands of shades of color and are referred to as "raster" files. The graphics are made of many rows of pixels and each pixel can have its own shade. These files have a fixed resolution when created or scanned and cannot be enlarged without losing detail. When enlarging a

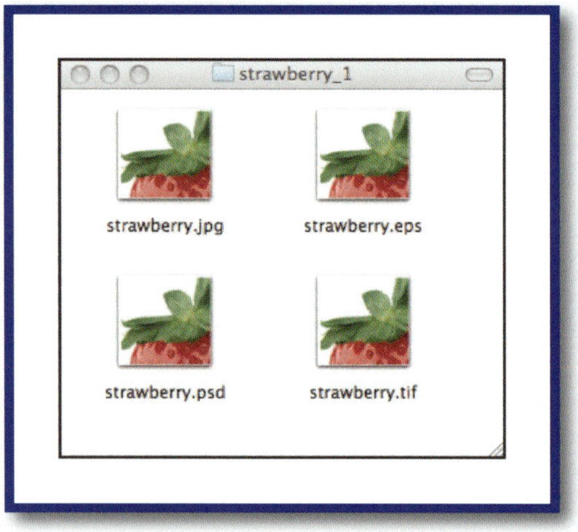

7.1 File Names: *File names should be short but descriptive.*

previously captured image, check with the prepress provider for input on maximum enlargement without significant loss of image detail.

3. Page Layout Programs: Page layout programs provide an assembly environment where all kinds of elements can be combined. These programs, such as InDesign and QuarkXPress, are generally superior for dealing with multiple scanned images and volumes of text in multipage documents, but are less adept at accurate placement of elements relative to a template. Therefore, *FIRST* does not recommend using page layout programs for package design.

If documents with placed (nested) images are imported into another document, the final RIP may not find the nested elements. For this reason, *FIRST* does not recommend placing files with nested images in page layout document. If it is necessary to deliver embedded or nested files, always send the original file with the job in case editing is required.

7.0 DOCUMENT STRUCTURE

7.1 Naming Conventions

When the design process is in the early stage, identifying a common naming convention is in the best interest of all parties and is vital in assuring smooth production. Many times the product being developed is part of a larger project or product line. Before the design and production files are created, check with the customer (CPC) to identify if they have already developed a common naming convention to be used by all suppliers.

File Names

File names should be short but descriptive. On some systems, file names may be truncated to the first eight characters when RIPed. Some systems cannot handle characters such as asterisks, spaces, or punctuation, so *FIRST* recommends never using these characters when naming files and/or document elements.

Naming for Image Replacement

In some production environments, low-resolution versions of images (FPOs) are used through various stages of concept, design, and approval. These low-resolution images are smaller in size and faster to process than the actual high-resolution version. With the correct and agreed upon naming convention, the FPO can be linked to the high-resolution file and replaced automatically during the output stage. Determine with the prepress provider if image replacement procedures will be used and what naming procedures are to be followed.

7.2 Document Size

Designs must be built to actual size. If the art is too big to proof in one piece, it will be necessary to tile the proof. All proofs should be made to the final size (100%) of the printed product.

7.3 Working in Layers

FIRST supports the use of layers to organize a file. Additionally, *FIRST* recommends putting the template on one layer, marks on another and design elements and copy on different layers. Some workflows may require that colors be pre-separated; layers are an ideal way to organize these separations. Separate layers can also be used for variations in designs, such as special price banners or line extensions. This makes certain that the underlying graphics are identical in content, placement and prepress execution. This can also be helpful in jobs with common colors (cylinders or plates shared between two similar designs).

When documenting the file, give the layers meaningful names. Put notes, instructions, color mixes and other documentation on a layer, or include them on a separate annotation layer with the art. Creating an annotation layer assures these important instructions will not be lost as the file moves through the production chain.

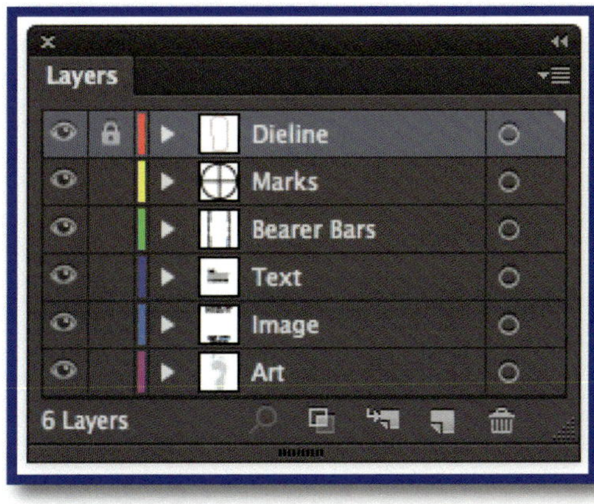

7.3 Working in Layers: *Use layers for variations in designs, such as special price banners, line extensions, etc. This makes certain that the underlying graphics are identical in content, placement, and prepress execution.*

7.4 Auto-Traced/Revectorized Art

Much of the fine-tuning of designs to achieve printability, die matching and cross matching occurs during the prepress stage of production. To eliminate repeating these changes on each new revision of a base design, it is recommended to send all changes made during the prepress phase back to the designer and/or customer (CPC) to be incorporated into the base design.

Some high-end systems can now convert completed files back to Mac format as Illustrator files. Such files should be used with extreme caution. Auto-tracing features ask a program to make decisions about placing nodes or points. These automatic choices are not the most efficient choices and produce complex files with too many nodes that can slow or stop file processing. In addition, the files are so massive they require large amounts of RAM to open.

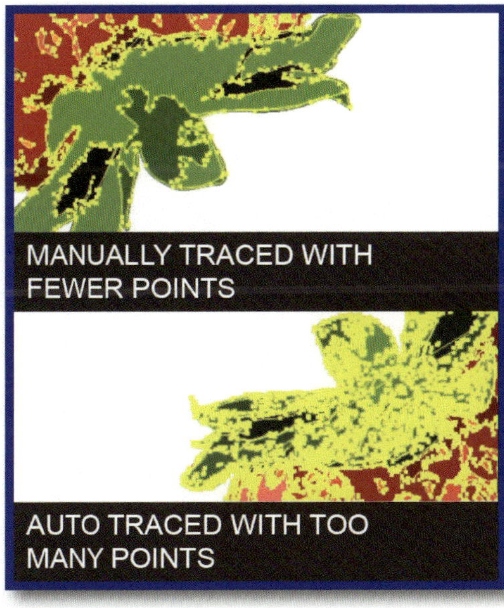

7.4 Auto-Traced or Revectorized Art: *Auto-tracing features ask a program to make decisions about placing nodes or points. These automatic choices are not the most efficient choices, producing complex files with too many nodes that can slow or stop file processing.*

Revectorized Files

Files that were created on a Mac, converted to a high-end system, and then converted back to a Mac are called "revectorized." If possible, these files should not be used. If these files are used, they should be simplified as much as possible. When a RIP

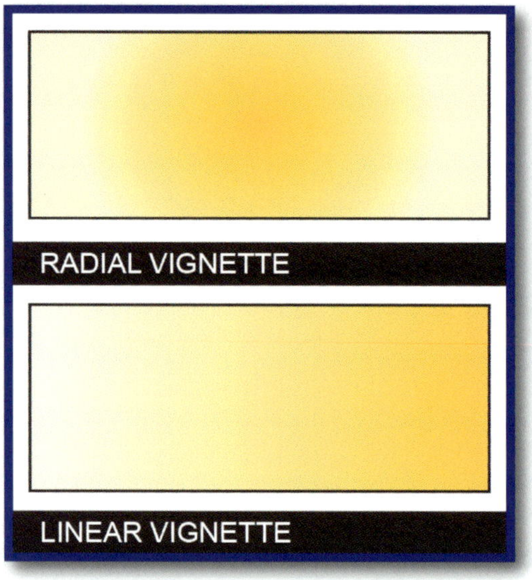

7.5a Radial & Linear Vignettes: *A holding line around a vignette protects the smallest highlight dots and helps to prevent hard edges and dirty print.*

7.5b Building a Vignette: *There are several approaches to building a smooth vignette as well as multiple problems in creating vignettes.*

converted the file to raster, the RIP decided which pixels to turn on, using the PostScript information sent by the application. Now another program has processed it, making more decisions about where to place nodes, making this a third-generation image. Some change is inevitable; in the best case, it may be in the range of 0.001" (0.025mm). For best results, use this image for position and move or adjust the original art to fit.

Recreate the art whenever possible; that is, redraw the elements in the program to create new elements that are native to the program. This solves the file size issue and produces elements that are easily incorporated into future designs and changes.

7.5 Blends/Vignettes/Gradients

The terms blend, vignette, gradient, fade-away, fountain and graduated tint are used interchangeably. *FIRST* uses the term vignette for clarity.

Building a Vignette

There are several approaches to building a smooth vignette as well as multiple problems in creating vignettes. Some of the approaches concern the way they print, others concern the way they are specified in software programs. Vignettes are subject to unpleasant banding (steps where tints do not transition smoothly) or dropping off (leaving a hard edge). Upgrades in software have resulted in higher quality vignettes. Although the algorithms used to create vignettes have improved, they still require skill and careful planning. A thorough understanding of current software applications and the printer's capabilities are required to create a printable vignette. Generally, the prepress provider is best equipped to create the vignette contained in the final production file.

Some of the primary considerations when building a vignette include:

Blending One Spot Color Into Another: When blending one spot color into another spot color, two final files should be produced; a file for creating a comprehensive proof (color comp) and a file for production. The production file must contain two separate vignettes, one for each color. Mark up a proof with instructions for how the vignette is to be created in addition to including instructions on the annotation layer. For example, "100% to 20% yellow overprinting 40% to 80% navy." There is no easy way to create one file that shows this effect and prints the correct tints except with process colors. Another solution is to substitute process colors for custom colors (ie. the magenta channel might print as red, the cyan as reflex blue, the yellow as gold and the black as green, etc.).

Blending A Spot Color Into White: When creating a vignette of a spot color fading to white, specify the minimum dot percentage of the spot color on the lighter end of the vignette. One technique is to use the same spot color for both ends of the vignette. One end should be set to the full tint value while the other end should be set to the printer's minimum dot size in the same color.

Trapping Vignettes: Vignettes are difficult to trap. The lighter color should trap into the darker color, but that relationship changes in a vignette. When placing type or graphics over a vignette, be aware that when the necessary trapping is applied, undesirable results may occur.

RIPing Vignettes: Designs that use multiple vignettes will take longer to process. To facilitate processing, consider using a raster program for the continuous tone image, the part of the design that looks like a picture. Use vector files for type and other elements that need hard, clear edges or very fine detail. Some processors will RIP vignettes from drawing programs to a continuous tone and add noise to prevent banding. This allows the prepress provider to separate the art, but requires more time to RIP.

Factors Influencing Banding

Many factors that influence banding in a vignette relate to the construction of the vignette. There is a mathematical relationship between the length, range and the number of steps in a vignette. The length refers to the physical length of the vignette and the range refers to the difference in color across or down the vignette. (ie. a vignette of 30% to 50% has a range of 20%).

- The longer the vignette, the more likely it is to show banding
- The shorter the range of the vignette, the more likely it is to show banding
- The fewer steps used, the greater the potential for banding
- Banding is more visible with darker inks
- Lower screen rulings are less likely to show banding

Higher output resolutions may also help reduce banding that may appear on some low-resolution printers and computer monitors. Professional film and direct-to-plate output devices usually run at a resolution of at least 1,200dpi which also helps minimize banding. If objectionable banding is observed when creating the file, make a notation on the annotation layer of the file, transferring the final inspection responsibility to the party outputting the file.

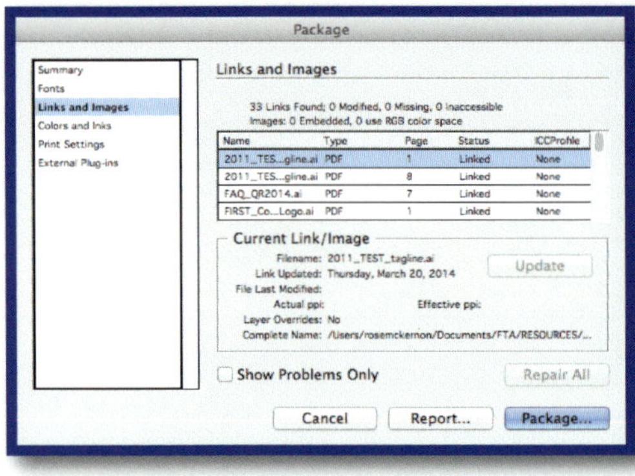

7.6 Imported Images: *After placing an image, do not rename the files. File names are a critical reference link between the document and the image file.*

7.8 Scan Resolution: *FIRST recommends images remain in RGB format for delivery to prepress.*

Factors Influencing Hard Edges & Dirty Print

To avoid hard edges and dirty print, it is important to maintain the printer's minimum dot and not fade to zero. The printer specifies the minimum dot used along the edge of any vignette. The lightest area of the vignette should adjoin a holding line or the edge of a graphic window; this will ensure that hard edges or dirty print do not appear across the vignette when the dot fades to the printer's minimum. When vignettes are made of more than one color, all colors must stop at the same place in order to prevent rainbowing and dirty print throughout the vignette.

7.6 Imported Images – Follow the Links

File names are a critical reference link between the document and the image file. After placing an image, do not rename the file. All images placed in the document must travel with the document for output. Most layout programs treat imported images as electronic "pickups" and refer back (by following the link) to the image file at output. Always make certain that all links are updated properly before sending files. If an imported image is modified, always update it in the final document to make sure that it has not shifted position.

FIRST recommends working with the appropriate packaging application. Problems, such as nested files, can be encountered when working outside of those applications. In many programs, it is an option to embed the placed image data with the EPS file. This is not recommended because some editing may be required downstream. Sending the native application files enables future changes.

7.7 Electronic Whiteout

Do not cover up unwanted elements with a white box. The RIP will still process unwanted elements. Files that are designed in drawing programs can use masking, clipping, or compound paths instead.

7.8 Image Capture Quality – Scanning Considerations

Optimizing scanner variables when capturing the original image is critical to achieving the desired printed result.

Scan Resolution

All scanners capture RGB data. Although some scanners can use hardware and/or software to translate the scanned data to CMYK, *FIRST* recommends capturing and supplying the image in the original RGB format to protect against data loss. Entry-level scanners generally are not adequate for production scans. Such devices use interpolation to achieve production resolution or size and real detail cannot be interpolated.

Image Sharpness/Resolution

The most important scanning factor is optical resolution. A scan at 100% scale should have a minimum sampling of 1.5-2 times over the final halftone line screen. Fine detail images may be sampled at up to 3 times the output line screen. For example, an image that will print with a 200 line screen may need a scan resolution of 300 to 600 pixels per inch, depending on the detail required in the image. If the image is enlarged, it will lower the effective resolution. The objective is to scan images at a high enough resolution to capture enough data to achieve the desired detail at the reproduction size.

Image Enlargement

Enlarging a scanned image will reduce the effective resolution of the image, and can compromise the image appearance. If possible, scan the original at the correct size and resolution; if rescanning is not possible, some enlargement may be acceptable depending on the scanned resolution. Adobe Photoshop is able to enlarge images using interpolation, a mathematical process of creating new pixels. Depending on the image, some interpolation may be tolerable. Whenever possible, it is always preferable to rescan the original image at the desired resolution.

Line Art

Theoretically, line art should be scanned at the same resolution as the output device. However, minimal improvement is visually apparent on most line art subjects scanned above 1,000 pixels per inch. Scaling will degrade quality; the best solution is to redraw line art in an illustration program. This also makes the file size smaller.

7.9 Scaling & Resizing

It is best to place images at the desired reproduction size and resolution, or larger. If upscaling is required, it should be done in Adobe Photoshop and not in the artwork layout. When upscaling an image, be careful to ensure the image resolution does not fall below the calculated resolution value, typically twice the halftone frequency.

7.10 Color Space

Images in a design file (whether captured or created) should remain in their native RGB color space for conversion in prepress to the color space described by the printer profile. Moving the image to any color space other than that of the final printer will result in unnecessary loss of color and detail accuracy. Refer to Section 14.4 for more detailed information on Color Management.

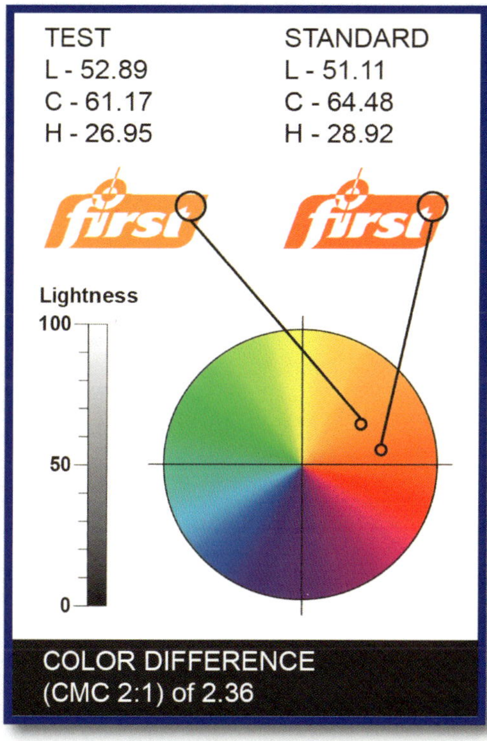

7.10 Color Management System: *Color Management Systems (CMS) translate from one gamut to another, allowing the proof to more accurately mimic the printing process.*

8.2 PDF: *Portable Document Format (PDF) is used to transport graphically rich content. It is typically used in direct-to-plate technologies.*

8.0 FILE FORMATS AND USAGE

Before using a new version of software, check with all parties downstream that will have to open and work with the electronic file to ensure compatibility. In newer versions, it is possible to save documents in older formats.

8.1 Specified Formats

The primary specified formats for vector images are: .ai (Illustrator native) and .eps. For continuous tone (raster) images the primary specified formats are: .psd (Photoshop native) or .tif. Refer to Section 8.2 for delivery of images via PDF.

There are numerous types of electronic file formats that can be generated from drawing, photo editing and page layout programs that should not be used. BMP, PICT and JPEG files generally lack detail due to the way their data is compressed. If other file formats must be used, it is imperative that all parties agree which file formats are to be created, exchanged and archived throughout the project. The prepress provider is in the best position to describe the advantages and disadvantages of each format for a specific purpose.

8.2 Portable Document Format (PDF)

PDF is an imaging file format used to transport graphically rich content. It is commonly used in direct-to-plate and digital proofing technologies. The “creator” of the file (designer, ad agency, prepress provider) must produce a file that meets the minimum imaging requirements of the “receiver” (prepress provider, printer). PDF/X is a PDF file with restrictions intended to facilitate the transfer of files from “creator” to “receiver”.

A PDF/X is a collection of standards defining a number of conformance levels, all of them targeted at ensuring predictable and consistent printing in a professional print environment:

All of these standards are published as parts of ISO 15930, under the general title Graphic Technology — Prepress digital data exchange using PDF:

- **Part 1:** Complete exchange using CMYK data (PDF/X 1 and PDF/X 1a)
- **Part 3:** Complete exchange suitable for colour-managed workflows (PDF/X 3)
- **Part 4:** Complete exchange of CMYK and spot colour printing data using PDF 1.4 (PDF/X 1a)
- **Part 5:** Partial exchange of printing data using PDF 1.4 (PDF/X 2)
- **Part 6:** Complete exchange of printing data suitable for colour-managed workflows using PDF 1.4 (PDF/X 3)

Part 7: Complete exchange of printing data (PDF/X-4) and partial exchange of printing data with external profile reference (PDF/X-4p) using PDF 1.6

Part 8: Partial exchange of printing data using PDF 1.6 (PDF/X 5)

As of this writing Part 8 is not supported by all software vendors. *FIRST* looks to the Ghent PDF Workgroup (GWG) as the international group establishing packaging PDF specifications.

While the GWG Packaging Specification is largely PDF/X compliant, there are deviations from this rule for applications that are packaging specific. Section 11.0 summarizes the rules for an ISO 15930-7:2006 compliant PDF file and also identifies the GWG Packaging Specification 2012 exceptions specific for flexography. For additional information, refer to the Ghent PDF Workgroup contact information in Appendix A.

8.3 Clip Art: *Clip art may come in the form of low-resolution PICTs, better-performing TIFFs, or as well-built EPS images.*

8.3 Clip Art

Clip art may come in the form of low-resolution PICTs, better-performing TIFFs, or as well-built object-oriented EPS images. Be sure to ask about the file format of the clip art being used to confirm the appropriate level of quality. If the image is a scan, identify the scanning resolution. If it was scanned at 72 pixels per inch, the clip art piece will be suitable only for display on a monitor and printing to a low-resolution printer.

The selected image may be one of several on a clip art page. Remember that masking out all the other images does not remove the images; they will all be processed. Save individual images under a new name and import the single image into the document.

8.4 Creating and Identifying FPO Images: *If an image is not properly identified as a "for position only" image, it may not be replaced.*

8.4 FPO Continuous Tone Images

Whenever possible, a FPO (for position only) continuous tone (CT) image should be created from actual high-resolution data with correct cropping and rotation. Otherwise, the high-resolution image will need to be manually placed. The letters "FPO" must be placed into the live image area because the file will go through many channels before being output and if not properly identified as a "for position only" image, it may not be replaced.

8.5 Special Effects

When editing low-resolution raster files to produce special effects, document the steps used. The effects of most functions change with a change in resolution. It would be difficult to reproduce the same result with the high-resolution image

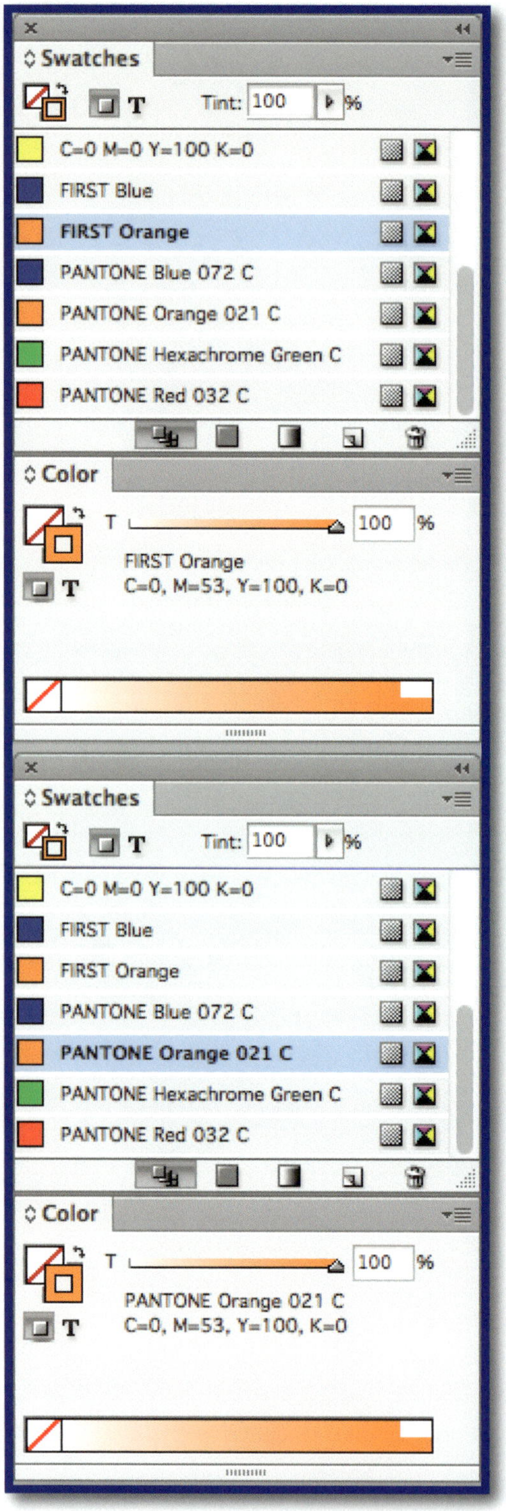

9.1 Images with Custom Colors: *Custom colors used in a placed image must have the same name as the corresponding custom colors in the final design file.*

without the documented information. Even with instructions, it is difficult to recreate several complicated special effects. The "action" sets within Adobe's Creative Suite allow the creator of the low-resolution file to record each edit step, in sequence, used to create the file. The "action" set can then be saved and shared with the user that will be creating the high-resolution original.

8.6 Image Substitution – Automatic Image Replacement

Using Low-Resolution Files for Automatic Image Replacement

A low-to-medium resolution file may be provided to the designer for automatic image replacement. These files contain links to full-resolution files on the prepress provider's system. It is important not to rename the file; the file name is the link back to the high-resolution image.

This method allows the designer to move, crop, or resize (within limitations) the APR/OPI image as if it were the live high-resolution image. It places the control of exact positioning in the hands of the designer. Resizing of low-resolution images must be employed with extreme caution. The high-resolution file will be scaled by the same factor. Enlarging the file will reduce its effective resolution significantly and reproduce an image that will not be pleasing due to loss of detail.

Specific recommendations on working with images for automatic placement may vary based on the workflow of the individual designer and prepress provider. The designer and prepress provider should agree on the procedures for using automatic image replacement.

9.0 PREFLIGHT OF FINAL DESIGN PRIOR TO RELEASE

Preflight is required by *FIRST*. The process entails documenting, collecting and testing files prior to release to another vendor in the production process. The preflight requirement was designed to ensure all components of a design have been supplied and received as intended. The designer should keep an electronic back-up of all released files for safety.

9.1 Documenting the Design

Revised Art

Revised files should be renamed with a revision number or date. Do not rely on the operating system modified date because each time the file is opened the date changes. Keep the old file name the same except for the revision number or date.

Images with Custom Colors

The custom colors used in a placed image must have the same name as the corresponding custom colors in the final design file. This applies to images pasted in as well. Otherwise, the two colors will not output as one color separation. Many programs will now import colors from placed images into their palettes, but the artwork must then be edited in the file to use these same colors.

TIP: For the placed raster file to output with the line art in the composite file, custom colors must be edited to the corresponding CMYK inks.

Design Report

The final design may seem very simple to the designer, but it can be difficult to decipher when someone in the production process starts to work with it. To make the design flow smoothly through production, details must be provided on how it was developed and the expected end result. Some programs have report features to list details about a file, others use comment layers within the file itself. The following list identifies what information should be included in the design report:

- Final file name(s)
- All placed full resolution and FPO images
- Mechanical name (die drawing used to build the design) including the date and source of the template
- Application/version of files
- Fonts used
- Colors used (CMYK, PMS, Custom)
- Common and/or base layers
- Instructions for vignettes or effects

9.2 Release to Prepress

Files to be released to prepress must be supplied in their entirety including all supporting files (linked/embedded high-resolution images, fonts, etc.). The most reliable way to verify all necessary supporting files have been supplied is to:

- Copy the files to another computer, or copy files to a different directory or volume on the local machine
- Open document with all supplied fonts loaded, and all supplied images linked
- Output full color, full size proof (tile if necessary)
  Note: Another option is to convert all type to outlines and print a PDF document, which is essentially the same as creating a PDF from a distilled postscript file
- Verify all content against the approved comps

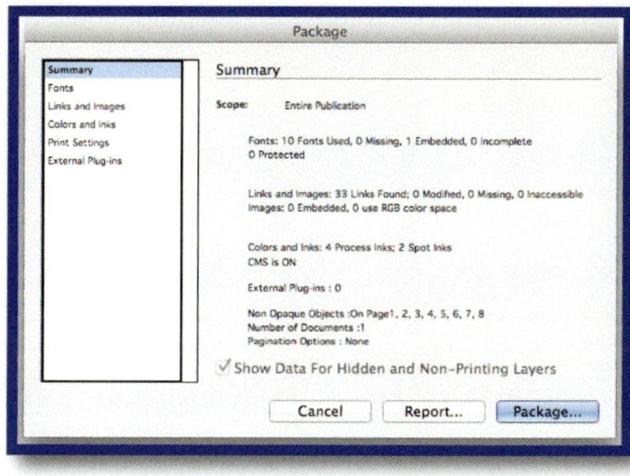

9.2 Release to Prepress: *Supply files to prepress in their entirety including all supporting files.*

Materials to Be Sent With the Job:

1. Final files, including all supporting high-resolution images, fonts and mechanicals (templates). When sending multiple designs, file-manage each design folder to house the relevant working design file and all applicable supports.
2. Full color, full size, hardcopy proof or a PDF printed out of native application (which is essentially the same as creating a PDF from a distilled postscript file).
3. Documentation described in Section 9.1 Design Report.
4. A printout of the disk directory if supplying files via disk.

Verify the file transfer method with the recipient. Many different options are available for file transfer, refer to Section 5.7 for more information. Additionally, if using data compression (.sea, .zip), check with the prepress vendor to determine compatibility.

DOWNLOAD *FIRST* 5.1 Extras Referenced in this Section at:
http://www.flexography.org/FIRST_extras

10.0 PREPRESS INTRODUCTION

10.1 Overview

FIRST is created to facilitate communication among all participants involved in the design, preparation and printing of flexographic materials. Prepress is responsible for transforming the product design into production files/films/plates ready to be flexographically reproduced on press. The ability of prepress to apply their knowledge of the printing and converting process to the current design directly impacts the quality of the printed piece, the efficiency of the press run and the overall time-to-market of the new product. The Prepress Section is intended to assist the prepress provider in understanding the flexographic print considerations necessary to produce quality graphic files in today's environment of compressed timeframes.

10.2 Responsibility

As packaging graphics continue to increase in complexity and production timelines continue to compress, the clear assignment of responsibilities is necessary to ensure a quality printed product in a timely manner. The assignment of responsibilities requires planning and collaboration among all involved parties. A more detailed review of the Package Development Process and the roles of and responsibilities of each party are provided in Section 1.4 The Package Development Process.

Consumer Product Company: Ultimately, the customer defines expectations and therefore must drive the collaboration process. The customer determines the effort expended to reach satisfaction. The CPC must facilitate communications between the supply chain parties: designer, prepress provider and printer.

Designer: The designer must work with both the prepress provider and the printer to understand the capability of the printing/converting process being utilized. Based on the print capability, the designer must provide a design concept that will enable the printer to meet the expectations of the customer (CPC). The earlier in the design development process the prepress provider and printer are involved, the better the team is able to determine specific capabilities and ensure the final product meets the customer's design objectives.

Prepress Provider: The prepress provider must work with the printer to understand the capability of the printing/converting process being utilized. In order to create a printable design, the prepress provider is responsible for providing the designer with accurate and timely information regarding print capabilities at

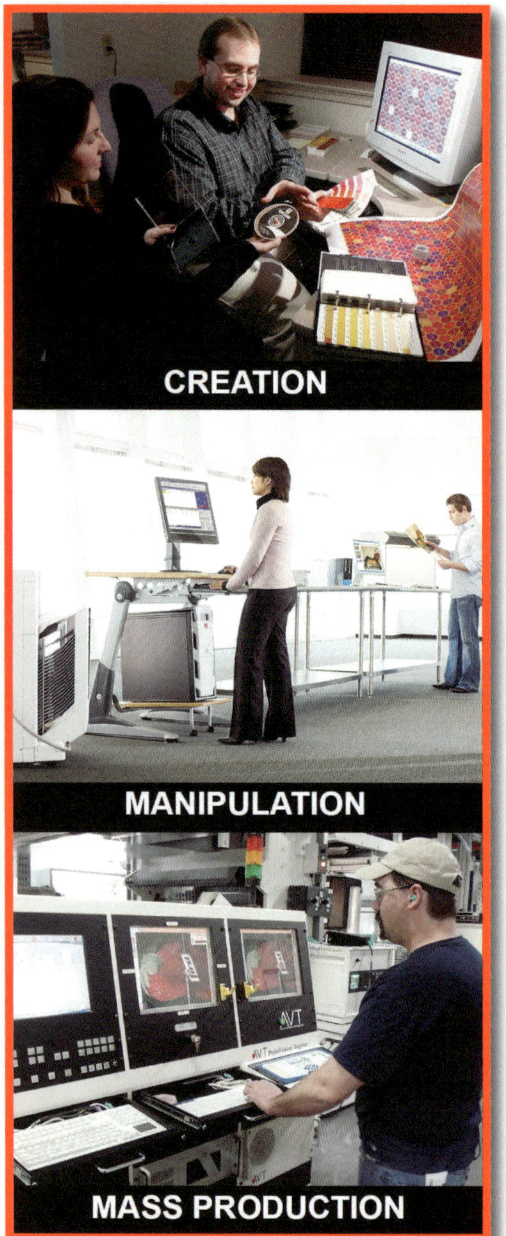

10.2 Package Development Responsibilities: *In short, the designer creates the image, the prepress provider manipulates the image, and the printer mass produces the image.*

the beginning of the design phase. Based on the print capability, the prepress provider produces appropriate films/files/plates that will enable the printer to meet the expectations of the customer (CPC). They must document the controls that ensure the consistency and accuracy of the supplied media (films/files/plates). Additionally, the prepress provider is responsible for producing a contract proof calibrated to predict the printed result. The prepress provider must provide the printer the ability to objectively confirm the accuracy of the prepress work and the printing process. This can be accomplished through the use of agreed-upon control targets.

Printer: The printer is responsible for consistently reproducing the graphic design to the satisfaction of the customer (CPC). They must utilize and document the process controls necessary to ensure that accuracy and consistency are achieved. They must work with the other parties and suppliers to define the capability of the printing process. At the beginning of the design phase, the printer is responsible for providing the designer with accurate and timely information regarding process capabilities. This information facilitates the creation of a printable design.

11.0 ELECTRONIC/DIGITAL FILES

FIRST specifications support the work presently underway within the Committee for Graphic Arts Technology Standards (CGATS), the International Standards Organization (ISO), Technical Committee for Graphics Technology (TC 130) and the Ghent PDF Workgroup. These four groups are working to develop a standard format for digital data exchange. CGATS is the accredited standards development committee within the American National Standards Institute (ANSI). ISO TC-130 has a similar function in the area of international standards development. The Ghent PDF Workgroup specifies scripts to create a PDF that is "correct" for printing. There are several principles that must be followed when dealing with digital data files to ensure a successful exchange. Reference Appendix A to obtain contact information for these organizations. Section 8.0 has additional information on the exchange of digital data.

11.1 File Formats

Digital files exchanged between the originator and the prepress provider must be in a format that can be written, read and manipulated easily by all parties using the files. Digital files must contain data that will ultimately produce film, color proofs and plates that conform to *FIRST* specifications. Standard file formats include TIFF, EPS and PDF. *FIRST* recommends working in native file formats such as ".ai" when using Adobe Illustrator. For global file transfer ISO15930 PDF/X is the requirement for file transfer (all parts).

11.1.1 TIFF/IT & 1-BIT Files

Prepress files can be exchanged as CMYK raster files using the P1 compliance level of the TIFF/IT file format as defined in ISO 12639:2004 (Graphic technology – Prepress digital data exchange – Tag image file format for image technology – TIFF/IT). The use of 1-BIT TIFF(s) is common in digital platemaking and digital proofing applications. A 1-BIT TIFF is essentially a digital negative created by the same RIP (Raster Image Processor) that would be used to create film for an imagesetter/platesetter. Normally, 1-BIT TIFF(s) are written as individual colors in the same way a film negative is used to describe individual colors. 1-BIT TIFF(s) contain only black-and-white (positive or negative) information. All screening information is inherent to the file. This information is used directly by the platemaking device to image a digital plate. Refer to Section 17.2 for information on file transfer of 1-BIT TIFF(s) for digitally-imaged and laser-engraved plates.

While software is available to edit 1-BIT TIFF(s), *FIRST* does not support this practice. Changes should be made to the original file and re-RIPed. Typically, final package (FP), continuous tone (CT), and line work (LW) files will be required. Incorporate all of the logical parameters for final graphic file output (UCR/GCR, gray balance, register marks, etc.) into the file. The original Photoshop files of unflattened images should be submitted to the prepress provider. Other information required when sending a 1-BIT TIFF file should include: specification of the resolution (dpi), compression method, positive or negative, right or wrong reading, line screen, screen angles, curve information and distortion. Files containing specialized screening methods will require additional custom specifications based on the screening technology. Agreement on specifications must be made between the sender and the receiver for a successful process.

Table 11.1.1 illustrates how compressing the 1-BIT TIFF significantly reduces the file size and therefore the transfer time. In this example, the uncompressed file required 4 hours to transfer while the compressed file transferred in 10 minutes (using ace format). Therefore, 1-BIT TIFF file should never be transferred uncompressed.

1-BIT TIFF File Compression *Example: 8 Color Job, 2400 Resolution; 12.5" x 20"*		
Format	Size (bytes)	Percent (%)
tif	1,322,194,256	100
len	474,145,262	36
lzw tif	100,765,228	8
zip	73,041,637	6
ace	36,073,032	3

Table 11.1.1

11.1.2 PDF/X: *FIRST* Recommended File Transfer

PDF is an imaging file format used to transport graphically rich content. It is commonly used in computer-to-plate and digital proofing technologies. The "creator" of the file (designer, ad agency, prepress provider) must produce a file that meets the minimum imaging requirements of the "receiver" (prepress provider, printer). PDF/X is a PDF file with restrictions intended to facilitate the transfer of files from "creator" to "receiver".

The standard PDF format for printing had been PDF/X1a:2001 until 2009. This file format was used for transfer of files across all printing types. PDF/X1a:2001 required postscript to create the file and processing to rip the file. In 2000, Adobe gave the PDF format to ISO to become an international standard. The PDF's created were made with each vendor's postscript processing, rendering different results in most systems. The postscript processing was the limiting factor to early PDF's and PDF/X1a. Postscript engines had to continuously add patches and fixes to adapt to tools in the creative software. Postscript could not consistently manage color, address blending modes, overprints and fonts were incomplete. If text was changed depending on the system it could reformat, have wrong characters and could flow differently from system to system.

Over the last 15 years, ISO and most rip manufacturers have worked together to create the updated PDF standards. ISO 12647-7, GWG PDF Workgroup and *FIRST* recommends using PDF/X-4:2010 for the files format. PDF/X-4 is known for complete blind transfer (all inclusive). PDF/X-4 rips are not required to use PostScript processing. Most manufacturers have integrated the Adobe PDF engine into their systems. Processing is now PDF to PDF, without the need for PostScript processing. Utilizing the tools available in PDF/X-4 color, blending modes and overprints can be managed correctly. The entire font library is required however, most of the limitations from postscript processing are corrected.

ISO 15930-7:2006 (Graphic technology — Prepress digital data exchange using PDF Part — 7: Complete exchange of printing data (PDF/X-4) and partial exchange of printing data with external ICC profile reference (PDF/X-4p) using PDF 1.6) is both an application standard and a file format standard. It defines a set of conditions to govern the creation, viewing and imaging of PDF/X-4 files. Content and structure of PDF/X-4 files are limited to ensure imaging integrity. The objective is to create a blind exchange of graphically rich content ready for imaging. Note: Be careful not to use Adobe CS4 or CS5 PDF/X-4 as the correct settings from Adobe are only in CS6 and Adobe Creative Cloud listed as PDF/X-4:2010.

Ghent Workgroup Packaging Specifications – 2012
The Ghent PDF Workgroup (GWG) is an international group comprised of graphic arts users, associations and developers whose goal is to establish and disseminate process specifications for best practices in graphic arts workflows. The Ghent Workgroup specifications establish a minimum threshold. If a file fails to meet these minimum specifications, a warning flag will appear. Even if a file conforms to Ghent specifications, flexographic segment-specific capabilities must also be considered.

While the GWG Packaging Specification is largely PDF/X compliant, there are packaging specific deviations from this rule. This section summarizes the rules for an ISO 15930-7:2006 compliant PDF file and identifies the GWG Packaging Specification-2012 exceptions specific for flexography. *FIRST* supports the GWG Packaging Specification-2012. The ISO 15930-7 standard and the Ghent Workgroup Packaging specifications are ever evolving. While the specifications outlined are current, consult the Ghent Workgroup website periodically for updates. For additional information, refer to the Ghent PDF Workgroup contact information in Appendix A.

11.1.2.1 Terminology & Guidelines

"Shall" = Mandatory requirement. Requirement MUST be followed when creating a PDF file. A preflight violation will result in a preflight error.

"Should" = Suggested but not required. Requirement SHOULD be followed when creating a PDF file. A preflight violation will result in an informational message.

"The preflight report should mention" = Informational only. Requirement may be ignored when creating a PDF file. A preflight violation will result in an informational message.

11.1.2.2 PDF/X Compliancy Requirements

Outlined below are the actual requirements a PDF file must conform to in order to be compliant. The Ghent Workgroup Packaging Specification-v2 flexographic specific exceptions are noted. The deviations are clearly marked in the title as [PDF/X Exception].

File/Document Creation Requirements

PDF Version [PDF/X Exception]: PDF/X imposes the rule that the PDF version of a compliant file must be 1.3 or lower. This rule is relaxed. The preflight report should mention the PDF version if it is 1.3 or less.

File Encoding & Compression: To minimize file size, the data in a PDF file must be compressed where possible, compliant with the PDF/X standard. Compression should not be used on metadata inside the PDF file, so that such metadata is readily accessible. Object compression as introduced by version 1.5 of the PDF format must not be used in a PDF file. ASCII encoding should not be used.

Document Title [PDF/X Exception]: PDF/X imposes rules on the document creation and modification dates. These rules are not imposed in the Ghent Workgroup Specifications.

Document Dates [PDF/X Exception]: PDF/X imposes rules on the document creation and modification dates. These rules are not imposed in the Ghent Workgroup Specifications.

Trap Flag [PDF/X Exception]: In PDF/X documents, the state of the "trapped" flag must indicate that either the document is not trapped or that it is trapped. This demand is not imposed in the Ghent Workgroup Specifications.

Use of Adobe PDFWriter: A PDF file must not be created with the Adobe PDFWriter product.

Page Set-Up Requirements

Page Boxes [PDF/X Exception]: PDF/X imposes rules on the different page boxes in PDF/X documents. These rules are not imposed in the Ghent Workgroup Specifications.

Number of Pages: A PDF file shall not contain more than one page.

Use of Empty Pages: A PDF file shall not contain pages that are completely empty (have no PDF objects). Objects that are completely outside the trim box (and thus will not appear on the finished page) are not taken into account.

Page Scaling [PDF/X Exception]: A PDF file should not use the page-scaling factor introduced in PDF 1.6. This ensures that a page will be printed with the same scale factor as it is displayed in Adobe Acrobat or Adobe Reader.

Text/Font Requirements

OpenType Fonts [PDF/X Exception]: The use of embedded fonts, which are embedded as OpenType is allowed.

True Type Fonts: The preflight report should mention all occurrences of True Type fonts in the PDF file.

Type 3 Fonts: The preflight report should mention all occurrences of Type 3 fonts in the PDF file.

Multiple Master Fonts: A PDF file shall not use Multiple Master Fonts. It shall not contain a uninstantiated Multiple Master Font, or instances of a Multiple Master Font.

Composite Fonts (Double Byte): A composite font is the official name used in the PDF reference manual for the type of fonts typically used, but not exclusively, for the representation of large character sets such as Japanese, Chinese, etc. If a PDF file contains any composite fonts, this information should be included in the preflight report.

City Font (Macintosh System Fonts): If a PDF file contains any font named "Athens", "Chicago", "Geneva", "London", "New York", "San Francisco", "Venice", "Monaco", etc. (typical Macintosh system fonts), this information should be mentioned in the preflight report.

Font Style: If a PDF file contains any font that has been styled with any characteristic (such as bold, italic, or outline) that is not an inherent characteristic of said font (artificial style applied), this information should be included in the preflight report.

Text: For the "Packaging Design" specification, the PDF should not contain text smaller than 4 point. For the "Packaging Flexo" specification, the PDF should not contain text smaller than 5 point. Refer to Section 12.2 for segment specific minimum type size specifications.

Text - 2 or More Colors: For the "Packaging Flexo" specification, text which is colored with two or more printing colors should be no smaller than 14 point. For the "Packaging Design" specification, text that is colored with two or more printing colors should be no smaller than 8 point. Refer to Section 12.2 for more guidance on producing multicolor text.

Black Text: A PDF file should not use black text smaller than a certain point size that is set to knockout. For the "Packaging Flexo" specification, such text should not be smaller than 12 point. Refer to Section 12.2.3 for additional information on Process Reverse/Knockout Text.

White Text: A PDF file shall not contain white text set to overprint.

Invisible Text: If a PDF file contains text which uses neither stroke nor fill, this information should be included in the preflight report.

Line Art Requirements

Line Art: For the "Packaging Flexo" specification, all line art should not have a line weight smaller than 0.30 point. For the "Packaging Design" specification, all line art should not have a line weight smaller than 0.15 point. Refer to Section 12.2 for segment specific Minimum Rule Widths.

Line Art - 2 or More Colors: For the "Packaging Flexo" specification, all line art which is colored with two or more printing colors should not have a line weight smaller than 1.50 point. For the "Packaging Design" specification, all line art that is colored with two or more printing colors should not have a line weight smaller than 0.30 point. Refer to Section 12.2 for additional information.

Invisible Line Art: If a PDF file contains line art, which uses neither stroke nor fill, this information should be included in the preflight report.

File Size Requirements

Image Resolution: The resolution of color and grayscale images shall not be below 240dpi for the "Packaging Flexo" specification and 72dpi for the "Packaging Design" specification. In addition, no image resolution should be above 450dpi for "Packaging Flexo" specification. The "Packaging Design" has a maximum resolution of 150dpi. Resolution of 1-BIT images (either regular images or image masks) shall not be below 2000dpi or above 3600dpi.

16-bit Images [PDF/X Exception]: For the "Packaging Flexo" specification, the PDF shall not contain images that use 16 bits per channel. For the "Packaging Design" specification, no such restrictions are imposed.

Image Compression: The section "File Encoding & Compression" imposes general restrictions on compression. This section imposes additional requirements on the compression of images. For both the "Packaging Design" and "Packaging Flexo" specifications, all color or grayscale images should be compressed using ZIP compression. All 1-BIT images should be compressed using ZIP or CCITT compression.

General Requirements

Layers [PDF/X Exception]: Layers (optional content) is explicitly allowed. Any use of it should be mentioned in the preflight report.

JavaScript [PDF/X Exception]: The use of JavaScript is allowed.

Actions [PDF/X Exception]: The use of Actions is allowed.

Trapnet Annotations [PDF/X Exception]: PDF/X imposes rules on Trapnet annotations; these rules are not imposed in the Ghent Workgroup Specifications.

PDF/X Version Key [PDF/X Exception]: PDF/X imposes rules on the PDF/X version key; these restrictions are not imposed in the Ghent Workgroup Specifications.

Pre-separated Pages [PDF/X Exception]: PDF/X imposes restrictions on pre-separated pages; these restrictions are not imposed in the Ghent Workgroup Specifications.

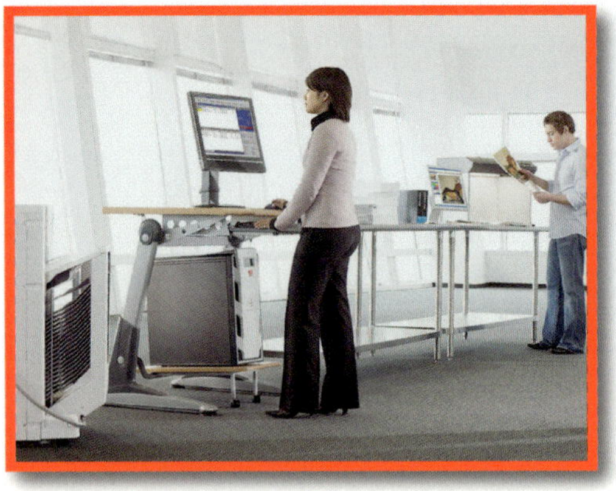

11.2 File Exchange: *The provider of digital files must identify the software programs and type fonts used to create the original files.*

Unknown Objects [PDF/X Exceptions]: PDF/X imposes rules on unknown objects in the page contents; these restrictions are not imposed in the Ghent Workgroup Specifications.

Annotations [PDF/X Exception]: A PDF file shall not contain annotations that are set to print; all other rules on annotations, as imposed by the PDF/X standard, are not imposed by these specifications. For the "Packaging Flexo" specification, the preflight report should mention the use of annotations of the following types: Text, Link, FreeText, Line, Square, Circle, Polygon, Polyline, Highlight, Underline, Squiggly, Strike-out, Stamp, Caret, Ink, and Popup. All other annotation types are allowed. The "Packaging Design" specification does not impose any restrictions on the type of annotations used.

11.2 File Exchange

The provider of digital files must identify the software programs (brand name and version identification) and type fonts (style, manufacturer and version) used to create the original files. Refer to Section 4.1.8 for type font transfer guidelines. Include any other pertinent information that may assist the receiving party in opening, accessing, compensating and outputting the files (such as identifying the data compression convention used). *FIRST* recommends supplying two files. One file with type outlined to show design intent and a second file that is a live text document enabling the prepress supplier to access the copy.

Prior to sending data, all participants must agree on the method of delivery of digital files (CD-DVD, jump drives, electronic transmission, etc.). All potential recipients of digital files must provide their requirements with respect to standard data formats, proprietary file formats (brand and version) and ability to perform any desired edits or modifications.

An electronic or hard copy job jacket should accompany all digital files. All business information originating from the customer (information encoded in a bar code, type of bar code symbol required, product identification number, SKU number, or other packaging identification codes) should be transmitted with digital files in either digital or hard copy format.

Unless remote proofing is implemented and robustly controlled, a profiled contract proof must be supplied with all digital files, regardless of delivery process. This proof should include all of the required sign-off documentation. Refer to Section 16.0 for proofing information.

Hardware & Software Considerations
Many design applications are available to the designer, and all can be integrated into a design-to-prepress workflow. Packaging files supplied today typically originate from Adobe Photoshop for continuous-tone images and Adobe Illustrator for linework elements. When the designer and prepress supplier use the same version of the application, with the same fonts and plug-ins, there is less opportunity for error.

Many packaging prepress suppliers require extra tools to manipulate files. There are numerous proprietary image-manipulation applications in use today from a variety of suppliers that bring packaging-specific solutions to prepress problems in both Macintosh and PC environments. Prior to the transfer of files to a proprietary system, the prepress provider should outline the fonts to avoid the possibility of text reflow.

After a digital asset has been manipulated, customers often require the file to be returned for future use. Therefore, backward compatibility is important in any prepress workflow.

Ghent PDF workgroup has developed settings for testing in different industry sectors. Your workflow and rips should process the test files from the GWG test suite to determine compatibility and settings required for your system and files. To obtain these files vist: http://www.gwg.org/download/test-suites/.

CxF Spot Color Communication
ISO 17972 is the new standard for "Color Exchange Format" for communicating color data across systems and users. The standard is broken up into 4 parts.

- ISO 17972-1 Graphic technology - Colour data exchange format (CxF/X) - Part 1: Relationship to CxF3
- ISO 17972-2 Graphic technology - Colour data exchange format (CxF/X) - Part 2: Scanner target data (CxF/X-2)
- ISO 17972-3 Graphic technology - Colour data exchange format (CxF/X) - Part 3: Output target data (CxF/X-3)
- ISO 17972-4 Graphic technology - Colour data exchange format (CxF/X) - Part 4: Spot colour characterization data (CxF/X-4)

Color Exchange format was developed in the mid 90's, then given to ISO in 2006 to be used in a standard format to communicate color data between users and systems. CxF has been evolving for several years as the standards have been developed. Development on this standard has been done through the FTA with Ghent PDF Workgroup, ICC, and ISO to create a open format needed for the packaging industry.

12.1 Image Trapping: *Image trapping is accomplished through the use of chokes and spreads. This technique should be used when two colors are adjacent to each other whether the graphics are line or screen.*

12.2a Serif vs Sans Serif Fonts: *Serif fonts have decorative details at the end of each stroke where as Sans Serif fonts do not.*

ISO 17972-4 is used for spot color communication, passing color data from brand owners through design files to proofing systems, ink formulation and process control solutions. CxF X-4 is a custom resource that contains the spectral reflectance data of the color. Providing the spectral data provides the receiver with ink formulation required for brand consistency. While the ICC profile embedded in the PDF/X defines the CMYK images and senders intent, there is currently no way to define spot colors or include a 7 color ICC profile. To include color and colorant information, alignment between PDF/X and CxF has been integrated.

Several solutions are available to use CxF for placing ink data in the design file. The file can then be saved as PDF/X adding ink and color data in the PDF/X for delivery through the supply chain. The tools included in ISO 17972-4 are aligned with *FIRST*. If printing solids only a solid patch shall be communicated. If tints are built into design at least a solid and 50% shall be defined. Ideal communication of spot color should include 9 evenly spaced patches with a solid and substrate, over substrate and over black.

The full CxF/X-4 provides data needed for users working downstream in the supply chain, and provides the required tool for compliance to ISO 12647-6 Flexographic Printing Standard.

12.0 JOB ASSEMBLY

12.1 Image Trapping

Image trapping is accomplished through the use of chokes and spreads. This technique should be used when two colors are adjacent to each other, whether the graphics are line or screen. While the minimum amount of image trap varies by press, the following chart summarizes "typical" image trap tolerances. Print system variables that will influence image trap include: general press design & condition, gear wear, print cylinder TIR, parallelism, tension controls and substrate. Printers must first optimize print variables and then determine minimum trap during the press fingerprint trial. Refer to Section 1.3 for more information on print optimization and fingerprint trials.

12.2 Text & Graphic Elements

Fine Serifs, medium or small type and thin lines must comply with *FIRST* specifications. All text printed on a line deck must be printed in a single color. Combining tints with text and/or solids should be avoided. Determine specifications with the printer prior to producing final graphics. Minimum positive and reverse type size and rule width is determined during the press

Total Trap Tolerance: General Guidelines *Trap tolerance is print system dependent; determine minimum trap with print optimization and fingerprint trials (ref. 1.3.1 & 1.3.2)*				
Print Segment		**Color-to-Color**		**Printer Specific**
Wide Web	Preprint Linerboard	Total Trap	</= 0.0156” (1/64)	
			</= 0.3969mm	
	Combined Corrugated	Between Station	</= 0.0625” (1/16)	
			</= 1.5875mm	
		Through the Press	</= 0.125” (1/8)	
			</= 3.175mm	
	Folding Carton	Total Trap	</= 0.0156” (1/64)	
			</= 0.3969mm	
	Multiwall Bag	Total Trap	</= 0.0313” (1/32)	
			</= 0.7938mm	
	Film Products	Total Trap	</= 0.0156” (1/64)	
			</= 0.3969mm	
	Newsprint	Total Trap	</= 0.0156” (1/64)	
			</= 0.3969mm	
Narrow Web	Paper Products	Total Trap	</= 0.0156” (1/64)	
			</= 0.3969mm	
	Film Products	Total Trap	</= 0.0156” (1/64)	
			</= 0.3969mm	
	Envelope	Total Trap	</= 0.008” (1/125)	
			</= 0.2032mm	

Table 12.1

fingerprint trial. Various rule widths and type sizes of both Serif and Sans Serif type, in positive and reverse format, should be included in the fingerprint test design to identify minimum specifications for each print system. Variables such as substrate, anilox volume, ink metering system, mounting tape compression and ink viscosity will influence the minimum achievable type size for each condition.

Minimum type specifications are normally expressed as “points,” which relates to the height of the characters. 1 point = 1/72”. The minimum specifications differentiate between Serif fonts and Sans Serif fonts. Serif fonts have “non-structural details” at the end of each stroke. A common Serif font is Times Roman. Sans Serif fonts do not have these “non- structural details” at the end of the strokes. A common Sans Serif font is Helvetica.

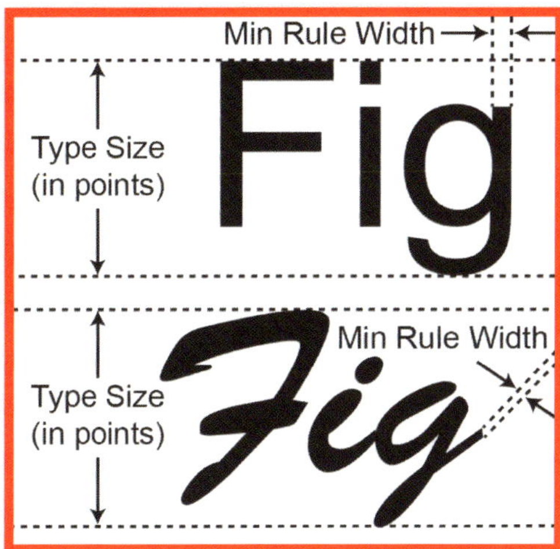

12.2b Minimum Rule Width: *Minimum rule width should be used when evaluating Kanji characters and some script type fonts with thin strokes.*

12.2.1 Line Color Type: *The heavier the solid ink coverage, the larger the type must be to prevent the type from filling in.*

12.2.2a Multicolor Text: *Text created out of more than one color will look unsightly due to the misregistration that is inherent in the printing process.*

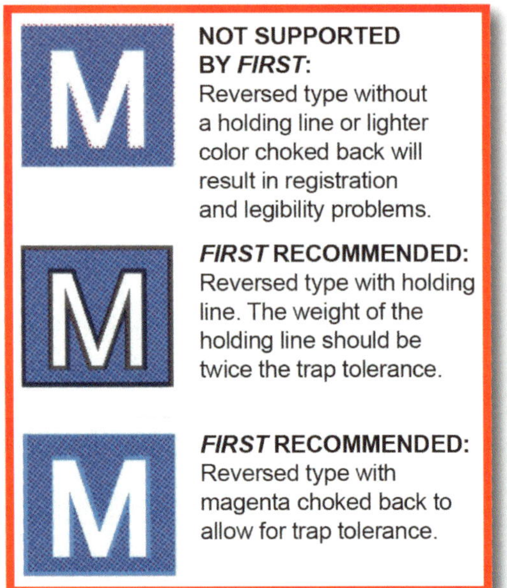

12.2.2b *FIRST* Recommendations

It is because of the "non-structural detail" at the end of each stroke that the minimum Serif type size is typically larger than the minimum Sans Serif type size for a given print segment. The decorative detail of Serif fonts will fill in faster with reverse print and plug or print dirty with positive print; thereby, necessitating a larger minimum type specification.

However, some fonts do not relate to these recommendations because of their unique design. In these cases, it is better to use the Minimum Rule specifications. Referencing the minimum rule ensures the thinnest lines of the font are not thinner than the minimum specifications. Kanji characters and some script type fonts are examples of fonts better suited to the Minimum Rule specification.

12.2.1 Line Color Type and Graphic Elements

Line colors are typically printed with heavier ink coverage than process colors. Avoid combining tints with text and/or solids, and ensure that text is created from a single color to minimize press registration issues.

12.2.2 Process Color Type and Graphics

When type is created with multiple colors, the dominant color should be used to hold the type shape. Generally, lighter color(s) will be choked back to eliminate any obvious misregistration that will likely occur in the printing process. Lighter colors used to create multicolor elements should be choked back by the minimum trapping requirement.

Keylines can be used; typically, a dark color is used to mask misregistration. The keyline thickness should be double the minimum trapping requirement. Care should be taken to preserve the designer's original intent where possible.

12.2.3 Process Reverse/Knockout Text

When type is reversed out of a color and has no holding line, the use of a single color outline will greatly enhance the reproduction capabilities of the process and the impact of the final printed product. Fine Serifs, medium and small type, and thin lines should be avoided when possible. Reverse copy printed on a line print deck should be limited to one color. Type and rule lines should be confined to sizes no less than those indicated in Table 12.2.3a and 12.2.3b.

Font graphics that are too small will not image onto the plate, and will not print. Therefore the Minimum Type Size and Minimum Rule Widths tables help the designer and prepress provider avoid

Minimum Type Size: General Guidelines *Minimum type size is print system dependent; determine minimum type size with press fingerprint (ref. 1.3.2)*										
Print Segment		**Substrate**	**Positive**		**Reverse**		**Printer Specific**			
							Positive		**Reverse**	
			Serif	Sans Serif	Serif	Sans Serif	Serif	Sans Serif	Serif	Sans Serif
Wide Web	Preprint Linerboard	All	8 pt	6 pt	10 pt	8 pt				
	Combined Corrugated	White Top	8 pt	6 pt	10 pt	8 pt				
		Coated Paper	6 pt	4 pt	8 pt	6 pt				
	Folding Carton	All	6 pt	4 pt	8 pt	6 pt				
	Multiwall Bag	Coated Paper	8 pt	6 pt	12 pt	10 pt				
		Uncoated Paper	10 pt	8 pt	18 pt	12 pt				
	Film Products	Polyester	8 pt	6 pt	12 pt	10 pt				
		Polypropylene, Polyethylene & Metallized	8 pt	6 pt	10 pt	8 pt				
	Newsprint	Uncoated Paper	10 pt	7 pt	11 pt	10 pt				
Narrow Web	Paper Products	All	6 pt	4 pt	8 pt	6 pt				
	Film Products	All	6 pt	4 pt	8 pt	6 pt				
	Envelope	All	6 pt	4 pt	8 pt	6 pt				

Table 12.2.3a

this issue. Because minimum type size and rule width are print system dependent, consult the printer to confirm.

12.2.4 Overprint Type

Overprint type consists of type that will print over a screened area or a line color without the use of a KO (knock-out) or trap. Overprint type is used when the letter widths are smaller than the minimum trap width, or where the design allows the printer to avoid unnecessary trapping issues. Overprint type is normally limited to a single color.

12.3 Vignettes/Gradations

The terms vignette, gradation, fade-away, fountain, degrade and graduated tint are now used interchangeably, though vignette was the term adopted first. Similar consideration is also given to the term drop shadows, even though this term refers to a different effect.

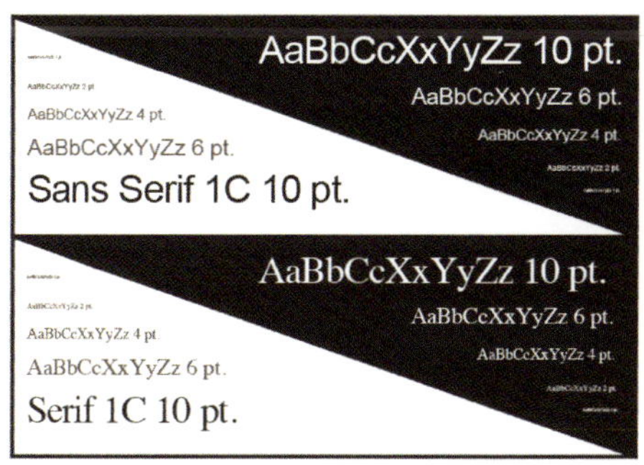

12.2.3a Minimum Type Size: *Using type sizes below the printer's minimum recommended size can result in type filling and is not supported by FIRST.*

Minimum Rule Width: General Guidelines *Minimum rule width is print system dependent; determine minimum rule width with press fingerprint (ref. 1.3.2)*						
Segment		**Substrate**	**Positive Rule**	**Reverse Rule**	**Printer Specific**	
					Positive Rule	**Reverse Rule**
Wide Web	Preprint Linerboard	All	0.010"	0.015"		
			0.254mm	0.38mm		
	Combined Corrugated	White Top	0.013"	0.020"		
			0.33mm	0.51mm		
		Coated Paper	0.007"	0.010"		
			0.18mm	0.254mm		
	Folding Carton	All	0.006"	0.008"		
			0.15mm	0.20mm		
	Multiwall Bag	Coated Paper	0.007"	0.010"		
			0.18mm	0.254mm		
		Uncoated Paper	0.013"	0.020"		
			0.33mm	0.51mm		
	Film Products	All	0.007"	0.013"		
			0.18mm	0.33mm		
	Newsprint	All	0.007"	0.013"		
			0.18mm	0.33mm		
Narrow Web	Paper Products	All	0.005"	0.010"		
			0.13mm	0.245mm		
	Film Products	All	0.004"	0.008"		
			0.10mm	0.20mm		
	Envelope	All	0.007"	0.010"		
			0.18mm	0.254mm		

Table 12.2.3b

12.2.3b Minimum Rule Width:
Determine the minimum positive and reverse rule width for the specific print conditions during the press fingerprint.

Vignettes

Vignettes are probably one of the most difficult elements to print with good quality and stability using flexographic printing. These elements will magnify any printing problems on press, and they are susceptible to unpleasant banding (places where the transition is not smooth or where it shows lines), or dropping off (what we usually call a "hard edge"). However, in speaking favorably about vignettes, the quality of most designs would be improved by adding a more contrasted and holistic design in comparison to a plain solid. Also for the printer, the benefit of printing vignettes instead of solid colors is that it is hard to find a place where it's possible to compare those solid colors against a Pantone number. The idea of matching a single color is fundamental, and for printing vignettes the color itself is not as relevant, but the use of vignettes will allow for more press variation than when using a spot color.

Vignettes in flexo won't correspond to a single standard. While one press can print a beautiful and smooth vignette in a high linescreen, another press wouldn't be able to repeat the same result. This means that in order to print good vignettes, it is always necessary to fingerprint the press for these types of elements.

Factors Influencing Banding

Many factors that influence banding in a vignette relate to the construction of the vignette. There is a mathematical relationship between the length, range and the number of steps in a vignette. The length refers to the physical length of the vignette. The range refers to the difference in color across or down the vignette. For example, a vignette of 30% to 50% has a range of 20%.

- The longer the vignette, the more likely it is to show banding
- The shorter the range of the vignette, the more likely it is to show banding
- The fewer steps used, the greater the potential for banding
- Banding is more visible with darker inks
- Lower screen rulings are less likely to show banding
- Higher output resolutions help to minimize banding

Factors Influencing Hard Edges & Dirty Print

Generally, to avoid hard edges and dirty print, it is important to maintain the printer's minimum dot and not fade to zero. However, screening and plating technology now exists that can allow printers to fade to zero without setting a minimum dot. An agreement on the specifications must be reached between the sender and receiver of files and plates. The printer specifies the minimum dot used along the edge of any vignette. The lightest area of the vignette should adjoin a holding line or the edge of a graphic window; this will ensure hard edges and dirty print do not appear across the vignette when the dot fades to the printer's minimum. When vignettes are made of more than one color, all colors must stop at the same place in order to prevent rainbowing and dirty print throughout the vignette.

12.3.1 Designing Vignettes

Vignettes can be designed in primarily 2 different ways:

1. Using a line-work editor (ie. Adobe Illustrator, etc.). In this case the vignette will always be a vector, depending on the program, and it will have different levels of grays.
2. Using an image editor (ie. Adobe Photoshop). In this case the vignette will be a raster image, and it can be managed as such.

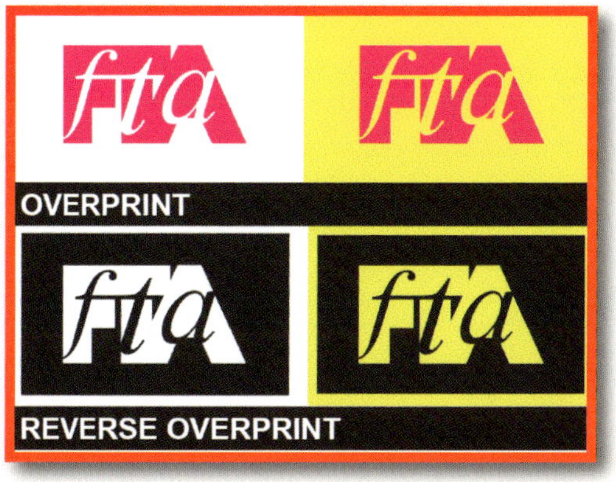

12.2.4 Overprint Type: *The top of the image illustrates an overprint. The magenta logo prints on top of the yellow background to produce a red overprint – no registration concerns. The bottom of the image illustrates a reverse overprint. The second down color (black) contains the reversed image. When black overprints yellow, the reversed image appears as yellow – again no registration concerns.*

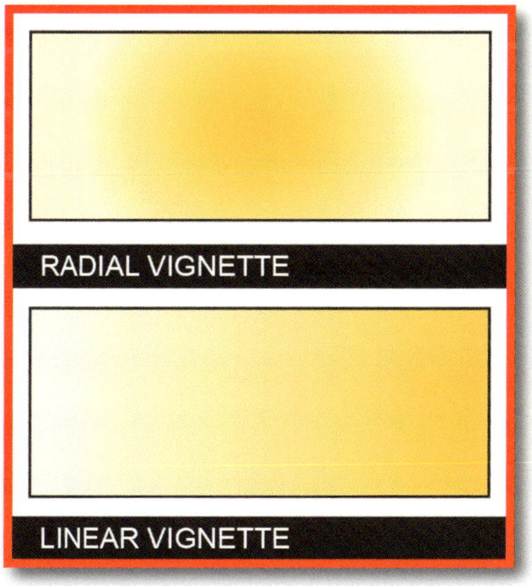

12.3 Vignettes: *The lightest area of the vignette should adjoin a holding line or the edge of a graphic window; this will ensure hard edges and dirty print do not appear across the vignette when the dot fades to the printer's minimum.*

In today's graphic industry there is not a common preference about which way works better. Some people prefer to work with these elements as images because they feel they can better control the effects and the minimum dots. Others think that a line-work vignette will provide more control on transitions.

It is necessary to understand what possible outcomes are implied in both cases, and we must include them in the vignette fingerprint chart.

Vignettes Best Design and Construction Practices

Single Color Vignettes

Banding and hard edges typically reflect printing issues, but there are still many things we can do in the design and construction of vignettes to reduce this problem. We need to understand that there is a relationship between the length of the vignette (how long it is) and the range of the vignette (how many steps your gradation will have; for example it may range from 30% to 70%, so in this case the range would be 40%). The third factor is the number of steps, or basically, determining how you want to distribute the vignette (ie. linear vignettes will have 2 steps).

The longer the vignette, the more likely it is to show banding or undesired mottling. If the range of the vignette is shorter, it is also more likely to show banding. Also, the fewer the steps used to distribute the vignette, the greater the potential for banding. Vignettes that cross the transition zone (mid-tones) are more likely to show banding than those that don't. Additionally:

- Banding is more visible with darker inks
- Lower screen rulings are less likely to show banding
- Higher output resolutions help to minimize banding

Vignettes that end in 0% are very likely to show hard edges, although some technologies will help you to improve this, although testing is still required. If you are creating a vignette that needs to end in 0% and you want to see a minimum dot, it is advisable to not push it too close to that minimum dot, or you will create a situation that is difficult to print. Creating minimum dots for vignettes is only a solution for when the area you need to cover is very small.

Two Color Vignettes

If you have to create a vignette where you need to combine two colors together, you need to be very careful about the colors you are mixing. Vignettes print better when the colors are closer in hue to one another than when they are far apart. If the colors are far apart in hue, you need to know that somewhere in the transition you may create a different color. If the colors are opposite (ie. cyan and orange, magenta and green, or yellow and blue) and you make them blend together, you will create a gray color in the middle of the vignette. It is easier to make transitions of colors that are very different in brightness, like a very light color and a very dark one. If you can include in the design a white space in the middle of both colors (so they don't mix together), then you will be more likely to hold the hue of both colors better. Don't make the transitions (from mid-tones to ¾ tones) to be coincident in the blend. Transition points are harder to print, and if you combine two inks it will become difficult to reproduce on press.

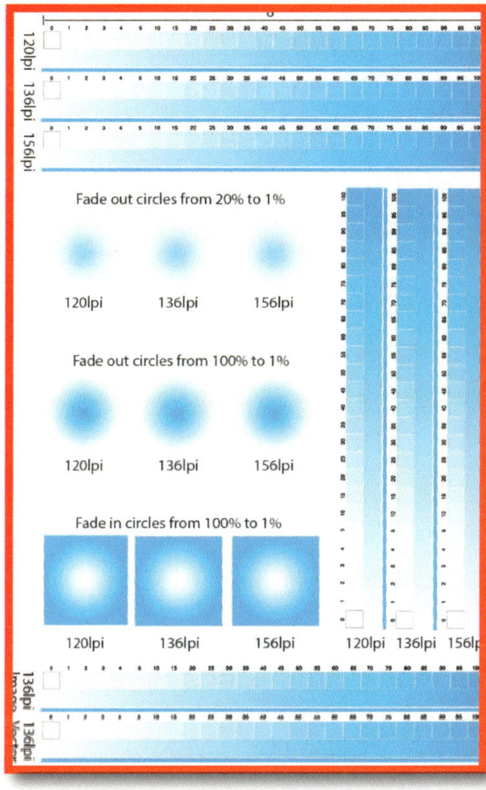

12.3.2 Vignette Fingerprint Test Form

12.3.2 Vignette Fingerprint

FIRST recommends running a fingerprint for vignettes. It is important to understand that vignettes are going to be used in different ways. Some vignettes require a spot color with a high density, others require a fine detail and in other cases the vignette will use process color inks. By analyzing these different conditions, we will know what settings we need to use, depending on the case in question. It is impossible to profile all the potential combinations, but we need to understand the most common ones.

In order to do this vignette fingerprint test you must define the following variables:

1. Working with your desired high linescreen, select 2 lower linescreens. (ie. using 156lpi linescreen, select 133lpi and 120lpi.)
2. Select 2 or 3 common anilox rolls. (ie. 1000lpi roll for 156lpi, select at least one 700lpi or 500lpi anilox roll with a higher BCM)
3. Select 2 different types of mounting tape; use a medium density and softer density.
4. Select your process inks; they could be cyan (as indicated in Image 12.3.2) and also one common spot color (darker colors work best). You may also want to test your process black, since it is very common to use this ink for drop shadows and shades.
5. Add any other variables to this test that you consider useful (ie. plate type, ink type, blend, etc.).

12.3.3a Shadows: *If the edge is dropping off to white, special screening may be used to lessen the appearance of a hard edge.*

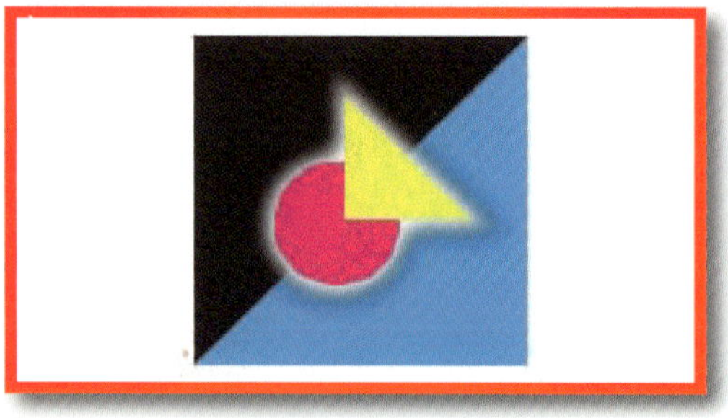

12.3.3b Blurs: *The lightest area of the effect should fade to the printers minimum dot percentage.*

After selecting the colors and variables make a table. For our example the table will consist of:

- Anilox rolls
- Mounting tape
- Ink

Image 12.3.2 is an example of the Vignette Fingerprint Test Form. Regarding the possible combinations, select one that makes the most sense. If you are only using 1000lpi anilox rolls for the cyan ink, then this is the one you need to test. The same goes for the mounting tape density.

There are 3 blocks of vignettes for different linescreens, and the notion is that these vignettes are generated by your preferred method (vector or raster). The size of the vignette must be at least 8", because a shorter vignette won't be enough to analyze mottling in the transition zone. On the right side of Image 12.3.2 there is a 1-1 comparison between the preferred method and the alternative method (vector vs. raster). In the middle of Image 12.3.2 there are radial vignettes. The purpose of these is to see how smooth the transitions are when approaching zero. The long transitions will show the hard edges, as well as inverse ones. The idea is to have an actual comparison between them for different job configurations.

12.3.3 Transparency/Effects

Transparency and effects are common ways to produce design and production enhancements to vector and raster elements within software applications. These techniques can often create poor or unexpected results if not constructed and applied properly. Output of effects are often difficult to predict, on screen these effects are live, final output requires flattening which can affect underlying object quality and ultimately design intent.

Design Factors to Consider when using Transparency and Effects

There are many ways to produce or introduce transparency into a design. Opacity, soft shadows, blurs, feather and glows are all examples of effects that may add transparency to a design. Most objects, groups, or even layers can have effects applied to them but it is important to only apply effects to objects that actually are needed to produce the intended look. Overuse of this method can add complications to output and in extreme cases failure.

The nature of many effects often resemble and have similar characteristics to vignettes. Refer to Section 12.3. In fact, vignettes can now be created using transparency rather than just a fade between one or more colors. A thorough understanding of current software applications and printer specifications are needed to create and print transparent effects. This is especially true when attempting to overcome factors that influence hard edges and dirty print. Special screening technologies along with careful planning can aid in the successful output of transparency effects.

Transparency and Effects Best Practices

1. Create effects or have software application resolution settings at >/=300dpi.
2. Vector effects are defined in terms of measurement. Scaling or change in resolution generally will not affect the design.
3. Raster effects are defined in pixel amounts. Scaling or change in resolution will have a direct impact on design.
4. Text or artwork not intended to be affected by the transparency effect, should be layered or stacked above the effect.
5. Effect color space recommended to be that of underlying objects.
6. Avoid fade to zero situations by maintaining printer's minimum dot specifications.

12.4 Bar Code Prepress Considerations

Formerly, the Uniform Code Council (UCC) was responsible for managing the bar code system in the USA. The UCC is now the GS1 US organization. GS1 US manages the GS1 system and assigns GS1 company prefixes to companies/organizations in the USA. The most common use of a GS1 assigned company prefix is the creation of UPCs (Universal Product Codes), which contain a 12-digit Global Trade Item Number (GTIN).

The GS1 US publishes the following electronic data interchange guidelines based on the ANSI ASC X12 standard:

- Industrial/Commercial EDI
- Uniform Communication Standard (UCS), used in the grocery industry
- VICS EDI, used in the general merchandise retail industry

The GS1 US is also the code manager for the United Nations Standard Products & Services Code (UNSPSC).The UNSPSC provides an open, global, multi-sector standard for classification of products and services. Applicable commodity codes are available on the UNSPSC website (www.unspsc.org).

VERSION 1 (21 x 21)

VERSION 2 (25 x 25)

VERSION 3 (29 x 29)

12.4 QR Codes: *Illustrated are some common examples of QR codes.*

For more information on design and print considerations for bar codes, refer to Sections 4.3 and 22.0 respectively. To learn more about correcting symbol dimensions to accommodate the addressable output resolution of the output device refer to GS1 US: "Guidelines for Producing Quality Symbols". The GS1 US & UNSPSC contact information is included in Appendix A. For more information regarding 2D Bar Codes reference Appendix G.

QR Codes Design

Unlike the traditional one dimensional barcode, known as UPC, which can be scanned by a uni-dimensional narrow beam of light, the QR code must be detected by a 2 dimensional digital image sensor and then processed and analyzed by an specific software. The processor works by locating the three squares in the corners to give the dimension to the code.

It is important to have the dimension and print respect the interspaces between the code and apply the same considerations descripted for the UPC code.

12.4.1 Bar Code Specifications

Bar code print specifications are produced by combining three types of related specifications:

1. **Application Standards** are published by accredited standards organizations. Bar codes are used in many different applications; for example, one bar code application involves bar coding products for retail checkout lanes while another application is for bar coding shipments for conveyor lane routing in distribution centers. The specifications for bar codes used in these two applications are different because the conditions for scanning the bar codes are very different. Accredited standards organizations provide specifications in the form of guidelines and standards to assist in:
   - Selecting the bar code type to be used
   - Structuring the data inside the bar code
   - Defining the printed human-readable information that is inside the bar code
   - Selecting bar code size within the acceptable range
   - Understanding where the bar code should be placed on the printed product
   - Defining the minimum print quality requirements
2. ***FIRST* Print Specifications** prescribe a minimal level of capability for all compliant printers. These specifications fall within the acceptable limits of the appropriate application standard for the bar code being printed and will assist in:
   - Determining the minimum size for a bar code depending on the printing press and substrate

- Identifying the preferred bar code orientation given the direction the web or sheet will travel

3. **Job Specifications** should be published for film or plate output. This type of specification should assist in:
   - Identifying optimum film/plate output resolution
   - Determining bar width reduction (BWR) required by the specified print conditions

12.4.2 Prepress Provider Responsibilities

The designer, prepress provider and printer all share responsibility for producing quality bar code symbols. The printer will be held responsible for the final verification of the bar code symbol as they are the final step in the process. See Section 22.0 for more details. Prepress providers play a critical role in ensuring that a bar code conforms to all applicable Application Standards and *FIRST* Print Specifications. When creating a production ready bar code, the prepress provider must consider the printability of the design. Variables such as symbol contrast, size and orientation must be evaluated. In addition, the prepress provider is responsible for working with the printer to determine the minimum printable symbol size and the appropriate bar width reduction. Section 1.3.1 outlines an optimization press trial for determining symbol size and corresponding bar width reduction. Refer to Sections 4.3 and 22.0 for additional information on design and print considerations, respectively.

9 01234 56789 8

UPC - A

9 012345 678906

EAN - 13

901234567890

GS1 - 128

901234567890

Code 39

12.4.2a Bar Code Symbology: *The type of bar code depends on many factors including where it will be scanned and how it will be printed. Prepress providers should consult with the customer to determine the appropriate symbology.*

1. Creating Appropriate Symbology
Bar Code Formats

For prepress operations that design bar codes internally, there is software that can produce a full range of bar code types. Some of the common bar code types printed by flexographic printers include:

- UPC – Version A and Version E (including add-on and composite component)
- GS1-128 (formerly known as UCC/EAN-128)
- EAN 8 (including composite component)
- EAN-13 (including add-on and composite component)
- ITF (Interleaved 2-of-5)
- Code 128 (full ASCII character set supported)
- Code 25 (also known as Interleaved 2 of 5, supported with and without check digit)
- Code 93 (full ASCII character set supported)
- Code 39 (supported with and without check code)
- MSI (including option to display data)
- JAN 13 (variation of EAN 13 used in Japan)
- JAN 8 (variation of EAN 8 used in Japan)
- Plessey (hexadecimal character set)

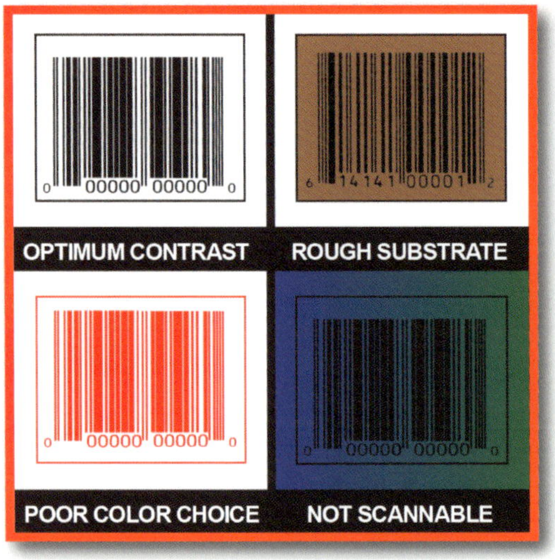

12.4.2b Symbol Contrast: *Using a background color other than white or a bar color other than black reduces symbol contrast and therefore, scanability. Colors should be tested and verified both at the proofing stage and on press.*

- Telepen (including compressed numeric mode)
- 2D Codes
- Codabar (both USS and Traditional format supported)
- USPS 4CB (United States Postal Service Intelligent Mail Barcode)

Encoding Accuracy
It is the responsibility of the supplier providing final files/films/plates to ensure the bar code is properly encoded. A good manufacturing practice for each phase of the production process is to confirm the human-readable text associated with the bar code, matches the information encoded within the symbol. A spreadsheet for calculating check digits can be downloaded from www.flexography.org/FIRST_extras.

2. Separating for Printability & Symbol Contrast
Printability Considerations
Bar code symbols should not be placed on a plate used to print a large solid ink coverage because these plates require extra impression and a higher volume of ink.

Bar codes require bars with sharp edges in order for the scanner to correctly read the code. The bars and the background of a bar code must each be printed in only one color on a single print station. Scanning accuracy is reduced if the symbol bars are printed in two separate print stations due to variation in register. The supplier providing final compensated art must ensure that the bar code's bars and background colors are of the correct color and location on the specified layers.

Symbol Contrast Considerations
The optimum bar code color combination is opaque black ink for the bars and opaque white substrate or ink for the background. Alternate colors can be used with caution:

- Opaque black, dark blue, or green ink for the bars
- White, red, orange, pink, peach, or yellow ink for the background

Using a background color other than white or a bar color other than black reduces symbol contrast and, therefore, scanability. Colors should be tested and verified both at the proofing stage and on press.

It is important to remember that colors with acceptable ANSI/ISO Symbol Contrast on an opaque substrate may be completely unacceptable on an opaque substrate of another color or on a translucent or transparent substrate. When printing on a

Minimum Bar Code Magnification: General Guidelines *Bar code magnification is print system dependent; determine optimum magnification with press fingerprint (ref 1.3.2)*			
Print Segment		**Magnification** *(Machine Direction)*	**Printer Specific Magnification** *(Machine Direction)*
Wide Web	Preprint Linerboard	100%	
	Combined Corrugated *(flute dependent)*	UPC: 110% - 200% ITF -14: 100%	
	Folding Carton	100%	
	Multiwall Bag	115%	
	Film Products	100%	
Narrow Web	Paper Products	80%	
	Film Products	100%	

Table 12.4.2

transparent substrate or colored substrate, a solid light-colored (white is optimum) background with maximum reflectance is recommended in the area where the bar code is to be located. Each ink color should be evaluated for use on a different substrate.

3. Determining Symbol Size
Minimum Magnification/Size

Bar code symbol sizes are specified in an acceptable range for the optics found in the scanning system and vary with symbology. For example, EAN/UPC symbols that are scanned by omni-directional point-of-sale scanners have a fixed relationship between height and width and are specified in a range of magnifications (80%-200%) around a nominal size. Other methods of specifying a symbol's size include providing the area reserved for the symbol (including its quiet zones) or providing the symbol's narrow element width (X-dimension). Printing a bar code below the minimum size specified by the applicable symbol specification, and printing constraints specified by the printer, is not supported by *FIRST*. Table 12.4.2 is only applicable to symbols that are printed with the bars running in the machine direction.

Because the symbol dimensions are slightly adjusted before output, it is necessary to specify the addressable imaging resolution whenever bar code symbols are designed. This is because bar codes, unlike typical graphic images, are machine-readable elements based on predictable decoding formulas. The bar code output size must be based on the resolution of the output device; most likely the film plotter or plate-imaging device. The GS1 refers to these adjustments as "Corrected Magnification" when applied to the symbol's overall size and "Corrected BWR" when applied to the symbol's bar width reduction (BWR). These adjustments should also apply to the special EAN/UPC symbol characters 1, 2, 7 and 8 when output is at a higher resolution. Digital bar code files should only be resized using the bar code design software package that created the symbol and accounts for output resolution.

Combined Corrugated – Bar Code Size Specifications

UPC Symbol: A 200% symbol is the maximum size specified by the GS1 system. Generally, this represents the minimum size recommended for direct printing on combined corrugated. However, many printers with newer presses, thinner printing plates, smaller flutes and improved white substrates are successfully printing in the 110-160% range. A minimum magnification of 150% is specified by the GS1 system for all containers that will be scanned on automated conveyor lines during the distribution stage.

ITF-14 Symbol: The nominal size specified by the GS1 for ITF-14 symbols carrying the GTIN-14 number is based on an X-dimension (narrow bar width) of 0.040" (1.0mm) and a height of 1.250" (31.8mm). Generally, the 100% specification is recommended as a minimum size for direct printing on combined corrugated. If the carton size prohibits the 1.250" (31.8mm) height (ie. the height restrictions of a tray), it may be preferable to print a slightly truncated symbol while leaving the X-dimension at 0.040" (1.0mm).

The current magnification range for ITF-14 symbols is between a minimum of 70% and a maximum of 120% based on GS1 system specifications. These specifications are changing in two ways:

1. If the container is large enough to be scanned on an automated conveyor line scanner during distribution, then the minimum height is fixed at 1.25" (31.8mm) regardless of the X-dimension used for the symbol.
2. To improve scanning efficiency, packages marked with ITF-14 symbols that are scanned in a conveyor-based environment should have a maximum X-dimension of 0.040" (1.0mm), not 0.048" (1.2mm). While current packages marked with ITF-14 symbols with X-dimensions between 0.040" (1.0mm) and 0.048" (1.2mm) remain acceptable based on historical GS1 system specifications, new packages should begin using 0.040" (1.0mm) X-dimensions as the maximum.

Symbol Truncation
Certain symbols have a fixed relationship between their height and width, while others have minimum heights specified. Bar code truncation is a reduction of a symbol's height below the application standard or symbol specification and is not supported by *FIRST*.

Creating Appropriate Size Quiet Zone
The quiet zone is the area, free of printing, that precedes the left bar and follows the right bar in a bar code symbol. The quiet zone allows scanners to detect when a bar code starts and stops. Quiet zones are based on multiples of the symbol's narrowest element width (X-dimension). Minimum quiet zone specifications depend on the symbol specified. For example, the UPC-A symbol requires a quiet zone of 9X on each side, while an ITF-14 symbol requires a 10X quiet zone on each side. USPS4CB codes require a minimum quiet zone of 0.040" (1.0mm) above and below the bar code and 0.125" (31.8mm) on each side of the bar code. Bar codes with the quiet zone omitted, obstructed, or too small are not supported by *FIRST*.

4. Applying Optimum Bar Width Reduction (BWR)
Bar widths increase in flexographic printing in a manner similar to dot gain/TVI. As the bar widths increase, the space between the bars decreases correspondingly. It is best to print bar codes under the same conditions as dots are printed, in order to minimize bar growth (plate material, mounting material, anilox roll). Just as a dot curve is typically applied to a process image to account for expected dot gain/TVI on press, a bar width reduction is typically applied to a bar code prior to output to account for the bar growth expected on press. The amount of BWR should be corrected at the symbol design stage for digital bar codes files. Different print conditions will result in varying bar growth during print. Refer to Section 1.3.2 for a press fingerprint method for bar codes or contact the printer for specific recommendations. Bar code designs that do not conform to the BWR specified by the printer are not supported by *FIRST*.

5. Ensuring Correct Orientation & Distortion
Orientation Considerations
FIRST strongly recommends the bars in a bar code, print parallel to the direction the web is moving through the press to avoid slurring. It is not recommended, but under certain situations the bars in a bar code must be placed in the transverse (across the web) direction. In these cases, the printer should be consulted. It may be necessary to use a larger symbol to meet the minimum print quality requirements specified by the appropriate application standard.

12.4.2c Symbol Truncation: *Symbol Truncation is a reduction of a symbol's height below the application standard and is not supported by FIRST.*

12.4.2d Quiet Zones: *Quiet zones allow scanners to detect when a bar code starts and stops. Quiet zones are based on multiples of the symbol's narrowest element width (X-dimension). Minimum quiet zone specifications depend on the symbol specified.*

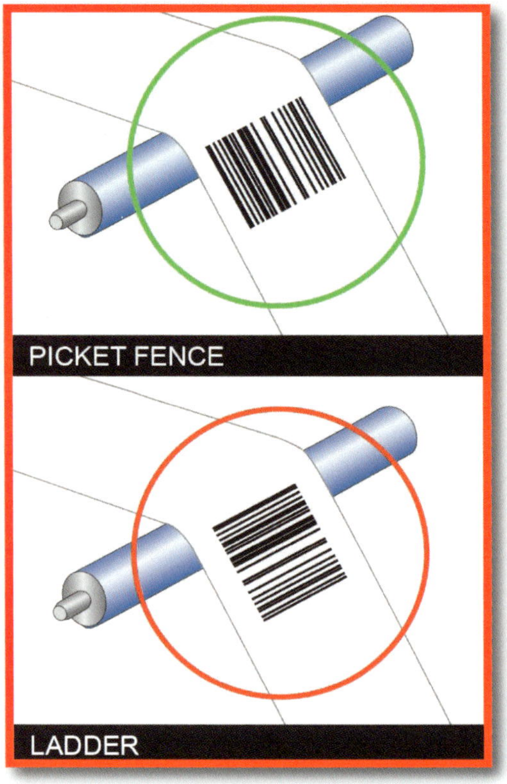

12.4.2e Bar Code Orientation: *Bar code orientation is critical. The top figure illustrates the symbols bars traveling in the machine direction, while the bottom figure illustrates the bars running across the press direction. If print slur occurs with the symbol in the picket fence orientation, the bars grow in length only and are still scanable. However, the symbol printed in the cross- web direction will cause the bars to grow in width, likely causing the code on the printed product to fail minimum grade specifications.*

Distortion Considerations
FIRST does not recommend printing bar codes in the cross direction because of slurring and other problems associated with this orientation. However, if placement in the cross direction is unavoidable and the design requires the bars in the bar code to be placed perpendicular to the direction the web is moving through the press. Distortion of the bar code is necessary to account for the plate cylinder circumference. This distortion will introduce pixel-rounding errors into the bar code unless the bar code design software accounts for this input variable in the design stage. If it does, follow the procedures provided by the software provider. If the software does not account for distortion when the symbol is created and distortion is unavoidable, outputting the file at the highest possible resolution (ie. 4,000dpi) is advised to avoid introducing rounding errors.

6. Documenting & Verifying Symbol Attributes

Documentation
For every film or plate that has a bar code image, the prepress provider should record and include on a CoA the following attributes:

- Bar code type
- Bar code size (magnification)
- Bar code bar width reduction (BWR)
- Output resolution used to generate the film or plate
- Human-readable content

Verification Method
Application: Retailers impose large fines and penalties for bar codes that do not scan properly. Most consumer product companies require ANSI grading of bar codes. The prepress supplier must include a bar code report scanned directly from the film supplied for platemaking regardless of who makes the plates. When an assembled print-ready file is supplied for imaging to the printer or plate provider, it must be accompanied by a written report identifying the human-readable number for the incorporated machine-readable code and the proper bar width reduction (BWR) applied.

Description: An ANSI/ISO based bar code verifier measures several parameters of the bar code to determine if it is scannable. The bar code parameters measured include: symbol contrast, edge contrast, minimum reflectance, edge determination, modulation, defects, decodability and quiet zones. An ANSI grade ("A" through "F") is given based on these parameters, which measures the quality of the symbol. The lowest scoring parameter determines the overall scan grade. For most applications, a 2.5 ("C") grade is the minimum acceptable grade. It is important to grade a bar code to ensure any properly functioning scanner in the marketplace will be able to read the bar code.

Industry Standard: While ANSI X3.182-1990 (Bar Code Print Quality Guideline) was the original bar code quality specification in the USA, the current international specifications are ISO 15416:2000 (Information technology – Automatic identification and data capture technique – Print Quality Test Specification – Linear symbols) and ISO 15415:2004 (Information technology – Automatic identification and data capture techniques – Bar code print quality test specification – Two-dimensional symbols). The methods outlined in these specifications are designed to assess a symbol's quality after it is printed. These methods do not replace the bar-width tolerance measurements taken in the pressroom for process control. These specifications require the use of an ANSI/ISO based verifier that can be used by the printer, prepress provider and consumer product company to assess the quality of the printed symbol and facilitate communication about the results between all parties. The minimum symbol grade required, as well as the verifier's measuring aperture and peak wavelength of light, are determined by industry application standards or symbol specifications depending on the symbol being analyzed and where it will ultimately be scanned. For example, EAN/UPC symbols are measured based on GS1 US "UPC Printed Symbol & Quality Specifications" and ISO 15420:2000 (Information technology – Automatic identification and data capture technique – Symbology Specification – EAN/UPC), which applies the ANSI/ISO method to the EAN/UPC scanning environment. Minimum bar code print quality specifications outside of those determined by the accredited standards organizations are not supported by *FIRST.*

12.4.2f Bar Code Verifier/Scanner

Calibration: To maintain instrument accuracy, bar code verifiers require periodic calibration. Some instruments require the operator to use the calibration patch that accompanies the instrument; other instruments may rely on periodic manufacturer participation in the maintenance process. *FIRST* recommends following the manufacturer's recommended calibration procedures and frequency.

In addition to routine calibration, an ANSI/ISO based bar code verifier must be validated by determining how well it agrees with a universally accepted and traceable conformance standard. Bar code verifiers should comply with ISO 15426-1:2006 (Information technology – Automatic identification and data capture technique – Bar code verifier conformance specification

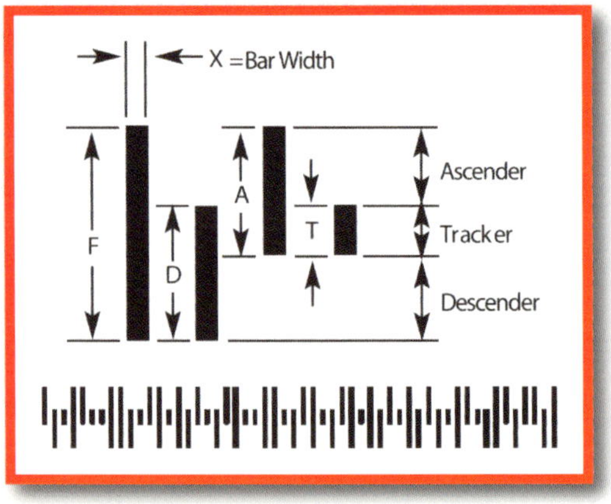

12.4.3 USPS CB4 Bar Code: *The Intelligent Mail Bar Code (CB4) is a 4-state bar code that consists of 65 bars.*

– Part 1: Linear symbols) or ISO 15426-2:2005 (Information technology – Automatic identification and data capture technique – Bar code verifier conformance specification – Part 2: Two-dimensional symbols) verifier conformance specifications. These specifications define test methods and minimum accuracy criteria of verifiers for both linear and 2-dimensional bar code symbols. Reference calibration standards against which bar code verifiers should be tested are also specified. The GS1 US Calibrated Conformance Standard, (a test card for EAN/UPC symbol verifiers) is a reference conformance standard traceable to NIST (National Institute of Standards and Technology). Refer to the Appendix A for ordering information.

12.4.3 USPS Intelligent Mail Bar Code

The information in this section was obtained from the United States Postal Service Intelligent Mail Bar Code specification USPS-B-3200C. For additional information, reference the USPS-B-3200C specification from the US Postal Service. Contact information is included in the Appendix A.

Description

The Intelligent Mail Bar Code (CB4), used by the United States Postal Service (USPS), is a 4-state bar code that consists of 65 bars. The bars are numbered from 1 (leftmost) to 65 (rightmost). Each bar exists in one of four possible states:

- Ascender (extends up)
- Descender (extends down)
- Tracker (has neither an ascender nor descender)
- Full (has both an ascender and descender)

Physical Dimensions

Quiet Zone

- Minimum 0.040" (1.02mm) above and below bar code
- Minimum 0.125" (3.18mm) on either side of bar code

Print Considerations

Background Reflectance: The area of the mail piece where the bar code is located shall be uniform in color. When measured with a Postal Service envelope reflectance meter or equivalent, the minimum reflectance from the background area shall be 50% in the red portions and 45% in the green portions of the optical spectrum.

Print Reflectance Difference (PRD): The PRD is the difference between light reflected from the printed bars in the bar code and the background, when measured with a Postal Service envelope reflectance meter or equivalent. The minimum PRD is 30% in the red and green portions of the optical spectrum.

Horizontal Dimensions: USPS CB4		
Parameter	Minimum	Maximum
Bar Width	0.015" (0.38mm)	0.025" (0.64mm)
Space Width	0.012" (0.31mm)	0.040" (1.02mm)
Overall Bar Code Width	20 bars per inch	24 bars per inch

Table 12.4.3a

Vertical Dimensions: USPS CB4		
Parameter	Minimum	Maximum
Overall Bar Code Width	0.134" (3.4mm)	0.23" (5.84mm)
"Full" Bar Height	0.134" (3.4mm)	0.23" (5.84mm)
"Ascender" Bar Height	0.088" (2.24mm)	0.155" (3.94mm)
"Descender" Bar Height	0.088" (2.24mm)	0.155" (3.94mm)
"Tracker" Bar Height	0.042" (1.07mm)	0.080" (2.03mm)

Table 12.4.3b

Quiet Zone Reflectance: Within the quiet zone around the bar code, background patterns, envelope insert "show through," and all other printing shall be limited to a maximum print contrast ratio (PCR) of 15% when measured in the red and green portions of the optical spectrum using a Postal Service envelope reflectance meter or equivalent.

Requirements for Human-Readable Information

Vertical Position: When human-readable information is required, it shall be printed immediately above or below the bar code but outside of the quiet zone. The human-readable information shall be at least 0.04" (1.02mm) above or below the bar code but not more than 0.50" (12.7mm) above or below the bar code. No other printing is allowed between the bar code and the human-readable information.

Horizontal Position: The human-readable information, when required, shall be printed so that the left edge of the leftmost digit aligns with the leftmost bar of the Intelligent Mail Bar Code.

Content: When human-readable information is required, it shall consist of the 20-digit tracking code and the 5-, 9-, or 11-digit routing code, if present. The tracking code shall include a space between each data field. When the bar code contains a routing code, the 5-digit ZIP code, the 4-digit add-on and

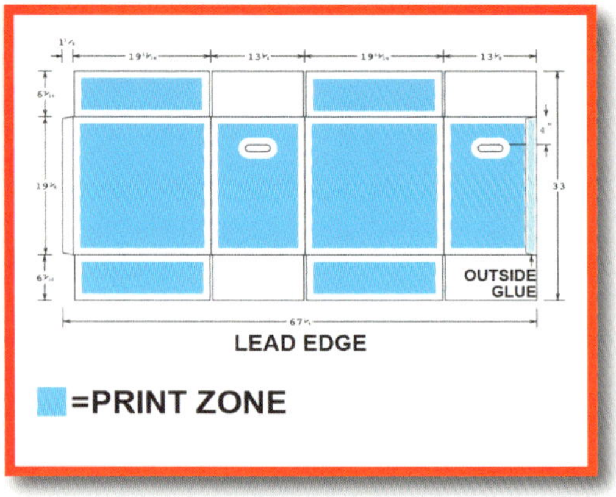

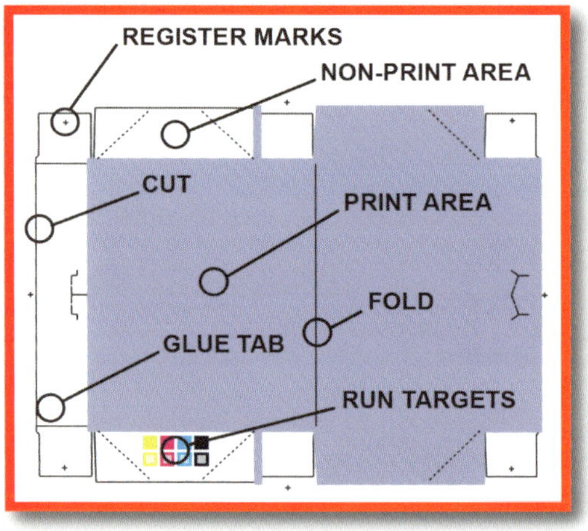

12.5 Template Layout: *The template must be provided by the printer. Include all pertinent information that must be considered during the design and prepress phase.*

the remaining 2 digits shall be separated with a space between data fields.

Font Specification: The human-readable information, when required, shall be printed using a Sans Serif font.

12.5 Template Layout

Template layout refers to a key line, mechanical, die strike, full-scale drawing or die vinyl. Usually, a separate layer is used to identify various elements on the template. All contract proofs must include the template layout. Film overlays on the proof or blueprints are acceptable.

The template layout must be provided to the prepress provider to ensure all job elements are correctly positioned, as required by the finished product, before beginning final job assembly. It is the prepress provider's responsibility to bring to the attention of the customer and design firm any known violations prior to preparing the final file/films/plates. The design firm, in conjunction with the consumer product company, should indicate areas where control targets may be placed to allow for production process control. When the prepress provider places the control targets, it must ensure the non-print areas of the layout are not violated. Refer to the Section 3.2.1 for additional template layout specifications. The file format for exchange must be agreed to by all parties.

The template layout should include, as applicable:

- Overall dimensions
- Cuts/Folds/Scores
- Glue zones
- Quiet zones
- Nonprint areas
- Inside vs. Outside
- Varnish areas
- Print direction
- Flute direction (corrugated)
- Copy limits
- Seal area
- Eyemarks
- Printer's marks (microdots, registration targets, color control bars, impression targets, etc.)

12.6 Eye Marks

Both the packaging manufacturer and the customer's product packing line use eye marks. At the packaging manufacturer, the eye mark communicates the cut-length of the package to the converting equipment. At the customer's packing operation, the eye mark can be used to stop the flow of filling material.

Specifications for eye mark size, location and colors vary greatly and depend on the requirements of the converting equipment. It is the responsibility of the consumer product company and the printer to provide machine specifications, blueprint, or die layouts for exact eye-mark location to the designer or prepress provider to ensure that the printed material will correctly trigger the electric eye. The eye mark lane (running in the machine direction), should be kept clear of any patterns or copy to allow for uninterrupted visibility by the electric eye. An "electric eye," using infrared technology, reads the eye mark. Therefore, it is recommended to use the darkest color on a job, preferably black, dark blue, or dark green for the eye mark. Colors that consist of red (brown, pink, peach) should be avoided. It is important for the eye mark to print solid. Print defects such as mottling and pinholes may hinder the ability of the eye mark to trigger the electric eye. If the eye mark requires more than one color, the underprinted colors must be reduced by the trap tolerance. Reference Section 12.1 Image Trap Tolerance, for print segment specifications.

12.6 Eye Mark Specifications: *Eye Mark Specifications for size, location, and color vary greatly and depend on the requirements of the converting equipment.*

12.7 Process Control Test Elements

Application: If repeatability and consistency are important to the customer, then space must be allocated on the sheet, web, or printed product for appropriate process control test elements. Measuring at set-up and throughout the run enables the printer to produce repeatable, consistent and accurate results on every job. The test elements used to measure the print characteristics outlined in Sections 12.8 Line Color and 12.9 Process Color, can be used for the print optimization and fingerprint trials as well as on every "live" job to facilitate process control. The test elements included will vary based on the process variable being optimized, the print characteristics pertinent to the job being printed and space constraints. Generally, all of the test elements discussed in Sections 12.8 and 12.9 should be included in a fingerprint trial to provide the most comprehensive information on the current process capability. Using similar test elements on the fingerprint trial as well as on "live" production jobs enables the printer to verify current print conditions and flag any changes since the press was last fingerprinted.

Placement: In order for the printer to deliver the desired print results, the customer and design team must include key test elements in the product design. Some packaging lends itself to placing test elements under flaps, in a glue zone, or on the waste matrix; other packaging requires the test elements to remain visible on the finished package. Therefore, each print application should determine where to place the individual elements to be monitored throughout the production run.

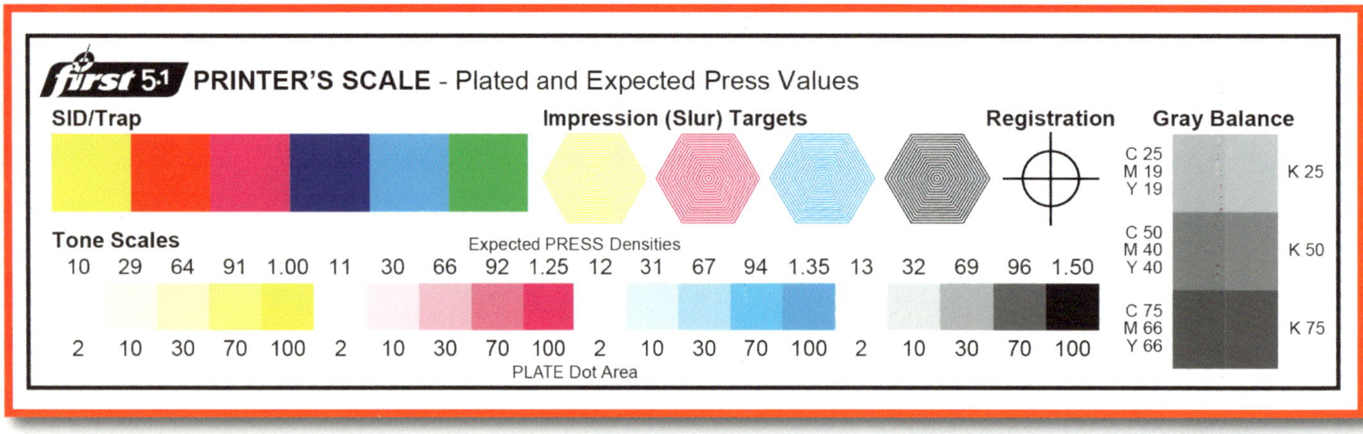

12.7 Example Control Target

Format: Sections 12.8 and 12.9 describe the key print characteristics for both line and process work and the test element used to measure each characteristic. Previous editions of *FIRST* have supplied the *FIRST* control target. All of the test elements discussed throughout *FIRST* are supplied for construction into a suitable control target, optimization or fingerprint test design for each print application. The test elements are available to all members and nonmembers through the FTA as an electronic file and are included in the *FIRST* Extras Download (www.flexography.org/FIRST_extras). Sample run targets are also included for review but should be considered only as potential working examples of what could be used.

Test Element Construction

Size: In general, when measuring with a spectrophotometer or densitometer, the largest aperture that the test element can support is the best. Larger apertures result in a larger measurement area and therefore less measurement variability. Different print applications may have unique requirements. *FIRST* does not recommend using a smaller test element because this requires a smaller aperture; a smaller aperture translates to increased measurement variability. That said, densitometers and spectrophotometers are available with smaller apertures, and are a better choice than not measuring at all. ANSI/CGATS.5 2003 (Graphic Technology – Spectral Measurement and Colorimetric Computation for Graphic Arts Images) provides the minimum and recommended aperture sizes (and therefore test element size) specified by line screen listed in Table 12.7. While these guidelines are useful, the print application must also be considered. The minimum acceptable aperture may be larger for some print applications. The designer and prepress provider should confirm test element size with the printer.

CGATS.5 Densitometer & Spectrophotometer Aperture Size					
Screen Frequency		Round Sampling Aperture (mm)		Non-Round Sampling Aperture (mm)	
Lines per inch	Lines per cm	Minimum	Recommended	Minimum	Recommended
65.0	26.0	3.8	5.7	11.3	25.5
85.0	33.0	3.0	4.5	7.1	15.9
100.0	39.0	2.6	3.9	5.3	11.9
120.0	47.0	2.1	3.2	3.5	7.8
133.0	52.0	1.9	2.9	2.8	6.4
150.0	59.0	1.7	2.6	2.3	5.1
175.0	69.0	1.4	2.1	1.5	3.5
200.0	79.0	1.3	2.0	1.3	3.0

Table 12.7

Imaging: All test elements must be imaged at the same time and with the same care and accuracy as the live job. The test elements must be imaged at the same line screen, angle, dot shape, etc. as the actual image. Surprinting, plate slugs and plate build up of the control target are not accurate representations of the live image area and therefore are not acceptable.

Special attention must be given to imaging tone scales. Refer to Section 12.9.2 for a detailed explanation of the type of tone scales required on press trials and production runs.

Process Control Test Elements

FIRST recognizes certain press configurations (narrow web) and product types (poly bags, envelopes, newspapers) may not have large enough trim areas or glue zones to maintain all recommended process control elements throughout the production run. On these products, the test elements used to verify density and at least one dot area (refer to Section 12.9.2) should be placed on the live area of the product to remain consistent throughout the press run. The more test elements included on production jobs, the better equipped the printer is to achieve the desired print result. Ideally, these five test elements should be on all process color jobs:

1. Registration: color-to-color and print-to-cut
2. SID/Trap
3. Tone scales
4. Impression: anilox-to-plate and plate-to-substrate
5. Gray balance

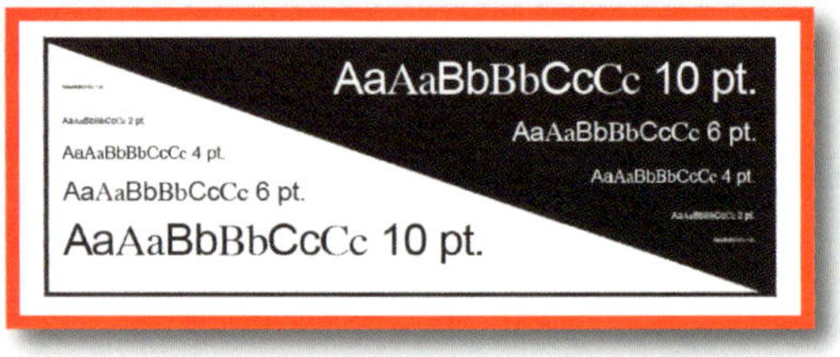

12.8.1 Type Element: *This test element incorporates both Serif and Sans Serif fonts in positive and reverse print.*

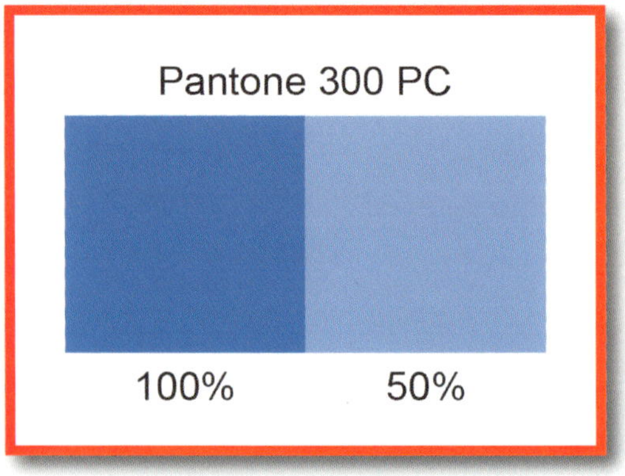

12.8.2 Spot Color Test Element

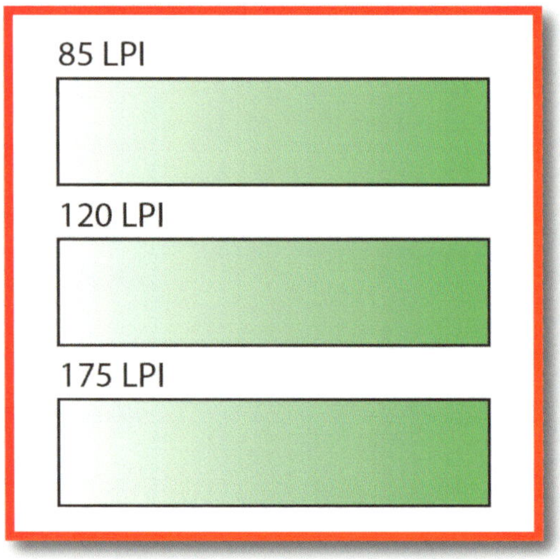

12.8.3 Vignette Test Element: *Include the line screens most appropriate for the print application being evaluated.*

12.8 Line Color: Print Characteristics Measured

When printing a non-process image (often referred to as a "line image"), the capability of the printing process to print positive and reverse type, achieve and maintain color matches, print clean vignettes and produce scannable bar codes must be identified, measured and controlled. If white ink is used to create a white background on either a transparent or colored substrate, then the opacity of the white ink is also critical to the printed outcome and therefore, must be measured and controlled during the printing process. This section reviews each of these print characteristics; Section 19.3 explains them in more detail.

12.8.1 Positive & Reverse Type Elements

Description: Minimum type specifications are normally expressed as "points," which relates to the height of the characters. The minimum specifications differentiate between Serif fonts and Sans Serif fonts. The decorative detail of Serif fonts will fill in faster with reverse print and plug, or print dirty with positive print, thereby necessitating a larger minimum type specification. Some fonts do not relate to Minimum Type recommendations because of their unique design. In these cases, it is better to use the Minimum Rule specifications, ensuring that the thinnest lines of the font are not thinner than the minimum specifications. Refer to Section 12.2 and 19.3.1 for more information on text & graphic elements.

Test Element: Minimum positive and reverse type size and rule width are determined during the optimization and fingerprint trials. Various rule widths and type sizes, of both Serif and Sans Serif type, in positive and reverse format, should be included in the optimization and fingerprint designs to identify minimum specifications for each print condition.

12.8.2 Custom/Spot/Line Colors

Description: Custom, spot and line colors are terms used interchangeably to describe non-process colors specified by the customer. Often they are brand colors. Spot colors may be specified as a PMS match (Pantone Matching System), GCMI match (Glass Container Manufacturers Institute), or the customer may provide a sample to match. Occasionally, the customer may approve the color on-press and the approved press sample or ink drawdown becomes the standard. Refer to Section 19.3.2 for additional information on custom/spot/line colors.

Test Element: *FIRST* recommends including a solid square of the custom color if the graphics do not provide a large enough solid area to measure with a spectrophotometer. If there are screens or vignettes of a custom color, include a tint patch as well to monitor dot gain/TVI during the press run, using either a spectrophotometer or densitometer depending on the color.

12.8.3 Blends/Vignettes/Gradations

Description: A halftone graphic (design element) that changes smoothly in tonal values from light to dark or vice-versa. It may or may not go all the way to a specular highlight (zero tone value). Also referred to as a gradient or blend.

Test Element: *FIRST* recommends printing vignettes in varying line screens on all print decks during the fingerprint trial. By printing a vignette in each color, the mechanical capability of each deck can be evaluated. Multiple line screens are used to identify the optimum line screen for a vignette given the set-up of the print deck. Typically, vignettes print with a coarser line screen than the 4 color process image in order to stay clean and print smooth at production speeds. Coarser line screens produce less dot gain/TVI and minimize all of the problems associated with dot gain/TVI when printing a vignette. Refer to Sections 12.3 and 19.3.3 for additional information on blends/vignettes/gradations.

12.8.4 Bar Code: Minimum Size & Bar Width Reduction

Description: While there are many variables that influence a bar code's ability to scan successfully, there are two variables in particular that the printer is responsible for specifying to the designer and prepress provider. The minimum size (or magnification) for a particular symbol and the corresponding bar width reduction (BWR) are unique to each print condition and are therefore, the printer's responsibility to identify and communicate upstream. For detailed information on variables influencing a bar code's ability to scan successfully, refer to Sections 4.3, 12.4 and 22.0.

Test Element: The minimum printable symbol magnification and optimum bar width reduction for each magnification can be determined via a print optimization trial or during the fingerprint trial depending on the scope of the test. Refer to Section 1.3.2 for more information.

1. Symbols should be printed with varying magnification and BWR combinations.
2. All symbols should be printed with bars running parallel to the machine direction. *FIRST* does not support running bar codes in the cross direction. The distortion inherent in the flexographic process will alter the bar width if printed in the cross direction and lead to unacceptable results.

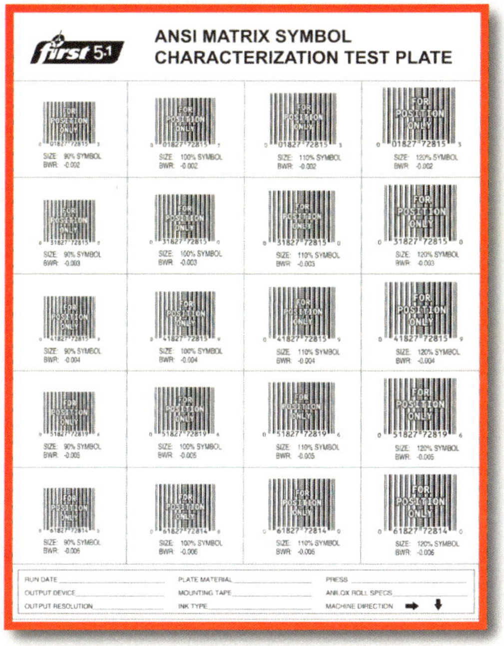

12.8.4 Sample Test Element: *Create a grid of the appropriate bar code. Vary the size of the symbol and the BWR in order to identify the optimum BWR for each magnification.*

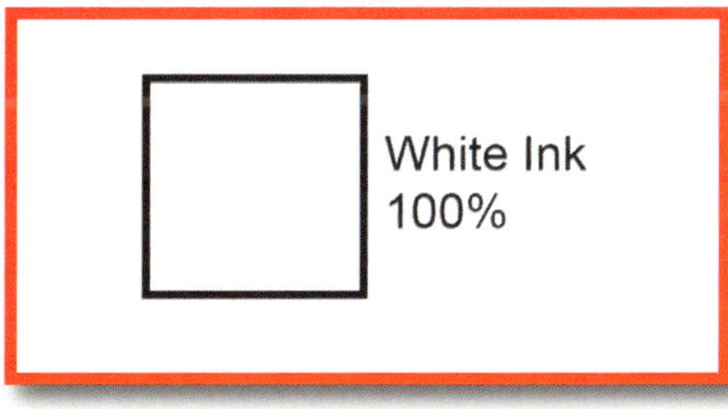

12.8.5 White Ink Test Element

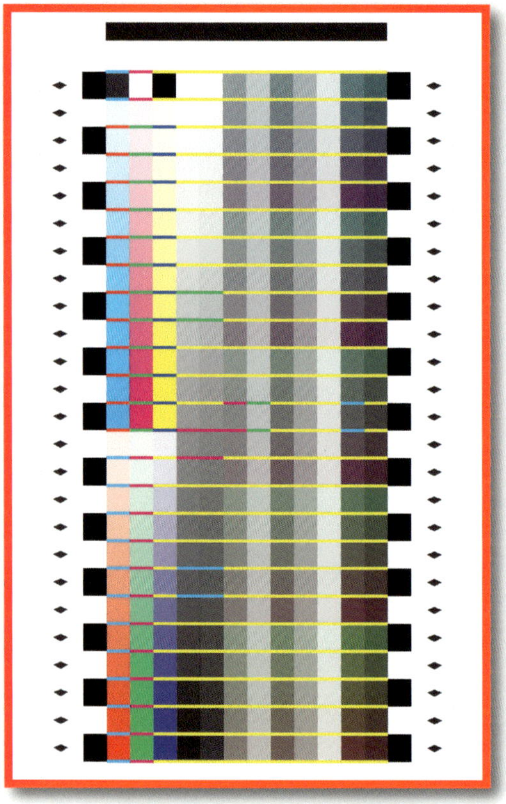

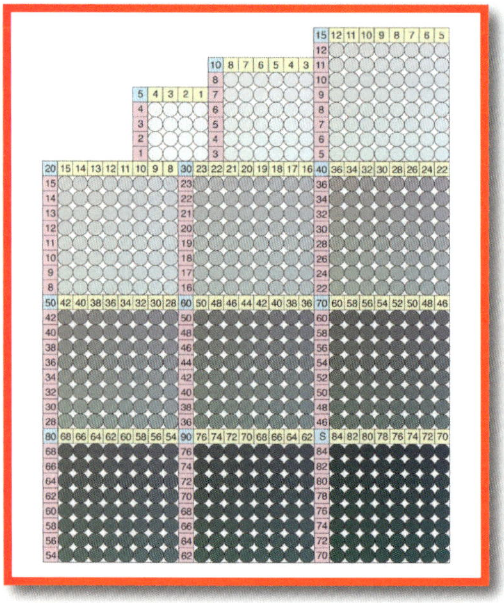

12.9.1a Gray Balance Test Elements: *The target on the top is the P2P25X target which is used in conjunction with the G7™ calibration technique. The target on the bottom is used to manually identify the combinations of CMY that produce a neutral gray throughout the tonal range.*

3. It is important to use a "live" symbol that can be scanned using an ANSI verifier. The printed symbol quality should be graded to determine the optimum BWR for each magnification as well as the minimum magnification.
4. The file containing the symbols must be processed under standard conditions (software, output device, output resolution, plate material, mounting tape, etc) and the press must be set-up and run under standard run conditions if the results are to accurately predict the printed outcome of live jobs.

12.8.5 Opacity of White Ink & Substrates

Description: The image appearance is influenced by the way light reflects from the substrate or the white ink under the printed image. When a custom color or a process color image must be printed on a clear substrate (film) or a non-white substrate (kraft paper) a background of printed white ink should be printed.

Test Element: A large, solid white patch should be included on the press optimization and/or fingerprint trials as well as on each production run. This allows the printer to use a spectrophotometer to monitor ink contamination, visually monitor pinholing and measure opacity using an opacity meter or densitometer. Refer to Section 19.3.5 for more information.

12.9 Process Color: Print Characteristics Measured

When printing a continuous tone image, the primary print characteristics to measure and control are: gray balance, density, dot gain/TVI and print contrast. Other variables for the printer to optimize and control include: print sequence, ink trap, registration, image slur and impression. The test elements for each of these variables are explained in this Section. Refer to Section 19.4 for a detailed explanation of each of these variables.

12.9.1 Gray Balance

Description: Gray Balance is the proper combination of cyan, magenta and yellow ink to produce a neutral gray as measured by a densitometer or spectrophotometer. The actual dot percentages required to produce a neutral gray must be determined during the fingerprint trial and press characterization. Refer to Sections 13.1, 14.3 and 19.4.2 for additional information on gray balance.

Test Element: There are two types of test elements used for gray balance: detailed formats which are used for the optimization, fingerprint and characterization trials and reduced formats, which are used for production runs.

1. **Optimization, Fingerprint and Characterization Trials:** These test elements are used to determine the optimal dot percentage of CMY to achieve gray balance across the tonal range. These elements are utilized in the optimization, fingerprint and characterization trials. They include highlight, quarter-tone, mid-tone, three-quarter-tone and shadow dot areas. Within each of these areas (blocks) there is a constant tint of cyan and varying tints of magenta and yellow. Using a spectrodensitometer, the combination of tint values that best achieves neutral gray can be determined. The size of the patches can be adjusted to conform to the space available on the printed web. However, each square must be slightly larger than the aperture used on the spectrodensitometer to allow for accurate measurement. This information may be used during the color separation process to set neutral gray control points for each process color image. The test element can take a variety of shapes and forms. Image 12.9.1a illustrates two examples of test elements.

2. **Production Run:** The reduced test element is used to control gray balance during production runs. Typically, there are gray balance patches for quarter-tone, mid-tone and three-quarter-tone dot areas. In Image 12.9.1c the patches on the left are black only and the patches on the right are the optimum combination of CMY to match the weight and neutrality of the adjacent black patch (as determined during the fingerprint and characterization trials). The size of the patches can be adjusted to conform to the space available on the printed web.

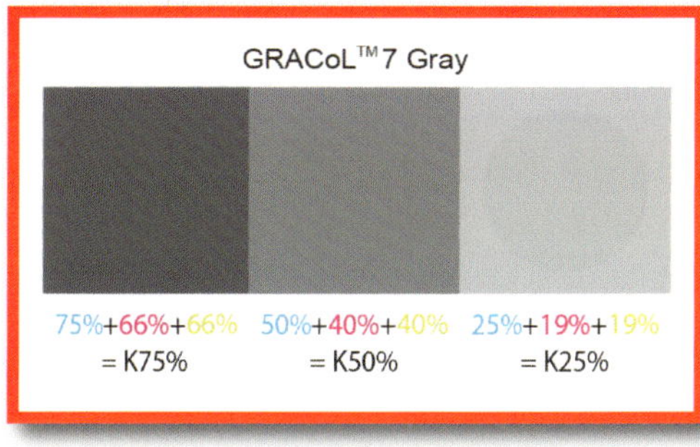

12.9.1b Gray Balance Run Target: *Determine the optimum percentages of CMY during the press fingerprint trial.*

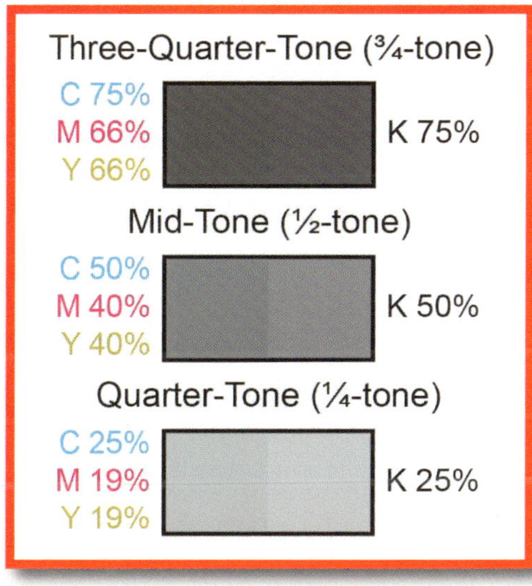

12.9.1c Gray Balance Patches: *The percentages of CMY will vary with each print condition. Determine the optimum percentages using the expanded test elements illustrated in 12.9.1a during the press fingerprint trial.*

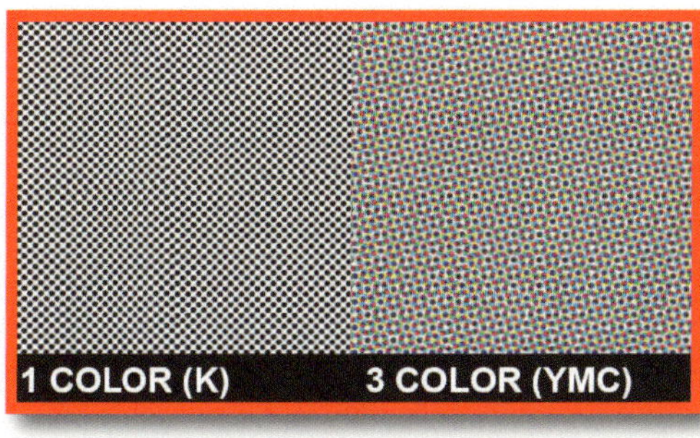

12.9.1d Magnified Gray Blance Production Run Target

12.9.2 Dot Area/Dot Gain/Tonal Value Increase

Description: If the printed image is to match the contract proof, the printer must achieve consistent dot reproduction. Measuring and controlling dot area/dot gain/TVI enables the printer to achieve similar results during production runs as realized during the press fingerprint and characterization trials. Refer to Section 19.4.3 for additional information on dot gain/TVI.

Test Element: Tone scales contain patches of flat tints from highlights through shadows. Tint patches slightly larger than the aperture on the color measurement device are required to measure dot area/dot gain/TVI. The target must be imaged onto the plate simultaneous to the "live" image. The selected tint values and number of patches will vary based on the information required and the space available. For direct-print corrugated, each patch should be 2 to 3 times the flute width to provide a stable measurement target.

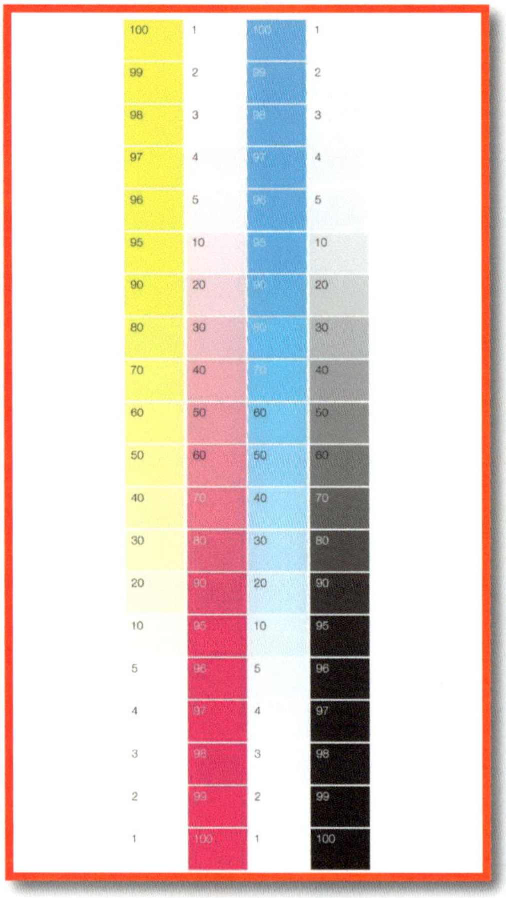

12.9.2a Dot Area Test Element: *Typically, multiple dot area scales, at varying line screens, are included on the fingerprint trial to identify the optimum line screen for the print condition.*

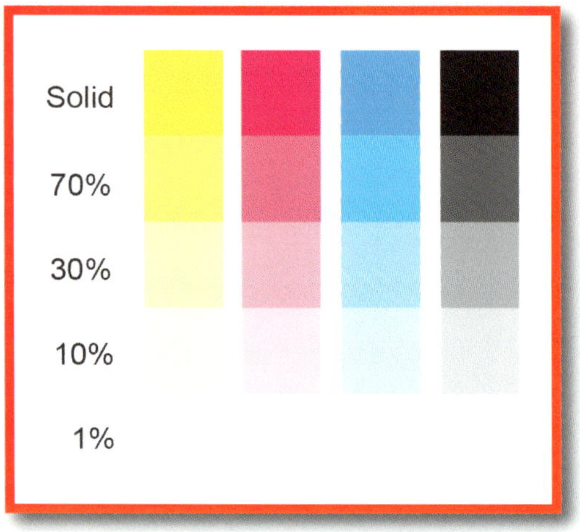

12.9.2b Dot Area Production Run Element

1. **Optimization, Fingerprint and Characterization Trials:** An extended tone scale containing patches from highlights through shadows for each process color should be included on all print optimization, fingerprint and characterization trials. These tint scales provide the raw data necessary to create dot gain/TVI and compensation curves. Typically, tints of at least 1%, 2%, 3%, 4%, 5%, 10%, 20%, 30%, 40%, 50%, 60%, 70%, 80%, 90%, 95% and 100% should be included. In addition to the extended tone scales, the Printer's Scale (Image 12.9.2c) should also be included on all press trials to allow for repeatable press set-up and run conditions.

2. **Production Run:**

Selecting Tint Patches: On production runs, it is critical to measure and control both dot gain/TVI and solid ink density. At a minimum for each color, a solid patch and the tint patch that is most sensitive to pressure adjustments as determined during the press fingerprint trial should be included on all production runs. Refer to Section 1.3.2 for more information on press fingerprint. Space permitting, it is desirable to include additional tint patches to allow for better control and monitoring of the printed image. Consult the printer to identify the appropriate tint patches.

Types of Tone Scales: *FIRST* recommends using tone scales to control both the platemaking process as well as the printing process. There are multiple approaches, described below, depending on the application one or more may be appropriate.

In flexographic platemaking, the final dot percentage on the finished plate is often not the same as the input digital dot percentage. Depending on the imaging and processing techniques, the halftone dots may shrink significantly. This can lead to confusion as to the size of the halftone dots on the finished plate where the labeled values and the actual values are different. Additionally, compensation curves and/or color management adjustments may modify the dot percentages that are sent to the imaging device thereby changing the values from the input digital values.

To facilitate process control, tone scales can be imaged in a couple of ways: linear, where the desired values are achieved on the printed plate; and adjusted, where the values are different from the input digital values due to a purposeful adjustment. These two approaches are explained more fully below. Regardless of the approach, if a patch is labeled with a value indicating its dot area, it should match the final plated dot area.

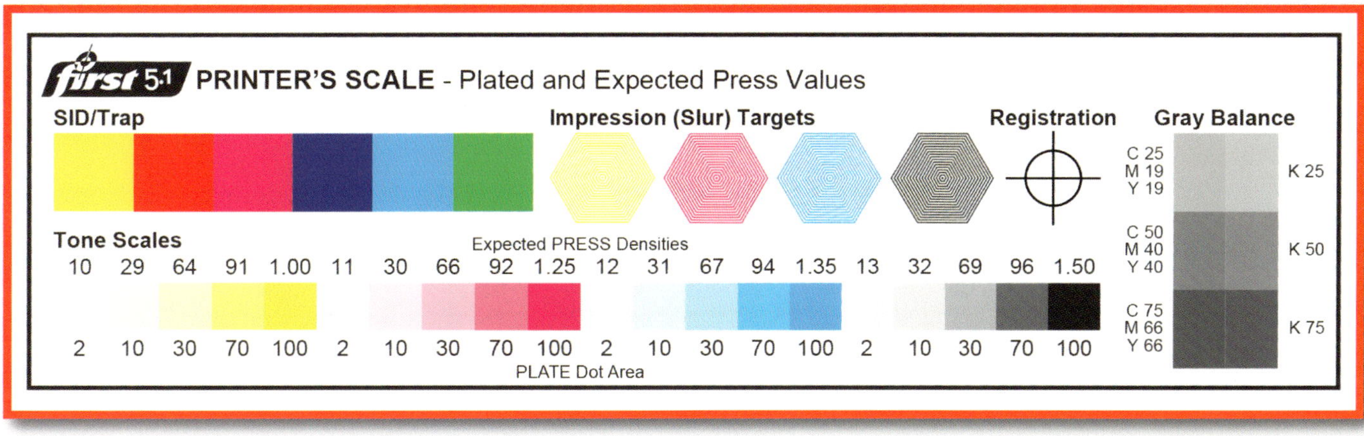

12.9.2c Printers Control Target

For the printer, it is vital to have a control bar (set of patches) imaged consistently with known values on every job in order to print in a consistent and repeatable manner. The values for the desired plate tint patches should be discussed and agreed upon with the printer.

1. **Printer (Linear) Scale:** The Printer Scale is a control scale used by printers to monitor and control press settings, components and materials. Using a liner scale is particularly useful for printers when multiple prepress providers and platemaking systems supply to an individual printer. In these situations, many varied compensation curves may be used during output. The Printer Scale must be labeled with known finished values rather than input (mask or film) values. For example, on the Printer Scale a tint labeled 30% must measure ~ 30% (within accepted tolerances) on the plate after exposure, processing and finishing. By providing "linear" identified values, the printer has the necessary tool to monitor and control dot reproduction regardless of the platemaking variables. The Printer Scale must be included on every optimization, fingerprint and characterization run to facilitate proper press set-up and run conditions.

2. **Prepress (Adjusted) Scale:** The Prepress Scale is created under the same conditions as the live image; any near neutral calibration curves, characterization curves, dot gain/TVI compensation, bump curve, etc. that are applied to the image should also be applied to the Prepress Scale. This scale represents the live image and is used by the prepress provider, platemaker and printer to confirm the appropriate compensation has been applied to the live image. The Prepress Scale can be printed

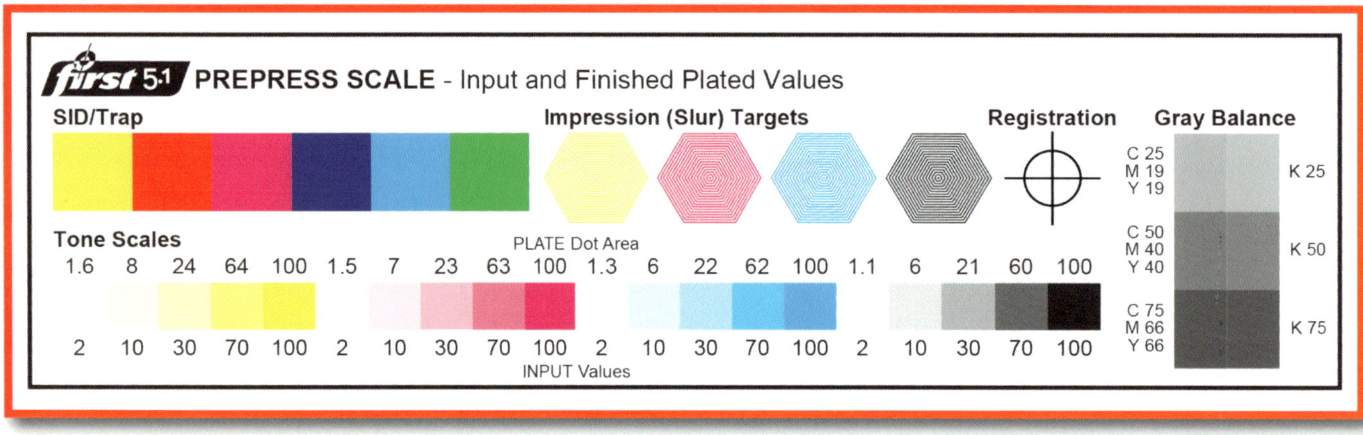

12.9.2d Prepress Control Target: *The critical difference between the Printer's and Prepress control target is the dot size placed in the file (and output onto the plate) for each dot area patch.*

with the job whether or not it's visible on the finished product. Regardless, the Prepress Scale must be plated in order to be measured on the finished plate as part of the plate's CoA. By verifying both the Printer and Prepress Scales on the plate, the platemaker can easily determine if the platemaking process is in control or not, as well as determine if the appropriate adjustment has been made to the live image.

The target values must have production tolerances placed around them for both the prepress/platemaking process and the printing process. Both scales must also be included on the contract proof.

Using both Printer and Prepress Scales provides an important "check and balance" mechanism. Mistakes and unacceptable variations in prepress/platemaking can be easily identified and corrected prior to going to press. When the print does not match the proof, the printer is able to quickly determine whether the problem is a printing problem or prepress problem. If the Printer Scale is printing within the agreed upon tolerances, then the printer can be confident the press is set-up correctly. If not, then adjustments to the press must be made to achieve the target range identified in the Printer Scale. The goal is to obtain the desired print result while minimizing press downtime and waste.

Labeling Tone Scales: Each type of scale must be clearly labeled in the file and on the plate to avoid confusion. For example, label "Printer" or "Linear" and "Prepress", "Profiled" or "Adjusted" next to the corresponding scales. With both tone scales, the printer may choose to have values for each patch placed in the file above or below the patch for clarity. There are two approaches for labeling tint patches. The printer may choose either labeling approach or both.

1. Label each patch with the plated tint value. *FIRST* recommends labeling the plated tint value on the Prepress Scale because it is a QC tool for the prepress/platemaking process.
2. Label each patch with the expected printed tint value. For example, if the 30% plated tint patch is expected to print as a 55% on press, the label for the 30% plated tint patch should be 55%. The expected printed tint value is determined during the press fingerprint and characterization trials.

Tone Scale Placement: Placement of tone scales is dependent on product design and layout. Ideally, a printer scale should be placed on both the operator and non-operator sides of the press so the press operator can control pressure settings across the press. *FIRST* recognizes this is not always possible. Typically, the Prepress Scale is located in a non-print area of the plate and removed prior to printing; therefore, only one is required on each plate.

12.9.3 Solid Ink Density

Description: The objective of each press run is to match the densities and dot gains/TVI established during the press fingerprint regardless of whether or not these densities conform to *FIRST* density specifications. Then by running to the press characterization, the printer can consistently match the contract proof, provided the prepress provider has applied the correct ICC profile. When printing all optimization, fingerprint and characterization trials the printer should set up and run under pre-established conditions, just as a production job. Refer to Section 19.4.4 for additional information on solid ink density.

Test Element: The solid ink density (SID) patches should remain at 100% regardless of the maximum screen value used in lieu of a solid in the image. Each patch must be slightly larger than the densitometer aperture to ensure accurate measurement. The target must be imaged onto the plate simultaneous to the "live" image.

12.9.4 Print Contrast

Description: Print contrast is calculated using the 70% tint patch and the 100% solid patch of each process color. It is expressed as a percentage and indicates the ability to hold shadow detail. The goal is to print the highest print contrast possible. Refer to Section 19.4.5 for additional information on print contrast.

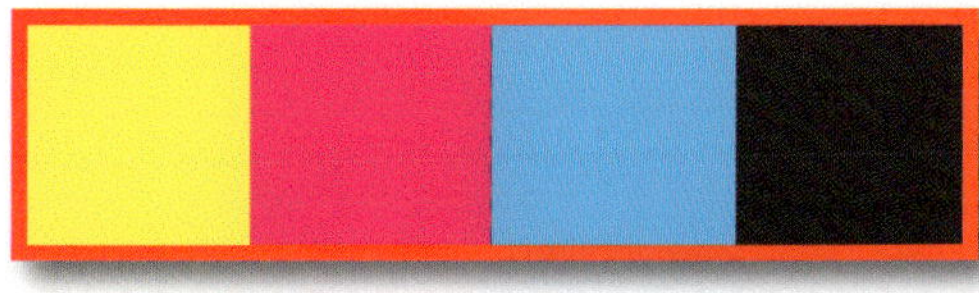

12.9.3 Solid Ink Density Test Element

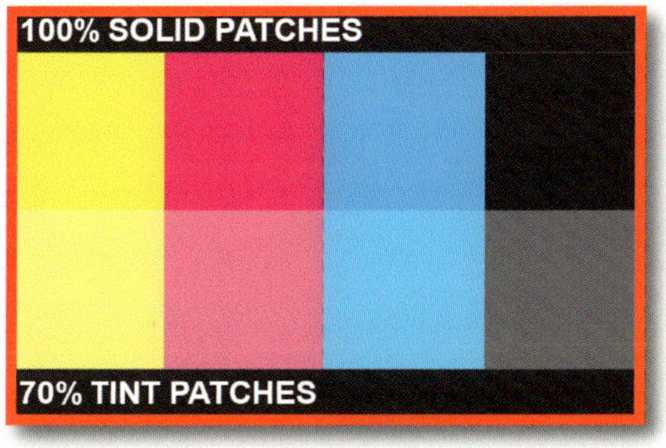

12.9.4 Print Contrast Test Element

Test Element: To measure print contrast, the 100% solid and 70% tint patches are required along with a spectrodensitometer. The patches must be slightly larger than the spectrodensitometer aperture to allow accurate measurement. The target must be imaged onto the plate simultaneous to the "live" image.

12.9.5 Ink Trap

Description: The ability of one ink to lay smoothly over the next is referred to as "ink trap". Process printing requires each color to overprint the previous color in order to produce desired secondary and tertiary colors. Generally, a trap of 80% or better is considered desirable. However, the achievable trap will vary based on press conditions. Refer to Section 19.4.6 for additional information on the variables influencing ink trap.

Test Element: Two and three color overprints in combinations of CM, MY, CY and CMY are used to evaluate ink trap. If a screen value less than 100% produces the highest density, the overprint patches should contain that screen value. Each color solid ink patch is also required in order to calculate ink trap percentage. The print sequence must be recorded. Each patch must be slightly larger than the densitometer aperture to ensure accurate measurement. The target must be imaged onto the plate simultaneous to the "live" image.

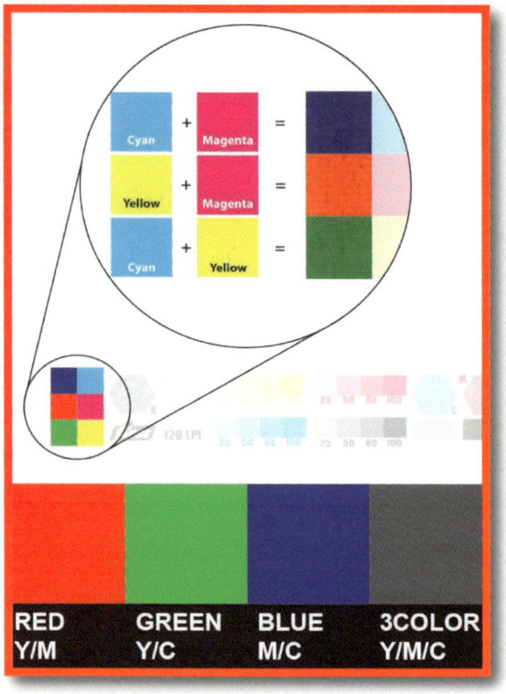

12.9.5 Ink Trap: *This graphic illustrates how process colors combine to produce the overprint colors red, green, and blue.*

12.9.6 Registration & Total Image Trap Tolerance

Description: Proper image register is necessary to prevent unwanted colors and misalignments. The three main uses of registration marks are to identify:

1. Position/location of the image on the substrate
2. Color-to-color registration
3. Square to the lead edge of the sheet (corrugated)

Refer to Print Section 19.4.7 for additional information on registration and total image trap tolerance.

Test Element: There are many different registration marks available for manual or automatic assessment. Register marks belong on the left and right side of the image area and must be imaged simultaneous to the "live" image. Registration targets are used for every color, both process and spot colors. When all marks are in register, the elements of the printed image should be in register as well. The prepress provider and printer must agree on the type of register mark for the print application before producing the job.

12.9.6a Traditional Registration Target

12.9.6b RIT Traffic Light Registration Target

When space allows, *FIRST* supports the use of the Railroad Track Target (RTT) register mark because it quantifies the exact amount of misregister without the use of a tape measure. This reduces the time required to achieve register. The RTT target can be created in millimeters or inches. Select the scale and tolerance that applies to the press. The size and tolerance should be determined during the fingerprint trial. If using inches, include the conversion table on the printed sheet. The objective of the RTT target is to place the triangle (of each color) on the centerline in both lateral & circumferential directions. The longer scale lines at the center of each target represent the "natural variation" of the press (3 sigma). As long as each color triangle remains between these lines, the press is operating in tolerance. The target can be easily adjusted for the gear pitch of the press, each line equaling the distance of 1 gear tooth. If the target does not fit on the web, it can be embedded in the bearer bars.

The RIT Traffic Light Target consists of 4/C cross hairs on either side of a CMY "traffic light". The traffic light element consists of a yellow, magenta and cyan solid circle butt-registered to a black rectangle. The butt-register makes it easy to see when the individual print decks are misaligned.

The RIT Visual Registration Scale contains registration scales measured in 0.001". The yellow, magenta and cyan are registered to the black print station. Each scale consists of black and one color (yellow, magenta, or cyan). When a color is out-of-register, the appropriate scale will show the out-of-register color on the scale (denoting how far out of register and in what direction). This makes it easy for the press operator to know how much to correct the print station. The test element could be placed in both a horizontal and vertical orientation on the printed web in order to measure registration accuracy both across and around the press sheet.

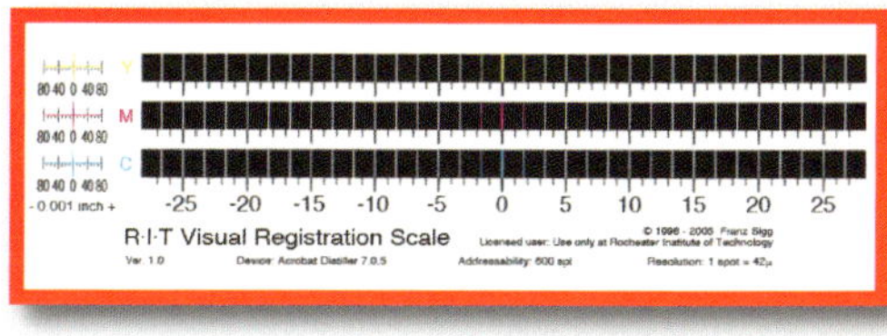

12.9.6c RIT Visual Registration Target

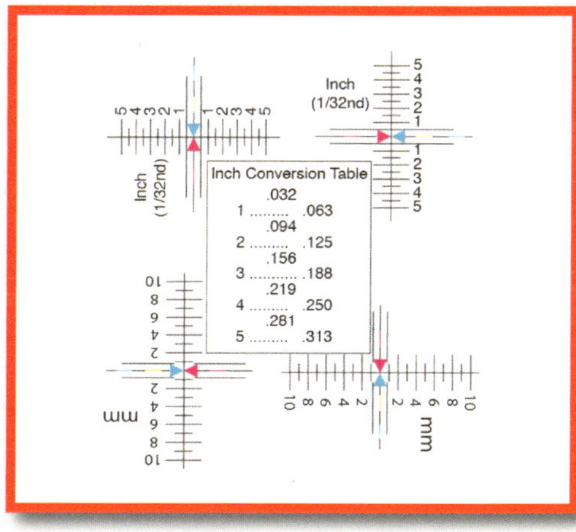

12.9.6d RTT Registration Target

12.9.7 Hexagon and Star/Flower Impression Targets

12.9.7 Image Slur & Impression

Description: Optimum (kiss) impression is the minimum necessary pressure to transfer ink from the anilox to the printing plate and from the plate to the substrate. Slur is a blurred image, caused by over impression and sometimes mechanical press problems. Slur and impression targets quickly and easily indicate when the condition exists. Refer to Section 19.4.8 for additional information.

Test Element: There are two types of impression targets, the hexagon and the star/flower target.

13.1 Near Neutral Gray

1. The hexagon target is produced in two resolutions. One is used for coarse screen rulings under 85lpi, and the other is used for 100lpi and finer. The hexagon shape radiates from the center out and becomes progressively larger. The relationship between the line weight and the distance between lines is critical. The line weight should be equal to 2/3 the distance between two lines. These scales must be built to the appropriate size. Reducing or enlarging the scales will alter the line weights. When an hourglass pattern appears in the hexagon, there is too much plate-to-substrate impression. The hexagon target can also indicate ink balance and drying issues on press.
2. The star/flower target contains "petals" growing in width from the center out. This target is useful for monitoring anilox-to-plate impression. Over impression of anilox-to-plate results in the center of the flower filling-in. Under impression results in the middle of the flower not printing. The size of the flower and the individual petals should be adjusted for anilox volume. Higher volume anilox rolls require a larger target. The appropriate size target should be determined during optimization or fingerprint trials.

13.0 COLOR SEPARATIONS

Reference ISO 12647-6 (Graphic Technology – Process Control for the Production of Halftone Colour Separations, Proofs and Production Prints – Part 6: Flexographic Printing) for additional information.

13.1 Gray Balance

Gray balance is a prerequisite for good color reproduction. It is a function of ink hue, ink film thickness and the percentage of dot area being printed. Color sequence, ink trapping, press characteristics and dot gain/TVI also influence gray balance. Refer to Section 19.4.2 for additional information on press variables influencing gray balance. The Near Neutral Calibration Process described in Section 14.3 is based on controlling gray balance across the tonal range. Table 13.1 illustrates sample CMY dot percentages and density values of the 3 color overprints compared to black.

If process inks were perfect color filters, with each color absorbing the proper third of the visible light spectrum, then cyan would absorb only red light, magenta would absorb only green light and yellow would only absorb blue light. Equal amounts of cyan, magenta and yellow ink could be printed and the eye would receive equal amounts of red, green and blue light. This result would be seen as a neutral gray. Pigments

Sample Gray Balance Densities & Dot Percentages					
Density	K Dot %	Black (K)	3-Color (CMY)	File C/M/Y Dot %	Density
0.25	25%			25/19/19%	0.22
0.54	50%			50/40/40%	0.50
0.90	75%			75/66/66%	0.90

Table 13.1

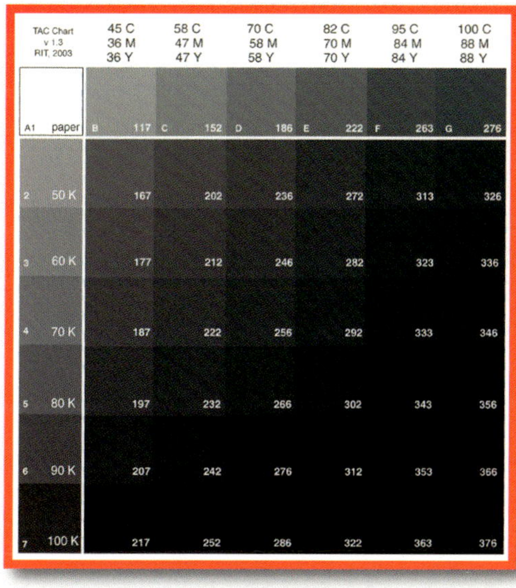

13.2 RIT TAC Test Element

that comprise the process colors are not as pure as the theory suggests. This contamination is known as "hue error." Because gray balance is achieved when equal portions of red, green and blue light are seen, the amount of CMY printed must be adjusted to account for the ink's hue error. Gray balance is not achieved by printing equal amounts of cyan, magenta and yellow; this combination would print a reddish-brown. Rather, the amount of yellow and magenta must be reduced in relation to cyan. There are a variety of targets that can be used to determine the amount of reduction (compensation) required. Refer to Section 12.9.1 to review common targets used.

When troubleshooting gray balance problems on press, some general guidelines to remember are:

1. If the ¼-tone CMY gray patch is not the same density, or neutrality, as the adjacent black tint, it is most likely due to an imbalance in dot gain/TVI. This is further reinforced if the ¾-tone CMY gray patch is neutral gray.
2. If the ¾-tone CMY gray patch is not the same density and hue as the adjacent black tint, it is most likely due to either: off-target ink densities, poor ink trapping, or shadow dot gain/TVI out of balance.

13.2 Total Area Coverage (TAC)

Also known as "Tone Value Sum" in ISO documentation, Total Area Coverage (TAC) or Total Ink Limit (TIL) is typically measured in the darkest shadow area of a process image by adding together the dot percentages of CMYK in the file or on the final films. For example: If C, M, Y and K each equal 75%, the total area coverage is 300%. If all four process colors were printed solid, TAC would be 400%. TAC is one of many parameters that can be fed into a color management system in order to build a profile used for separating images for a specific print system. Total area coverage is measured in the full color file prior to platemaking.

Total Area Coverage: General Guidelines *TAC is print system dependent; determine maximum TAC limits during press fingerprint trial (ref. 1.3.2)*					
Finish	Corrugated	Paper	Film	Newsprint	Printer Specific TAC Limits
Coated	260%	290%	N/A	N/A	
Uncoated	290%	320%	N/A	200 - 240%	
Film	N/A	N/A	300 - 340%	N/A	

Table 13.2

The objective of TAC is to extend the tonal range in the shadow region while minimizing ink film thickness. Higher TAC values may result in ink trapping and drying/curing problems, resulting in slower press speeds. Another consideration is cost. Applying additional ink, that does not result in a darker shadow, drives cost without improving quality.

The RIT Total Area Coverage (TAC) Test Element is a grid that contains increasing dot percentages of black going down each row and increasing dot percentages of CMY moving from left to right across each column. The TAC value is listed in the bottom right-hand corner of each patch. Evaluate this test element by visually, or densitometrically, assessing each patch to identify when increasing the TAC value does not result in increased darkness, or density. The square with the lowest TAC value and highest density is the optimum TAC value for the print condition being evaluated. Printing with total coverage higher than the optimal value may cause drying/curing issues forcing slower press speeds, and increase ink consumption and costs unnecessarily. The optimum TAC value will vary with the print condition (substrate, ink chemistry, anilox, mounting tape, dryer configuration, press speed, etc.).

13.3 UCR/GCR vs. Normal Separation: *The advantage of these separation techniques is that, by reducing the amount of CMY in the image, they make maintaining color balance on press easier. They also use less ink which minimizes potential ink trap/drying problems and cost. The potential disadvantage of these techniques is that, if not done well, the graphic can appear washed-out and lacking detail. When "over done" the quality of the printed image suffers.*

Because of the complexity of the variables involved, the Total Area Coverage chart included in *FIRST* is a ballpark reference only. *FIRST* recommends including a TAC test element on the fingerprint test form. Refer to Section 1.3.2 for more information.

13.3 Under Color Removal (UCR)

Under color removal is the balanced reduction of cyan, magenta and yellow in shadow areas. The addition of black in these reduced areas maintains the dark and near neutral shadows. UCR, used in traditional offset separations, is not always best suited for the flexographic printing process. Ideally, if the amount of color in the three process colors could be reduced while maintaining the shape and shadow detail of the image in all three colors, this would be the best application of under color removal for the flexographic printing process. UCR is done at the input/scanning or color correction stage and is measured prior to applying a compensation curve. Contact the printer for guidance.

13.4 Gray Component Replacement (GCR)

Gray component replacement is more easily defined as an unwanted color (cyan in red or magenta in green) being replaced by black, in part or in whole, as the graying component. The use of GCR is the responsibility of the prepress provider and the printer. It is recommended that GCR be restricted to a single unwanted color under normal conditions in the flexographic process. Reviewing the job with the printer or customer prior to prepress will help determine the best approach. The prepress provider and the printer should reach a consensus on the amount (percent) of GCR used in an image. When the printer must print line black on the same station as the process black, GCR should not be used. It is better to have a short black for the separation (skeleton black) so that the printer has more latitude in setting impression.

14.0 PROCESS COLOR CALIBRATION

There are three basic methods used to optimize process color separations and achieve the best possible match. Regardless of the method used, the objective is the same, the proof to the printed image should have a common visual appearance. The three basic methods are:

1. Dot Gain/TVI Curves based on Tone Scales
2. Tonal Curves based on Near Neutral Calibration (commonly referred to as Gray Balance)
3. CIELAB Color Management System

14.1 Process Color Calibration Techniques

The following paragraphs provide a brief overview of the three main process color calibration techniques. Each technique is explained in more detail in Sections 14.2, 14.3 and 14.4.

1. **Dot Gain/TVI Curves based on Tone Scales (14.2)**
   This method applies dot gain/TVI curves to the image prior to platemaking. Tone scales are printed during the press fingerprint trial and measured to determine the dot gain/TVI experienced on press. Prepress places the dot gain/TVI values into the prepress workflow software and applies the resulting curves to the output file. Each color separation (CMYK) is treated independently. This method has been utilized by the industry for many years and is well understood. The shortcoming of this approach is that it does not consider

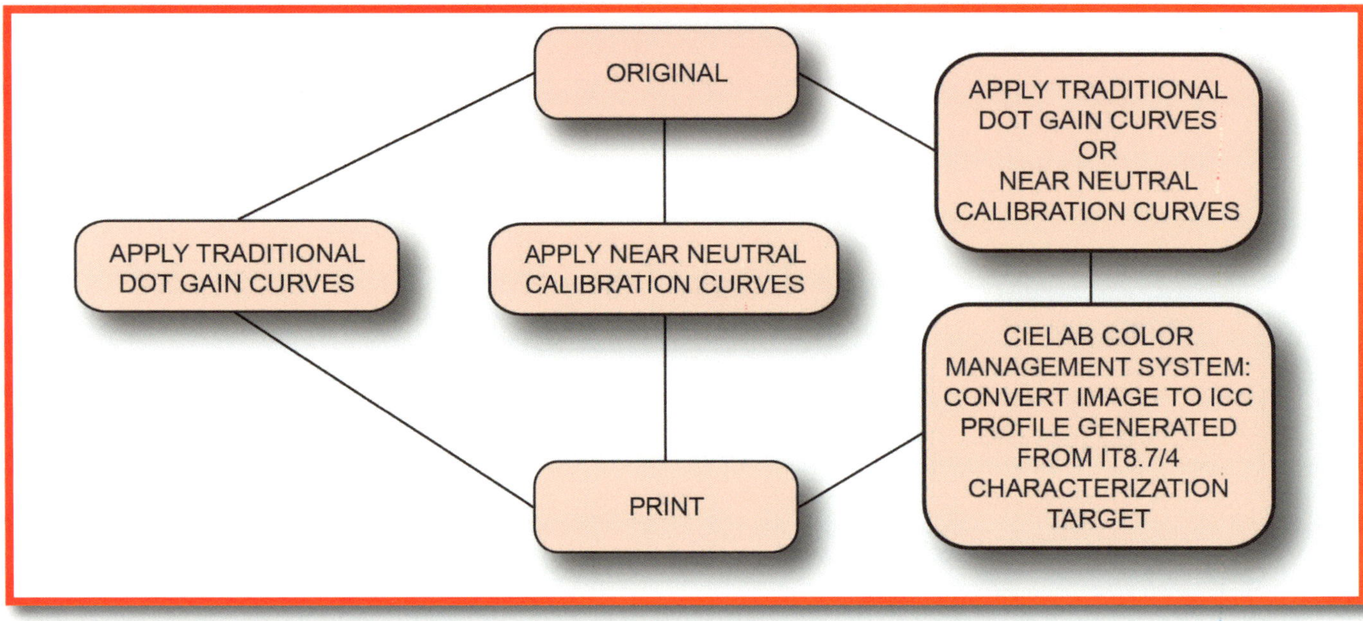

14.0 Process Color Calibration Techniques

gray balance, overprints, or differences in colorants and gamuts between the printing press, the proofing device and differences in substrate hue.

2. **Tonal Curves based on Near Neutral Calibration (14.3)** Near neutral calibration is based on an optimized tonal curve, the Neutral Print Density Curve (NPDC), derived from 3 color overprints (CMY) instead of single color dot gain/TVI curves. Matching a common tonal curve derived from 3 color overprints corrects for dot gain/TVI, ink trap and gray balance simultaneously. Therefore, NNC curves are used in place of single color dot gain/TVI curves. NNC does not correct for gamut or colorant differences. Some approaches to NNC, such as G7™, define the colorants using an international standard, such as ISO 12647-2. The primary advantage of NNC is improved color matches across presses and print methods.
3. **CIELAB Color Management System, CMS (14.4)** The CIELAB CMS documents the colors that can be produced by a given printing system with an ICC profile. The ICC profile provides the information to determine what input is necessary to create a specific color on press. The CMS approach optimizes the ability to match proof-to-press and predict achievable results. This approach results in an ICC profile that represents how an output device (such as the press or proofer) renders color under a given set of conditions. Printing a target such as the IT8.7/4 during the press characterization trial creates the ICC profile. Refer to Section 1.3.4 for more information. Each patch on the

IT8.7/4 test target is measured and plotted in CIELAB color space to develop the ICC profile. The color management system is the most comprehensive of the three approaches; it addresses dot gain/TVI, colorant differences and overprint characteristics.

14.2 Traditional Dot Gain/TVI Curves

A process color image consists of four separations printed on top of one another. In order to maximize the detail in the printed image, each separation (CMYK) must be optimized to achieve the full range of tones the printing process is capable of reproducing. Dot gain/TVI curves, derived from printed tone scales, represent the traditional method of optimizing each separation (CMYK).

The shortcoming of this approach is that it considers each separation (CMYK) independently. So while the dot gain/TVI experienced by each print deck is compensated for, gray balance, overprints, and color/gamut differences are not controlled. Along with traditional dot gain/TVI curves, the prepress provider may also use the traditional gray balance approach (identifying the best dot size of M and Y for a given C dot percentage) to achieve color balance within the image. Using the traditional method, it is more difficult and less efficient to accurately and consistently match a proof. Because the traditional approach does not consider and control for the actual color of the process inks (L*a*b*/L*C*h°), the results are usually inferior to the other two approaches, particularly when multiple print locations and/or presses are involved.

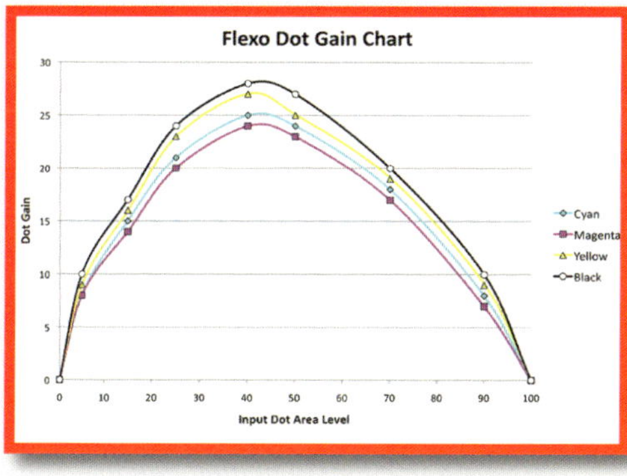

14.2 Dot Gain/TVI Curves: *Dot gain/TVI curves are used to reduce the size of the dot in the file, and therefore on the plate, to compensate for the expected dot gain on press – as measured during the press fingerprint.*

14.3 G7™ - Near Neutral Calibration (NNC)

G7™ - Near Neutral Calibration (NNC) is based on an optimized tonal curve, the neutral print density curve (NPDC), derived from 3 color overprints (CMY) instead of single color tone scales. The NPDC is a more evenly spaced tonal curve that represents ink on paper and a printing press. It is not proof manufacturer specific. In order to match the NPDC as closely as possible, an individual NNC curve is applied to each separation (CMY). For the black separation, a traditional dot gain/TVI curve is applied. Matching a common tonal curve the NPDC corrects for dot gain/TVI, ink trap and gray balance simultaneously. The NNC curves are applied to the image instead of traditional dot gain/TVI curves.

There are multiple approaches to NNC. While *FIRST* supports the concept of near neutral calibration, *FIRST* does not endorse any one product. In general, there are five steps in a near neutral calibration approach to process color separations.

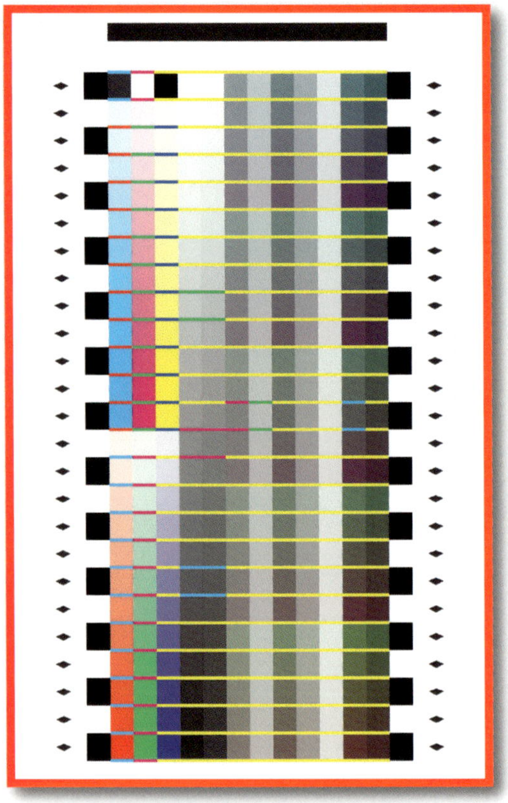

14.3 Near Neutral Calibration Target: *This test target is one example of a near neutral calibration target.*

Near Neutral Calibration – General Steps

1. Specify a mono-pigmented (preferable) ink set (CMYK) that satisfies the functional properties required by the printing process and finished product. The *FIRST* recommended process ink pigments identified in Sections 20.2.2 and 20.2.3 are flexographic ink chemistry specific recommendations. In addition, there are several ISO ink color specifications including ISO 12647-2. While ISO 12647-2 is a lithographic offset process ink specification, in this application it should not be regarded as such. Instead, it represents an idealized color specification that is the objective for all printing processes regardless of ink, substrate, or printing condition. The success of NNC is dependent on a consistent ink set. In most flexographic printing processes it will not be possible to match the ISO 12647-2 specification. Regardless of the specification targeted, it is critical for the printer to establish and identify a standard process color ink set that is used for all optimization, fingerprint and characterization trials as well as all live production runs. When attempting to match an ink color specification, focus first on each ink's color, or hue angle. Try to match the hue angle of the specified color as closely as possible (preferably within +/- 2°). Next, obtain as much chroma (ink strength/saturation) as possible without shifting the hue angle of the ink.
2. Once the ink set has been identified, print a gray balance/near neutral calibration target to collect the data necessary for near neutral calibration. A NNC target should be run during the fingerprint and characterization trials. It is important to print the target using the selected ink set under standard operating conditions. The printer should monitor and control tone reproduction using tone scales as explained in Section 12.9.2. Pull print samples throughout the press run. One approach to NNC, developed by IDEAlliance, is G7™. It utilizes the Plate-to-Plate (P2P) NNC target. Refer to Appendix A for IDEAlliance contact information, and Appendix C for a P2P target example.
3. Evaluate multiple samples; CGATS recommends measuring at least 6 random samples using a spectrophotometer. The samples used to evaluate NNC should be within the printer specified tolerances for solid ink density, dot gain/TVI, print contrast, and color (L*a*b*/L*C*h°). The data, derived from averaging the results of multiple samples, is used to create the NNC print curves.
4. Apply the NNC print curves to the image instead of traditional dot gain/TVI curves.
5. Run the press under the same conditions as the NNC test target was printed. Use a control target to monitor and

FIRST Recommended Process Ink Pigments							
Color	Chemistry	C.I. Ref	L	*a	*b	C	h°
Yellow	Solvent	Y14	90	- 4	111	111	92°
	Water	Y14	92	- 6	101	101	93°
	UV	Y13	92	- 5	83	83	93°
Magenta	Solvent	R57:1 & R52:1	46	72	9	72	7°
	Water		53	70	- 4	70	357°
	UV		50	76	- 1	76	359°
Cyan	Solvent	B15:3 & B15:4	56	- 41	- 47	62	229°
	Water		59	- 37	- 45	59	230°
	UV		56	- 38	- 45	59	230°
Black	Solvent	K7	21	1	3	3	N/A
	Water		16	0.8	1	1	N/A
	UV		16	0.2	5	5	N/A

Table 14.3a

ISO 12647-2 Specification *An idealized color specification - not an ink standard in this application.*					
Color	L	*a	*b	C	h°
Yellow	89	- 5	93	93	93°
Magenta	48	74	- 3	74	357°
Cyan	55	- 37	- 50	62	233°
Black	16	0	0	0	N/A

Table 14.3b

control density, dot gain/TVI, trap, print contrast and ink color. Refer to Section 12.9 for more information.

14.4 CIELAB Color Management System

What is a Color Management System?

Color Management Systems (CMS) are a collection of software/hardware tools that quantify and reconcile the color differences between various color output devices (such as monitors, scanners, proofers and printing presses) to help ensure consistent color throughout the color reproduction process. Typically, the available color gamut of the printing press is smaller than other devices further upstream in the process, like proofing systems and displays. A CMS will identify the colors outside the gamut of

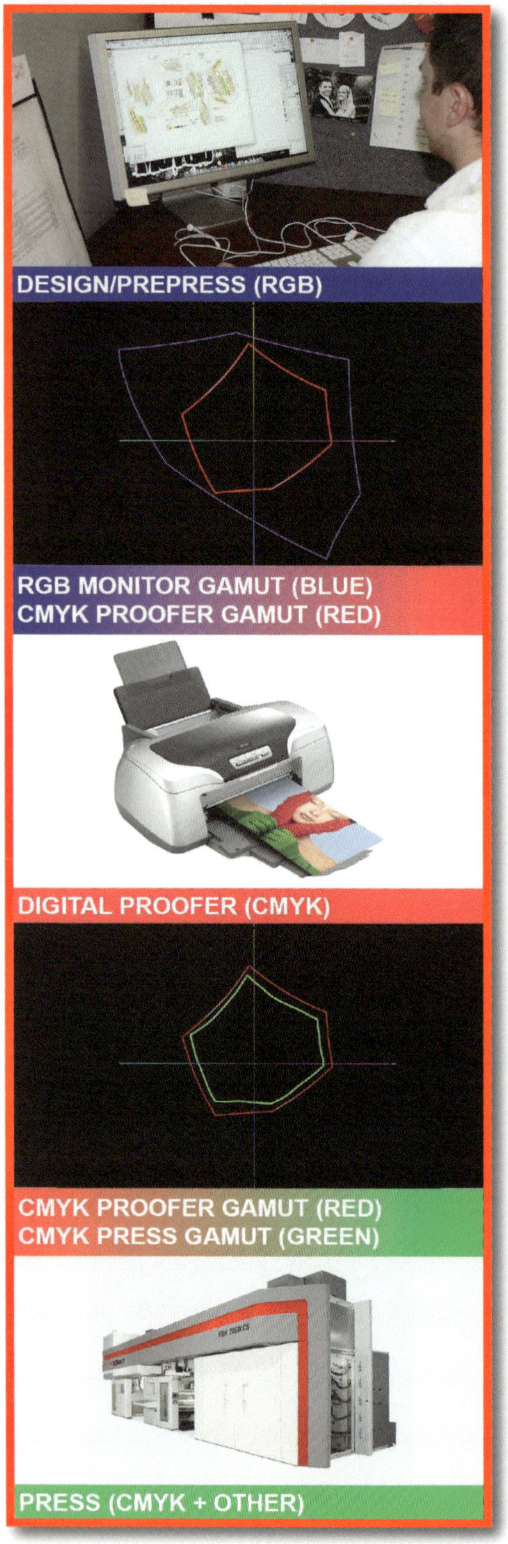

14.4a CMS Near Neutral Calibration Target: *This diagram illustrates the relationship between the various processes and the color management system.*

the printing press and limit, or map, the colors on larger-gamut device(s) (ie. proofer, display). Therefore, colors that cannot be reproduced on the press are not generated. This creates a realistic representation of the printed result at the proofing stage.

CIELAB-based color management uses the CIELAB color space to quantify color, independent of the device used to produce that color. Colors are numerically defined; this allows the use of mathematical models and algorithms that permit color conversion based on the measured color capabilities of an identified process. A device-independent color space enables conversion from one ingredient-based color space to another. Examples of ingredient-based color systems are RGB and CMYK. RGB color ingredients refer to phosphors or LED's in a monitor, dyes in a transparency, photomultiplier tubes in a scanner, or charged-coupled device (ccd) array processors in a digital camera. They can all have the same numerical RGB value, but not match because of the different ingredients. CMYK color ingredients include toners, inkjet ink or dyes in digital proofs, and different pigments in solvent, water and UV inks. The device-independent model produces the closest color match when moving from one ingredient-based system to another. Both input and output color ingredients are specified colorimetrically, using a spectrophotometer, and are easily compared for conversion between ingredient-based color systems.

How to Implement a Color Management System

The first stage of implementation calibrates the color reproduction process from computer monitors, digital cameras and scanners to proofing devices and printing presses. To calibrate, each device under standard operating conditions produces a target such as the IT8.7/4 characterization target. Each spot within the target is measured with a spectrophotometer. The CIELAB values of each spot are documented within the CMS software. The CMS software combines the CIELAB values (L*a*b*/L*C*h°) of every spot on the IT8.7/4 characterization target and produces a color gamut (ICC profile) for each device.

The second stage of implementation compares the ICC profiles of the various devices (computer monitors, digital cameras, scanners, proofing devices, printing presses). By identifying the areas of the CIELAB color space that are reproducible by all of the devices (typically the printing press produces the most limited color gamut), it is possible to transfer images from one device to another while achieving consistent and predictable results. This is accomplished by applying the appropriate ICC profile to the image for a given output device. For example, by comparing the reproducible color gamut (ICC profile) of a proofing device and a printing press, the resulting contract proof is able to accurately predict the printed result. This information affords the customer the option of altering the design, color selection, or materials if a result different from the contract proof is desired.

The final stage of implementation is to control the color reproduction process. In order for the ICC profiles (which are generated after the calibration stage) to be useful, they must reflect actual operating conditions. Therefore, all devices must be calibrated regularly based on the device manufacturer's recommendations. For a printing press, "regular calibration" includes routine cleaning and preventative maintenance as well as scheduled press rebuilds. Incoming consumables (ink, proofing dyes, substrates, etc.) must be specified and measured to ensure consistency. Finally, the device must be operated under controlled conditions reflective of the conditions under which it was characterized.

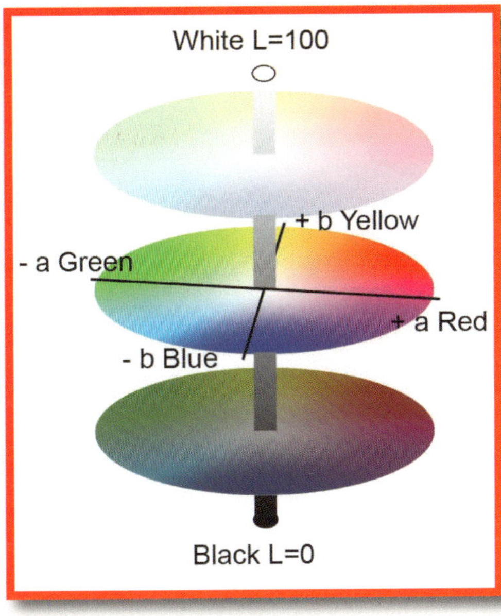

14.4b CIELAB: *The CIELAB color space numerically identifies colors, allowing for color communication and correlation between various devices.*

Benefits of Implementing a Color Management System
Some of the benefits of implementing a Color Management System in the color reproduction process include:

1. Prepress is better able to optimize the graphic file for the specific print conditions:
   - Opening overprints
   - Correcting gray balance
   - Increasing saturation of clean colors
2. The customer develops increased confidence that the final product will have a common visual appearance with the contract proof.
3. The customer is better able to make adjustments to the design to achieve the desired outcome prior to going to press.
4. The proofing system is better able to predict non-standard print conditions:
   - Before and after lamination
   - Non-standard ink sets
   - Unusual substrates

14.4.1 Calibrating The Color Management System

In order for the color management system (CMS) to be effective, a commitment to calibrate and characterize each production process and to maintain these systems must be made by every supplier contributing to the production of the printed product (designer, prepress provider, platemaker and printer). Targets such as the IT8/7.4 characterization target are used to establish the Color Management System (CMS).

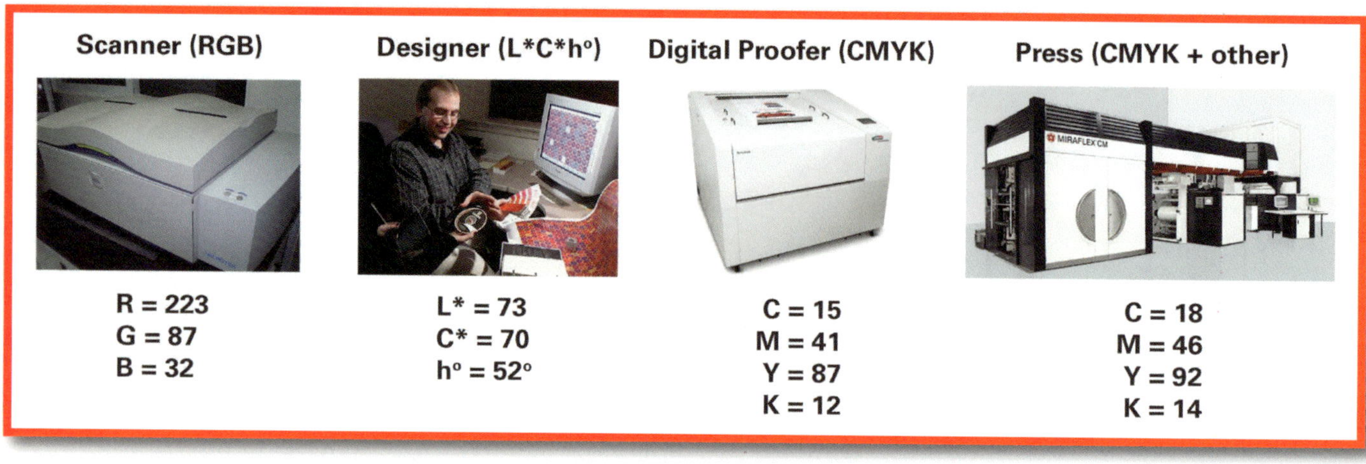

14.4.1a Calibrating the CMS: *This diagram illustrates different ingredient-based color systems.*

The process is divided into two steps:

1. Characterize the printing press to create an ICC profile.
2. Calibrate the color reproduction process from image capture to the printed result.

Step 1: Characterize the Press to Create an ICC Profile

The first step in calibrating the color management system, from a prepress perspective, is to output the IT8.7/4 characterization target direct-to-plate (or to film and make a plate). The plates must be made, mounted and printed under normal operating conditions and to *FIRST* specifications. Tone scales, as described in Section 12.9.2, must be included on all characterization plates. The result of this press run creates the press ICC profile (characterization). The accuracy of this press run is paramount to the successful implementation of color management. Prior to the press characterization run, the printer must establish aim points for press set-up and run conditions by completing the optimization and fingerprint trials detailed in Sections 1.3.1 and 1.3.2. A proof is not required for the characterization press run. All system components in the press must be identified and documented. Changes in anilox, plate, mounting tape, ink formulation, or other components will likely alter the results of the ICC profile. Production specifications should be noted in the characterization database and amended as production components change. CGATS TR-012 2003 (Graphic Technology – Color Reproduction and Process Control for Package Printing), summarized in Section 1.3, identifies conditions requiring a new characterization as well as process control methods. Refer to Section 1.3.4 for a detailed explanation of the press characterization process.

Averaging the measurements from multiple targets taken from the same characterization press run, will lead to a more accurate profile. CGATS recommends evaluating at least six samples that are within the aim points of the printing condition. This approach leads to more accurate results because it minimizes the "noise" that is typical of the printing process, caused by small defects in the printed target. Space permitting, *FIRST* recommends including multiple IT8.7/4 targets across the printed web, oriented in two directions with a 90° rotation in both the ordered and random formats. After a press profile has been created, it is a good idea to make a proof representing the press profile. With the highly structured form of the ordered (visual-layout) target, it is easy to identify areas that are inconsistent with the rest of the target.

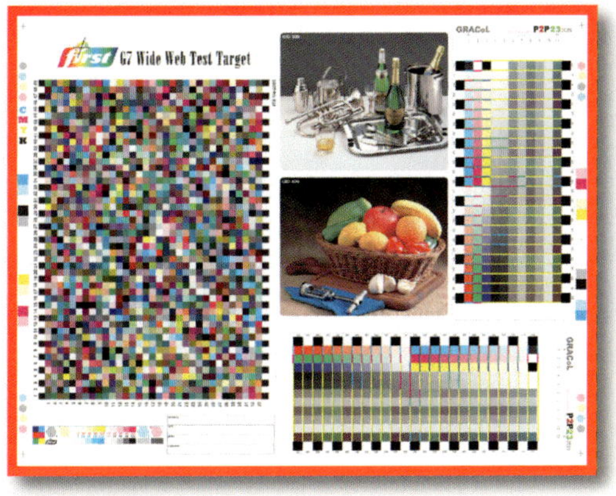

14.4.1b Characterization Press Sheet: *This is an example of a characterization press sheet that includes the IT8.7/4 ICC profile target and the G7™ P2P near neutral calibration target.*

On a "live" job, the image is converted using the ICC profile (generated from the IT8.7/4 characterization target) in the color management workflow. Because the IT8.7/4 target assumes a dot gain/TVI of roughly 17% at the midtones (50% dot area), the prepress provider will typically apply either traditional dot gain/TVI curves or near neutral calibration curves to each separation (CMYK) when the characterization file is sent to the platemaking equipment. For applicable platemaking processes, a bump curve must be incorporated with whichever curves (dot gain/TVI or NNC) are applied to the image.

It is common practice to apply either traditional dot gain/TVI curves or near neutral calibration curves to the IT8.7/4 target prior to the press characterization trial. If the prepress provider opts to compensate the IT8.7/4 press characterization target for dot gain/TVI, aim for the 50% tint in the electronic file to print ~67% on press (as measured using a reflection densitometer calculating dot percent using the Murray-Davies equation). When the 50% dot reproduces as 67% on press, an even distribution of color data is achieved which is thought to improve the accuracy of the mathematical modeling used by the color management software to calculate color values that are not specifically measured in the IT8.7/4 target. The prepress provider should estimate the proper dot gain/TVI compensation based on dot gain/TVI values measured by the printer during routine press runs.

In order to realize the benefits of a color management system, the printing process must first be in control. Gray balance, dot gain/TVI, density, ink trap and print contrast along with other print attributes must be routinely monitored and controlled. If it is unclear what the specifications are, or if they have been achieved, color management is not appropriate at this time. A

Steps to Creating a CMS Profile: Characterizing the Printing Process	
1) Printer	Select Inks (ref. 14.2, 21.2.2 & ISO 12647-2)
2) Prepress or Printer	Calibrate platemaking system (ref. 17.0)
a) Optional Printer	Print near neutral calibration target (ref.14.3)
b) Optional Prepress	Apply NNC data; neutral print density curves NPDC (ref. 14.3)
3) Printer	Print IT8.7/4 characterization target (ref. 19.4)
4) Prepress	Measure IT8.7/4 target and create color profile (ICC profile)
5) Prepress	Place control target on live job (ref. 12.7, 12.8, 12.9)
6) Prepress or Printer	Make plates using digital control strip and ICC profile (apply dot gain or NCC curves if applicable)
7) Printer	Set-up press to repeat target values achieved during press characterization (use control target)
8) Printer	Measure control target throughout production run and record data
9) Printer	Feed run data back to prepress provider (continuous communication loop)

Table 14.4.1

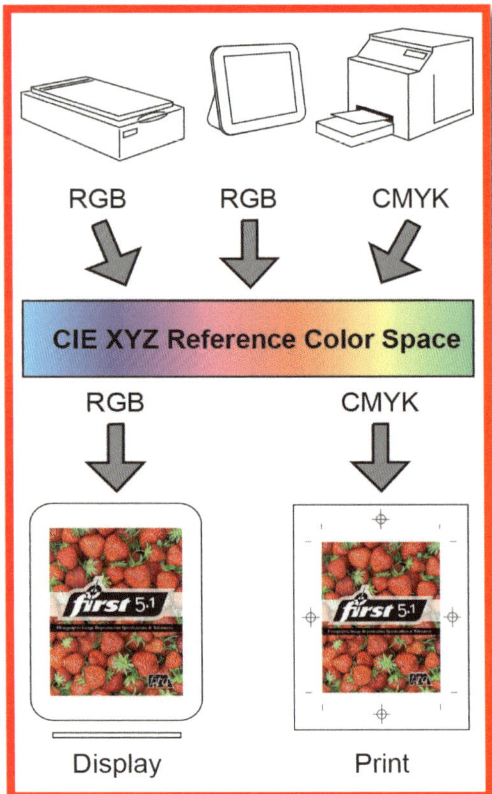

14.4.1c ICC Profiles: *In order to implement the CMS, the proofing system and RGB devices must be profiled and calibrated to the press.*

press characterization should not be attempted until the press has been optimized, refer to Sections 1.3.1 for more information.

Step 2: Calibrate the Color Reproduction Process from Image Capture to the Printed Result
The second step in calibrating the color management system is to calibrate the proofing system along with the RGB devices (computer monitors, digital cameras, scanners) and combine RGB profiles with the CMYK profiles from the proofing device and printing press. Monitor calibration is needed to make images match on screen across multiple software applications, or to simulate a press. This will enable a facsimile on screen to match the proof or press.

All digital proofs must contain a control target in order to verify the integrity of the proof. Include on all proofs the control target defined by ISO 12647-7:2007 (Graphic Technology – Process control for the production of half-tone color separations, proof and production prints – Part 7: Proofing processes working directly from digital data). The IDEAlliance 12647-7 Control Strip 2009 is a free control bar conforming to all of the requirements outlined in ISO 12647-7. Refer to Appendix A for contact information.

A "Certification of Result" should accompany all proofs. It should include, at a minimum, measurements of the proof density and the Delta E (ΔE) of each patch (as applicable). The operator, date, job number, customer and proofing device should

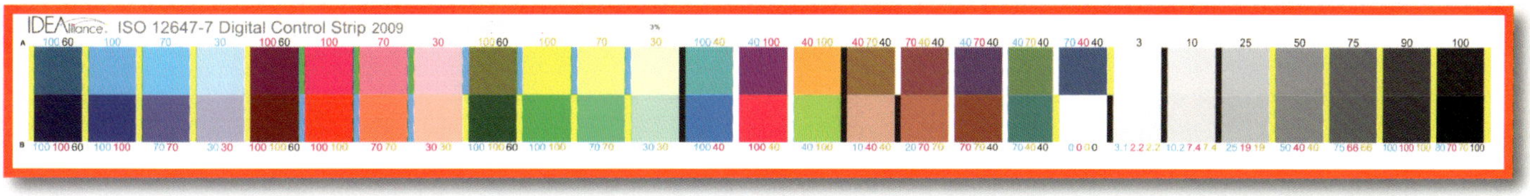

14.4.1d IDEAlliance 12647-7 Control Strip 2009: *This control strip is used to confirm the accuracy of the digital proof.*

also be included. As specified within ISO 12647-7, the proof-to-proof variability should be within a ΔE CIE of 1.5. All solid primary printing patches in the IDEAlliance 12647-7 control strip must achieve a ΔE CIE < 5.0 and have a hue angle within 3° of the ink hue specified in ISO 12647-2 specification. Refer to ink chart in Section 14.3. The maximum variation of any patch must be less than a ΔE CIE < 6.0 with the average of all patches achieving a ΔE CIE < 3.0. Do not evaluate dot gain/TVI when using inkjet proofing methods.

Proof-to-Press Match: Acceptable Difference

A Color Management System (CMS) uses two profiles, an input profile and an output profile. The input profile (also referred to as the Source profile) represents the colors desired, and the output profile (referred to as the Destination profile) represents the colors achieved on the device being used. For example, compare an image on a computer monitor and the same image printed on a desktop printer. The input profile is the profile of the monitor (an RGB profile), and the output profile is the profile of the desktop printer (a CMYK profile). Converting between the input profile and the output profile requires a CMM (Color Matching Module). The Color Matching Module (CMM), sometimes called the "Color Engine," is the part of the Color Management System software that maps one color gamut to another. A CMM uses device profiles and render intents to convert the Look-Up Tables (LUTs) between devices. The CMM does this by mapping the out-of-gamut colors into the range of colors that can be produced by the output device.

The CMM uses a set of proprietary algorithms to calculate the dot percentages required to produce the requested color. Those calculations are typically based on regression analysis of the data contained in the input and output profiles. Results can vary depending on which brand of engine is used. Consequently, there is always some error associated with the color management process. However, this error is far less than the error associated with any other color correction process, such as dot-gain compensation. For matching proof-to-press, this error can, and should, be quantified.

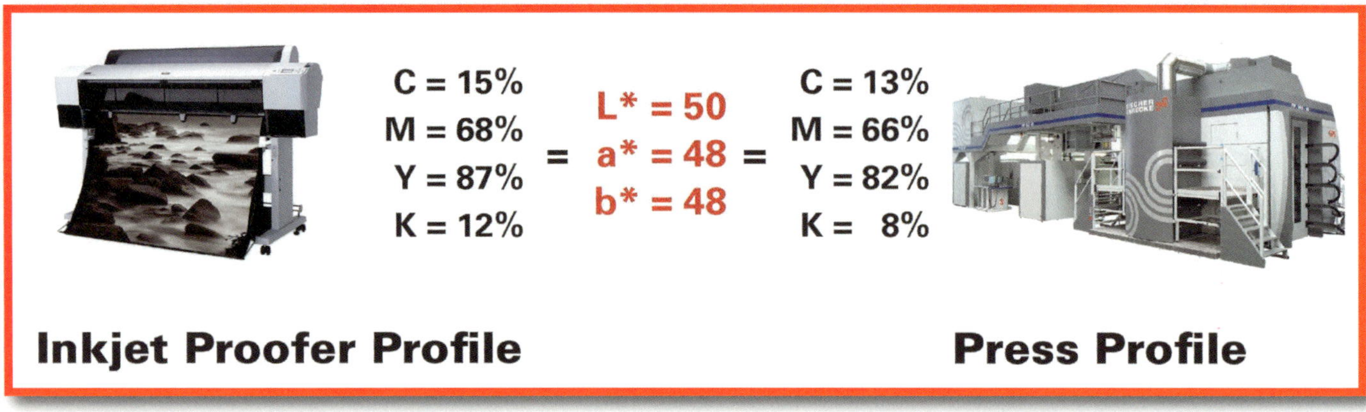

14.4.1e Proof-to-Press Profile: *This graphic illustrates how one area of an image might print on the proof in order to match the press. The different dot percentages are determined by profiling both devices and applying the ICC profiles to the image.*

To quantify the error within the color management process:

- Use the CMS software to reproduce the original target used to profile the printing condition
- Measure the resulting proof, color swatch by color swatch
- Compare it to the original press sample

When using the IT8.7/4 characterization target, there are 1,617 color comparisons (ΔE differences). With 1,617 color comparisons between the proof and the press, there will be some colors that are near perfect matches and some colors, that could be considered unacceptable matches (ΔE >3.0). An average ΔE < 2.0, based on CIE2000 tolerancing, is the acceptable difference for inkjet proofing systems. An average ΔE < 3.5, based on CIE2000 tolerancing, is the acceptable difference for thermal transfer systems. In either case, 95% of the data must have a ΔE < 6.0, based on CIE2000 tolerancing.

The acceptable difference depends on the technology used to create the proof. Conceptually this is counter intuitive, so it requires a detailed explanation. The two fundamentally different approaches to creating a proof are:

1. **Thermal Transfer Proofing Systems:** Match solid ink density and then use dot gain/TVI compensation to match individual tones. This approach produces smooth curves throughout and the final result is a smooth profile.
2. **Inkjet Proofing Systems:** Use color management software to map the color gamut of the press to the proofing device. Since this requires measurement of the 1,617 color patches, the profile reflects individual measurement error and press variation. The resulting profile is, most likely, not smooth.

The thermal transfer proof shows very smooth tonal gradations but, on average, exhibits a relatively large difference from the press sheet. An inkjet proof is very color accurate because it accurately mimics the inherent variations that were in that particular press sheet. Obviously measuring a large number of press sheets (20-30 press sheets) produces a much smoother profile; however, this is rarely done.

There are advantages and disadvantages to both systems such as, the ability to laminate to a specific substrate or the availability of white ink. Whichever technology is chosen depends on more than just color accuracy. Hence, there must be a larger acceptable difference for thermal transfer proofs.

14.4.2 The IT8.7/4 Characterization Target

FIRST supports using the industry standard, IT8.7/4 characterization target. This data set is the result of collaborative efforts of several standards committees, and represents the best technical approach to a standard target. The target can be formatted in a variety of ways to be read by various color measurement devices. *FIRST* recommends printing the target in the format that can be read by the color measurement device being used.

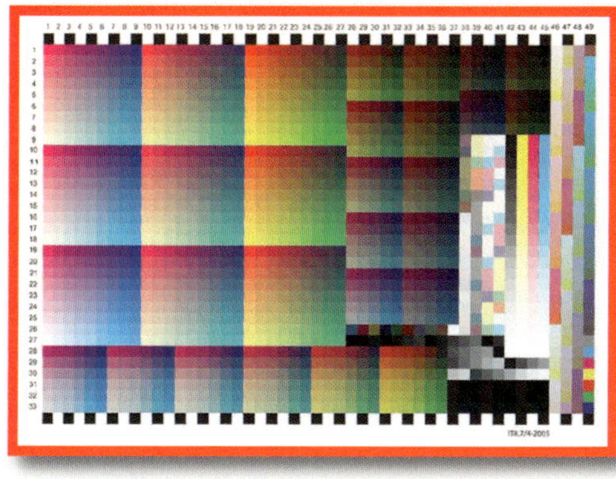

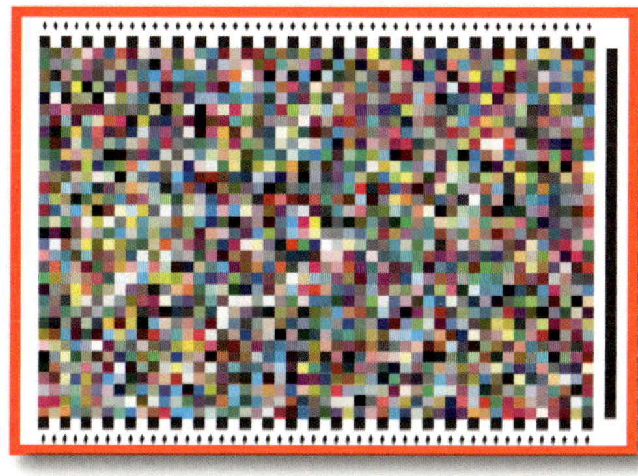

14.4.2 IT8.7/4 Targets: *There are two layouts of the IT8.7/4 target, the "visual-layout" (also referred to as the "ordered format") and a "random- layout" (bottom).*

There are two layouts of the IT8.7/4 target, the "visual-layout" (also referred to as the "ordered format") and a "random-layout". Both contain the same combinations of CMYK data. *FIRST* recommends utilizing the full print width of the web to incorporate multiple IT8.7/4 targets in both the random and visual layouts. Ideally, the target can also be oriented in two directions with a 90° rotation.

The random-layout target should be used for proof & print characterization. By randomizing the locations of the various CMYK combinations, ink takeout across the target area is more uniform. Errors introduced by proofing device or press anomalies will be randomized with respect to CIELAB color space.

The ordered, or visual, layout is useful for the initial "visual" analysis of the characterization and for those applications where manual measurement is required.

Refer to Section 16.4.2 for information on *FIRST* recommended spectrophotometer settings and color tolerancing methods.

IT8.7/4 Documentation

Color management systems from different vendors produce different ICC profiles. Even when reading the same test target, different programs will interpolate and expand the color data set according to custom settings and proprietary technology. Therefore, different programs using the same ICC profile may produce different results. Consistent results are best achieved by consistency in the workflow. Therefore, color characterization data should be archived. *FIRST* recommends maintaining ICC profile data in a database. ICC profile documentation should include, but not be limited to, the following:

- Company
- Date
- Project
- Measurement equipment settings
- Equipment: press, proofing device, computer monitor, scanner, CMS software
- Press conditions: anilox rolls, mounting tape, print sequence, plate material, etc.
- Materials: substrate, ink set, proofing dyes, etc.
    1. Spectral identification of ink sets
    2. Substrate characteristics: include flute profile for corrugated
- Prepress conditions when creating the IT8.7/4 target: dot gain/TVI compensation curve, GCR, UCR, bump curve, etc.

Device-Link Profiles

A device-link profile is a profile created from the printer-specific ICC profile (generated from the IT8.7/4 characterization target) and special instructions. The primary benefit of using a device-link profile, instead of the printer-specific profile, is that as an image is converted from RGB-to-CMYK or from CMYK-to-CMYK, single colors can remain single colors. For example, when utilizing a traditional ICC conversion, a "black only" shadow would be converted to a 3 color black shadow. A device-link profile enables the prepress provider to define the "black" separation, maintaining a "long black", "short black", or whatever is desired. The device-link profile gives the prepress provider greater control of the image as it is converted.

15.0 FINAL FILMS/FILES/DIGITAL MASK SPECIFICATIONS

15.1 Evaluating Physical Properties of Film Negatives

Refer to Section 17.3.1 for mask specifications of digitally imaged photopolymer plates.

Measurement Instrumentation – Transmission Densitometer

FIRST recommends measuring film negatives in accordance with ANSI/CGATS.9 1994 (Graphic Technology – Graphic Arts Transmission Densitometry Measurements – Terminology, Equations, Image Elements and Procedures). There are numerous definitions, equations and guidelines in CGATS.9 1994, some of them can be found in the Glossary section of *FIRST.*

The following information should be included when communicating transmission densitometric data:

- Conformance with, or deviations from, CGATS.9 1994
- Densitometer manufacturer and model
- Aperture size
- Density: Equation (absolute or relative); Channel (UV/Ortho/Visual/RGB)

All instruments must be calibrated in accordance with the manufacturer's recommended procedure prior to use. The transmission calibration standard must be traceable to a standard reference. The parameters listed in this section, along with the measurement results, should be included with the final films and contract proof as part of a Certificate of Analysis. Refer to Proof Compliance Cover Sheet/Label in Section 16.5.

Film Density

Traditional silver halide films that are developed, fixed and washed produce extremely dense, visually black images. These images absorb nearly all wavelengths of light from infrared (IR) through visible to ultraviolet (UV). Their visual density is a good indication of their actinic (wavelengths that expose photopolymer plates) density. They may be measured using the UV/Ortho/Visual channel of a densitometer with little risk of error. Dye-based heat or thermal imaging materials produce images that may have lower visual density but have high actinic density. Measuring with an Ortho or Visual channel on a densitometer may give false low readings.

The clear area (D-min) of the film should have a base orthochromatic density less than 0.05 and an ultraviolet density less than 0.10. The measurement of D-min using the ultraviolet channel on a transmission densitometer is important to ensure optimum proofing and plate exposures. The higher the D-min number, the more ultraviolet light is absorbed. Most proofing systems and photopolymer plates require ultraviolet light to accurately and consistently expose the image. Therefore, the lower the D-min value, the better. The black maximum density

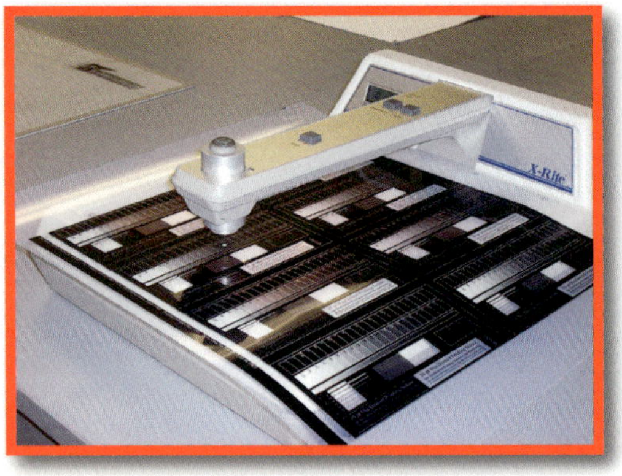

15.1 Film Density: *A transmission densitometer is used to measure film density (D-min & D-max).*

Platemaking Film D-min & D-max		
All Print Segments	D-min	D-max
Ortho Density	< 0.05	> 4.0
UV Density	< 0.10	> 4.0

Table 15.1a

(D-max) area of the film should have ultraviolet densities greater than 4.00.

Film Thickness

All film should be a consistent thickness and dimensionally stable. The recommended film thickness for photopolymer platemaking is 0.007" (0.18mm) for wide-web and 0.004" (0.10mm) for narrow-web printing applications.

Platemaking Film Thickness	
Wide Web (All Segments)	Narrow Web (All Segments)
0.007"	0.004"
0.18mm	0.10mm

Table 15.1b

Non-Traditional Film Types

Since digital imaging technology has been developed, the need and availability of photographic film in the marketplace has declined. However, some platemaking applications still require the use of film such as liquid photopolymer or specialty materials. Below is a listing of alternative film methods and potential uses.

Ink Jet Film: Inkjet films use inkjet printer technology to produce negatives. Required materials include inkjet receptive film/media, often available with a matte surface and special ink that includes UV light blocking additives. These systems generally include a RIP that separates the file into individual 1-BIT files for output. Inkjet film is often used for corrugated plate making. Most devices can output between 45-120lpi.

Thermal Film: Thermal image setters use heat sensitive films to produce negatives. There are a variety of sizes available and are often used in the corrugated or specialty industries. These devices do not require the chemistry or darkrooms of photographic films, and utilize resolutions of up to 1200dpi. Thermal image setters also utilize a RIP to separate files into color channels for output.

Laser Ablative Film: Laser ablative films are available for imaging on the same devices that image digital plates. The films have a thin carbon coating that the laser ablates away. File preparation is done similarly to digital plate preparation. Individual film sheets must be mounted on the drum of the device. Resolutions depended on imaging device capability with recommended line screens up to 133lpi or possibly 150lpi.

Film Negatives – Physical Specifications

- All film must be matte finish as specified by the raw plate material manufacturer
- Film must be supplied as one piece per color, identified by the color (Magenta, PMS 186 red, etc.)
- All films supplied for the flexographic printing process must be negative
- No etching or hand color corrections may be made on supplied film
- Opaquing may be done on the non-emulsion side of the film (Opaquing should be minimal and thin in consistency to preclude high spots in the film that may create contact and dot gain/TVI problems during platemaking)
- All extraneous marks, smaller than a minimum dot, should be opaqued or covered
- Excessive opaquing is unacceptable
- All films must be hard dot types or equivalent dye or inkjet materials; camera or soft-dot film is unacceptable
- The film must be free of kinks and scratches
- Visible fingerprints, bleaching, or colored markings are not acceptable when viewing the film's clear areas

Film Negative Emulsion

- Negatives prepared for surface print must be right reading emulsion up (RREU).
- Negatives prepared for reverse print must be right reading emulsion down (RRED)

15.2 Dot Characteristics for Film/Digital Masks

The shape of the prepress dot contributes to the dot gain/TVI characteristics of the resulting print. Therefore, when introducing imaging differences (such as equipment or dot type) different dot gain/TVI results should be expected. In a carefully controlled and calibrated process, it is important to test and understand the differences. When changing the shape of the dot, it is necessary to repeat the print characterization.

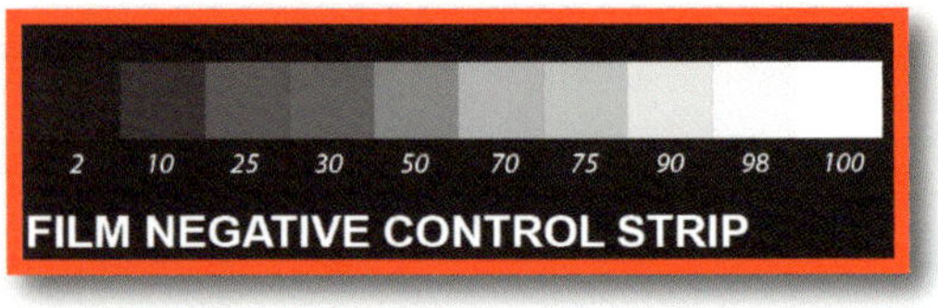

15.2a Film Negative Control Strip: *A control strip on the film negative can be measured with a transmission densitometer to confirm the imaged dot size.*

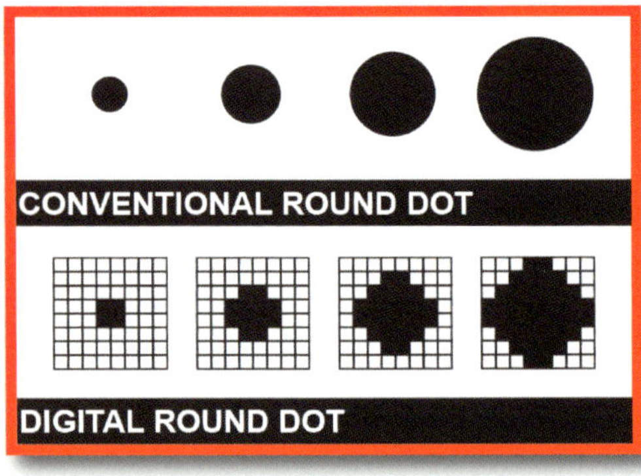

15.2b Dot Shape: *Dot gain is minimized when round dots are used, especially in vignettes.*

When TIFF/LEN files are used to make plates on more than one device, the platemaking systems must achieve the same results. Proper calibration helps to ensure that imagers, exposure frames and washout units are controlled to achieve a consistent result regardless of location.

Dot Accuracy

When imaging film or digital masks for the flexographic printing process, dot accuracy is extremely important, especially in the highlight tonal range. The verification of dot accuracy can be monitored only if the target dot percentage is known prior to duplication or imaging. With electronic imaging, output devices must be calibrated so that the output from the computer must match the dot percentage of the film/mask. If the films are duplicated or contacted, the scales should be used to verify contact exposure, film speed, and processing conditions. The dot percentage on the final platemaking film should conform to the tolerances in Table 15.2.

Dot Tolerance for Film and Digital Masks *All Print Segments*	
2%	+/- 0.5%
10%	+/- 0.75%
25%	+/- 1.0%
30%	+/- 1.0%
50%	+/- 2.0%
70%	+/- 2.0%
75%	+/- 2.0%

Table 15.2

Dot Shape

A round dot shape is best controlled in the flexographic printing process. Dot gain/TVI is minimized when round dots are used, especially in vignettes. A round dot is defined as a dot that is round in the highlights and round in the shadows.

15.3 Image Screening

Traditional Dot Shapes

Traditional screening incorporates halftone dots of a specific shape, organized in an amplitude-modulated (AM) pattern, where the frequency of dots remains constant, but the size of dots varies as the tonal value increases. While several dot shapes have become standard across the print industry, only two shapes are well suited for flexographic printing:

1. Circular (Euclidean): Dots maintain a round/circular shape throughout the tonal range. As dots converge, the space between dots appears diamond-shaped.
2. Fogra-Round: Dots appear slightly elliptical, resulting in two "touch points" where dots begin to converge. This helps even out the visual increase in intensity that occurs when dots converge all at once.

In circumstances where AM screening is not desired, frequency-modulated (FM) screening may be used. In FM screening, dot size remains constant, while spacing between dots varies. As tonal value increases, dots are placed closer together.

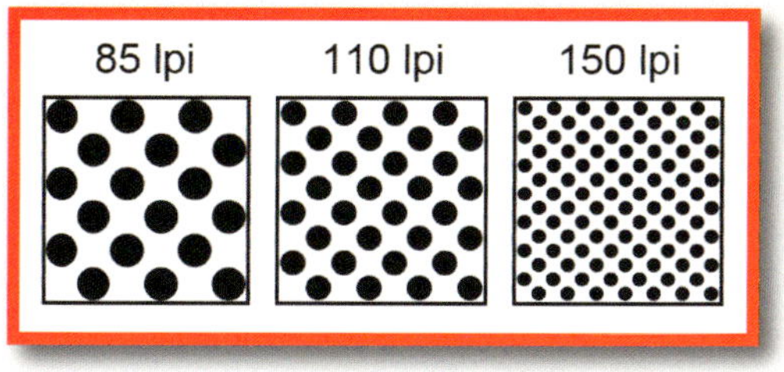

15.3a Line Screen: *The physical dot size of the same dot percentage (for example a 30%) decreases as line screen increases.*

Hybrid Screening
A hybrid screen is classified as such when the screen cannot be classified purely as either AM or FM screening. Most often, these screens utilize an AM screen in the mid-tones and transition into a variant of FM screening in the highlights and sometimes the shadows. The primary goal is to maintain the smoothness of AM screening through most of the tonal values, but achieve lower printed tone values in highlights with FM screening to reduce the likelihood of a harsh tone break where the highlights suddenly stop.

There are many types of hybrid screens from many different RIP vendors and are often designed for specific plating systems. It is important to choose the hybrid screen most compatible with your platemaking workflow and printing environment.

Plate Surface Modification
"Modification" of the printing plate surface, designed with the intent of improving ink transfer from plate surface to substrate, has been developed and implemented by several different plate and RIP manufacturers.

While not technically referred to as "screening," these applications are typically accomplished by applying a "patterning" effect to the plate surface, either in the plate manufacturing stage or through imaging a particular pattern at a higher frequency and resolution.

These plate surface modification techniques have demonstrated the capability to enhance ink laydown and reduce pinholing, particularly on film substrates, and result in higher perceived solid ink densities and improved opacity in white ink coverage.

15.3b Line Screen: *Higher line screens can print more image detail because there are more dots per square inch to create the detail. However, dot gain increases with higher line screens because the white space between the dots is less.*

Results can vary depending on substrate, ink system and anilox configuration. It is recommended to test and compare plate surface modification techniques prior to implementing in production. In most cases, it will be necessary to characterize and profile the print conditions with the application of plate surface modification.

Line Screen

Line Screen rulings vary based on imaging method, plate material, ink characteristics, and print conditions (such as press width, anilox engraving, metering system and substrate). Higher screen rulings may result in greater image detail, but may also narrow the tonal range that can be reproduced on the press. The printer must approve the line screen ruling for each job.

For any given line screen ruling, the RIP software may use the best mathematical equivalent to avoid moiré problems. Therefore, the specified line screen may be scaled up or down slightly to produce the best result when imaged using that supplier's equipment. For example, when 120lpi is specified, the actual screen ruling may be 116lpi or 126lpi depending on the manufacturer of the RIP.

Screen Angles

Screen angles and rulings should be selected to minimize moiré patterns. Some moiré is inevitable. All screen angles and dot shapes should be specified in the digital file before outputting film or plates (digital or laser engraved). Below are guidelines for selecting screen angles for a round dot:

- An aesthetically pleasing rosette pattern is achieved by selecting angles 30° apart. Therefore, cyan, magenta and black screen angles should be 30° apart.
- Yellow should be 15° between two of the colors (in most cases, cyan and black) because yellow is the most difficult to visually perceive.
- The nominal angle for viewing screens is 45°. However, it is necessary to avoid using the same angle as the anilox rolls. Usually, this is accomplished by rotating all color angles by approximately 7.5°.
- The dominant color, or the color containing the most information, should be placed closest to 45°. Typically, black is the dominant color; however, the dominant color varies with the image being printed. Screen angles for two-color or three-color printing should follow the same guidelines with the dominant color at the nominal angle.

- Apply these guidelines:
  1. Black is screened at 45° - 7.5° = 37.5° (assuming it is the dominant color)
  2. Magenta is screened at Black + 30° = 67.5°
  3. Cyan is screened at Black - 30° = 7.5°
  4. Yellow is screened at Black - 15° = 22.5°

As with line screen ruling, for any given screen angle, the RIP software will use the best mathematical equivalent to minimize moiré problems. Therefore, the specified screen angle may be scaled up or down slightly to produce the best print result when imaged using the supplier's equipment. For example, when a 67.5° angle is specified, the actual screen angle may be 66° or 68° (or some value in between) depending on the manufacturer of the RIP. This is considered normal practice in achieving the least noticeable moiré and the most pleasing rosette pattern.

15.3c Screen Angles: *Screen angles and rulings should be selected to minimize moiré patterns. Some moiré is inevitable.*

15.4 Registration Marks and Microdots

A printer-supplied template indicating the size and placement of registration marks supersedes the recommendations that follow.

All images must have registration marks indicating the center of the web direction as well as the centerline of the cut-off direction on one-up color. For jobs that are being die cut, final assembly of the product will hide the registration marks from the consumer. Marks that remain on the printed product need to be positioned so that, upon final assembly of the package, the registration marks will be hidden from the consumer, if possible.

Registration marks should be solid lines with each of the colors producing precisely the same mark in register with each other. When all colors are accurately registered using the incorporated marks, all elements of the job must be in register. For digitally imaged plates, a mounter's proof or stationary cameras during mounting are two ways to verify registration prior to press. Cross hair dimensions (line weight and thickness) are sensitive to anilox volume. As anilox volume increases, the cross hair dimensions should also increase (longer and thicker).

An alternative approach to registration utilizes the "Railroad Track Target" (RTT). The value of this target is its simplicity. The RTT allows the printer to achieve press registration without using a tape measure. The RTT is formatted in both millimeters and inches. The printer should select the appropriate format for the press based on the unit of measure used in gearing and press adjustments. Refer to Section 19.4.7 for a detailed explanation of how to use the RTT target.

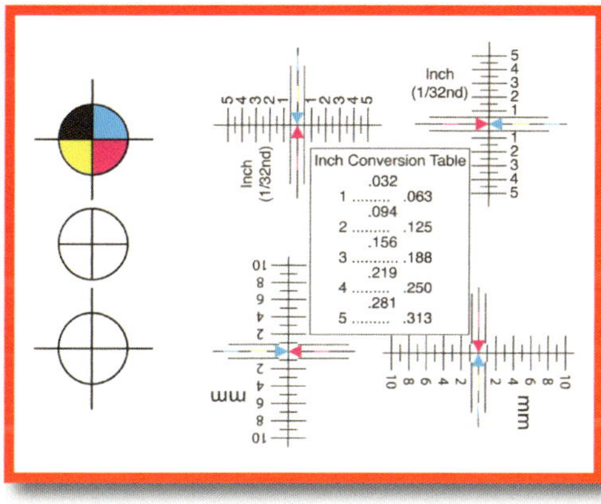

15.4a Registration Targets: *The printer should select either the millimeter or inch scale when using the RTT Target. Both are not required on a job.*

Line Screen (lpi & lpcm): General Guidelines
Line screen is print system dependent; determine optimum line screen during the press optimization and press fingerprint trials (ref 1.3.1 & 1.3.2)

Segment		Substrate	Conventional Plates	Digitally Imaged Photopolymer	Laser Engraved Rubber/ Cured Polymer
Wide Web	Preprint Linerboard	SBS Board	110 - 133 lpi (43 - 52 lpcm)	110 - 175 lpi (43 - 69 lpcm)	110 - 133 lpi (43 - 52 lpcm)
		Uncoated	100 - 133 lpi (39 - 52 lpcm)	100 - 133 lpi (39 - 52 lpcm)	100 - 120 lpi (39 - 47 lpcm)
	Combined Corrugated	All	55 - 110 lpi (22 - 43 lpcm)	55 - 150 lpi (22 - 59 lpcm)	55 - 110 lpi (22 - 43 lpcm)
	Folding Carton	SBS Board	120 - 150 lpi (47 - 59 lpcm)	120 - 175 lpi (47 - 69 lpcm)	110 - 133 lpi (43 - 52 lpcm)
		CRB Board	110 - 133 lpi (43 - 52 lpcm)	110 - 133 lpi (43 - 52 lpcm)	110 - 120 lpi (43 - 47 lpcm)
	Multiwall Bag	Coated Paper	75 - 120 lpi (30 - 47 lpcm)	75 - 120 lpi (30 - 47 lpcm)	75 - 110 lpi (30 - 43 lpcm)
		Uncoated Paper	65 - 85 lpi (26 - 33 lpcm)	65 - 100 lpi (26 - 39 lpcm)	65 - 100 lpi (26 - 39 lpcm)
	Newsprint	All	85 - 100 lpi (33 - 39 lpcm)	85 - 100 lpi (33 - 39 lpcm)	85 - 100 lpi (33 - 39 lpcm)
	Film Products	All	110 - 133 lpi (43 - 52 lpcm)	110 - 150 lpi (43 - 59 lpcm)	85 - 133 lpi (33 - 52 lpcm)
Narrow Web	Film Products	All	110 - 133 lpi (43 - 52 lpcm)	110 - 175 lpi (43 - 69 lpcm)	85 - 133 lpi (33 - 52 lpcm)
	Paper Products	Coated Paper	133 - 175 lpi (52 - 69 lpcm)	133 - 175 lpi (52 - 69 lpcm)	110 - 133 lpi (43 - 52 lpcm)
		Uncoated Paper	110 - 133 lpi (43 - 52 lpcm)	110 - 133 lpi (43 - 52 lpcm)	100 - 120 lpi (39 - 47 lpcm)
	Envelope	Coated Paper	133 - 175 lpi (52 - 69 lpcm)	133 - 175 lpi (52 - 69 lpcm)	N/A
		Uncoated Paper	85 - 133 lpi (33 - 52 lpcm)	85 - 133 lpi (33 - 52 lpcm)	N/A

Table 15.3

Step-and-Repeat Jobs – Registration Marks

Step-and-repeat jobs are handled differently than one-up jobs. The printer must confirm specific placement of register marks on stepped jobs. A registration mark equal in size, shape and character to the one-up job should be placed exactly in the center of the web direction. For example: if the total printed image area, including all of the various repeats across the web, equals 12" (30.5cm), the mark should be placed 6" (15.2cm) from either side.

Registration marks in the machine direction must be in the center of the repeat, taking into consideration any stagger requested by the printer. For example: if the printed product has

Register Mark Dimensions: Crosshairs			
Segment	Substrate	Length	Line Weight
Wide Web	Paper/Film Products	0.25"	0.01"
		6.35mm	0.25mm
Narrow Web	Paper/Film Products	0.125"	0.006"
		3.18mm	0.15mm
Corrugated	All Flute Profiles	0.25"	0.01"
		6.35mm	0.25mm

Table 15.4a

a measurement of 4" (10cm) and is being supplied to the printer one-up, then the center of this image is 2" (5cm). However, if the printer has requested a half stagger, the overall dimension will be 6" (15.2cm), making the center and the registration mark location now 3" (7.6cm).

15.4b Registration Marks on Step and Repeat Jobs: *This illustration indicates the proper positioning of registration marks on a step and repeat job. The marks are located in the center of the overall dimensions.*

Auto Registration Marks

Several electronically controlled press registration devices require the design firm, printer, and/or prepress provider to incorporate automatic registration marks into the final file. It is the printer's responsibility to provide the specific scale, type, and placement information to the prepress provider prior to generation of the final file. The printer should make a digital, vector file of the special target(s) required and provide to the design firm or prepress provider during the design review process.

Microdots

Microdots are used with video mounting devices. Microdots are placed on both the left and right side of the printed material and in the center of the web direction. They should be 0.010" (0.25mm) in diameter unless otherwise specified by the printer. Typical microdots will be freestanding dots that will print and remain on the printed sheet throughout the press run. When the job is printing in register, the dots will overprint each other and appear to be an almost perfect dot. With some designs, the printer may choose to knock out the microdot by obscuring it in a solid color. In the same area of the knocked-out color, use a positive print for each of the remaining colors to help hide the microdot.

It is recommended that an arrow of approximately 0.0625" (1.588mm) wide and 0.25" (6.35mm) long be placed approximately 0.125" (3.3mm) away from the microdot,

15.4c Microdots: *Typical microdots will be free standing dots that will print and remain on the printed sheet throughout the press run.*

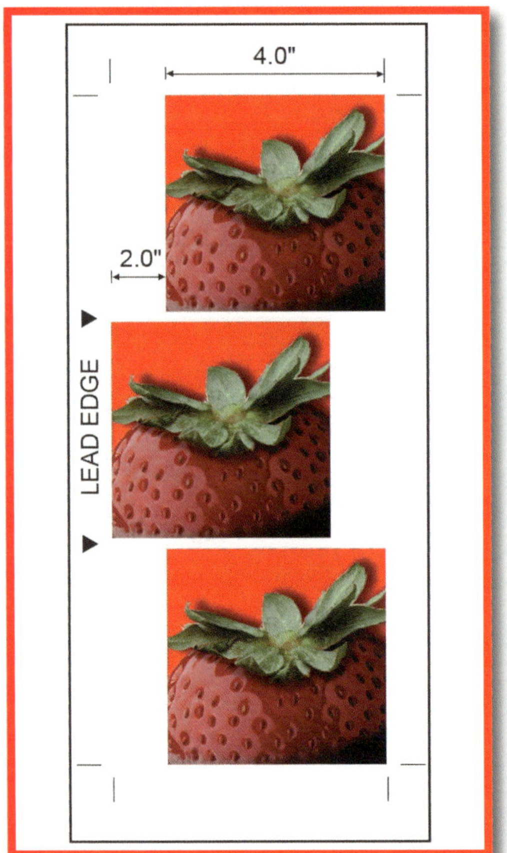

15.5 Image Stagger: *The graphic illustrates a job stepped three across and one around with a quarter stagger. The stagger achieves constant impression on all colors around the cylinder to minimize press bounce.*

indicating its placement, so the dot is not arbitrarily removed from the files (or opaqued out of the film). It is the responsibility of the platemaker to cover the arrow so it is not imaged with the plate and later printed. Only the microdot is imaged onto the plate, remains on the product throughout the press run, and is not cause for rejection.

Microdots on Step and Repeat Jobs

Microdots location on step and repeat jobs are based on the total image. Usually these dots are placed across the web on the centerline of the full stepped job. The size of the design and the stepped job will determine how many dots are needed. The microdots should be placed in areas that do not impact the design, but are in strategic positions for mounting. Microdots must be a minimum of 0.25" (6.4mm) inside the live printed area of the product. The printer should be consulted about specific placement that may be required for a particular press-mounting system.

Microdot Diameter	
All Print Segements	0.008" - 0.010" D
	0.20mm - 0.25mm D

Table 15.4b

15.5 Image Stagger

Because of the physical characteristics of printing from a raised plate, the technique of staggering the image is used to minimize press bounce. Within each print station, the goal is to have the plate constantly under impression. Of course, this is not always possible, but staggering does help maintain an even impression and minimize the bounce that may occur as the plates go onto impression. The printer may request a stagger of a third, a fourth, or half of the product's overall dimension, from top to bottom. Consult the printer for detailed instructions on how to stagger a stepped job. Also be aware that each printer, although printing similar images, may require different staggering techniques.

15.6 Calculating Distortion

When a photopolymer plate is laying flat, the top (Y) and bottom (X) of the plate are the same length (X=Y). However, when the plate is wrapped around a printing cylinder, the surface of the plate becomes stretched because the distance around the top of the plate is greater than the distance around the bottom of the plate ($Y_d > X_d$).

Since the plate is imaged flat and printed round, the original file must be reduced, or distorted, only in the direction it will be wrapped around the cylinder. Therefore, when the plate is wrapped around the cylinder, it will be the proper size. The percent distortion is simply the ratio X_d/Y_d. Where, X_d is the circumference of the inner circle and Y_d is the circumference of the outer circle. Circumference = $2\pi r$ (π = ~3.14159)

$$\% \text{ Distortion} = \frac{2\pi r}{2\pi R_2} = \frac{R_1}{R_2}$$

The values of R_1 and R_2 depend on the thickness of the plate (P), the thickness of the mounting tape (T), the cylinder radius (C) (radius equals half the diameter), and the thickness of the polyester backing (M) used to support the plate. The thickness of the polyester backing is important because it has a very high modulus of elasticity and will not stretch when wrapped around the cylinder. Consequently, R_1 is equal to the radius of the cylinder plus the mounting tape thickness plus the polyester backing thickness. R_2 is equal to the radius of the cylinder plus the mounting tape thickness plus the plate thickness.

$$R_1 = C + T + M$$
$$R_2 = C + T + P$$

Equation #1:

$$\% \text{ Distortion} = \frac{C + T + M}{C + T + P}$$

Often repeat length (R_L) is provided instead of cylinder diameter. The above equation can be converted into a new equation using repeat length (Y_d).

Equation #2:

$$\% \text{ Distortion} = \frac{[(R_1 \div 2\pi) + (M - P)]}{R_L \div 2\pi}$$

For example, a job with a repeat length of 11.5" using a 0.20" tape and a 0.067" plate will have a 96.61% distortion.

R_L = 11.5" M = 0.005" P = 0.067"

$$\% \text{ Distortion} = \frac{11.5 \div (2 \times 3.14159) + (0.005 - 0.0067)}{11.5 \div (2 \times 3.14159)}$$

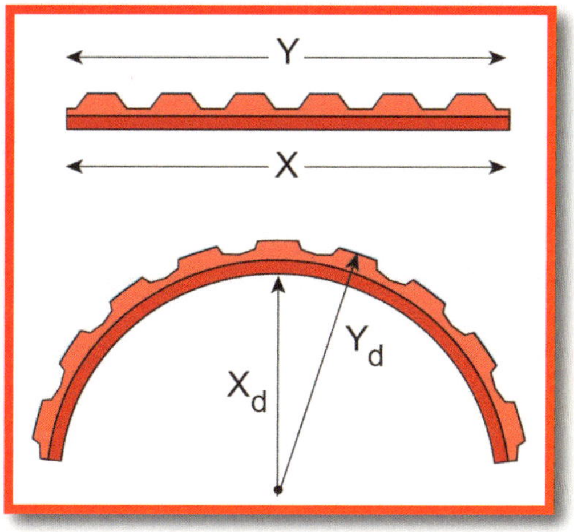

15.6 Calculating Distortion: *When a photopolymer plate is laying flat, the top of the plate and the bottom of the plate are the same length (X = Y). However, when the plate is wrapped around a printing cylinder, the surface of the plate becomes stretched because the distance around the top of the plate is greater than the distance around the bottom of the plate* ($Y_d > X_d$).

$$\% \text{ Distortion} = \frac{1.8302834 - 0.062}{1.8032834}$$

$$\% \text{ Distortion} = \frac{1.7682834}{1.8032834}$$

$$\% \text{ Distortion} = 0.9661254$$

$$\% \text{ Distortion} = 96.61\%$$

When calculating % distortion using repeat length, mounting tape thickness is not required for the calculation. However, it is necessary to confirm the correct tape thickness is used to account for the cylinder undercut.

For most plate material, the polyester backing is 0.005"; however, there are exceptions. Confirm the polyester backing thickness with the plate material manufacturer.

To confirm the proper distortion factor is applied to file/film or file/mask prior to plate making, the file/films or file/masks should indicate the distorted dimension, the distortion % and the final print dimensions. When using photopolymer plates or laser-engraved rubber, image size should be maintained at 100% across the press width and be distorted in the machine direction.

When using molded rubber plates, compensation in both machine and web direction is required. The printer should provide a through-the-press distortion factor. The platemaker must determine the amount of reduction required across the press/web.

Example: For 0.125" (3.2mm) rubber: To allow for the shrinkage that takes place in the dimension across the cylinder, make the image wider than the desired print dimension by 0.020" (0.51mm).

15.7 Final File/Film or File/Mask Inspection Attributes

Workflow tags, commonly referred to as "Smart Marks", can be used to confirm the films/files/masks are correct along with the "Certificate of Analysis (CoA)". The following attributes should be reviewed during QC of the final files and certified by the file/film/plate supplier.

Actual Dot Size: Include a control strip on the negative to confirm anticipated dot size. The control strip is gray scale with printer's minimum, 10%, 30%, 70% and 100%

(or shadow treatment values used in lieu of solid). Measure with a transmission densitometer or have the prepress provider supply a certificate of analysis (CoA).

Compensation Curve: Each printer has a specification based on the press fingerprint. Verify the correct compensation has been applied.

Bar Code Verification: No material should be accepted for print unless it is accompanied by a bar code report. The bar code report should be scanned directly from the film supplied for plate making or mounter's proof, regardless of who makes the plates. When a bar code is generated via software and the assembled file is supplied for imaging to the printer or plate provider, it must be accompanied by a written report. The report should identify the human-readable information.

16.0 COLOR PROOFS

16.1 Types of Proofs

All parties involved with a project must agree on the process and terminology used to evaluate and communicate color. Specifically, every proof created throughout the workflow should be clearly labeled to communicate:

- The purpose of the proof
- The system or device on which the proof was created
- Whether the output device was profiled, and if so, which profile was used
- The suitability of the proof for judging color

Concept Proof
This proof is common in the early creative stages of the project. It is used to capture input from all partners in the supply chain during design development and is also referred to as a "collaborative proof". This proof is not typically color profiled. Therefore, these proofs should be labeled "Not For Color".

Color Target Proof
The color target proof is often the selected "concept proof". It represents the ideal color intent of the designer and client, independent of the print process or the ability of an individual press to achieve that color. Some of the color in this proof may not be achieved in the final print. To avoid rework costs and unachievable expectations downstream, it is helpful, when possible, to produce this proof based on the known or expected capabilities and color gamut of the anticipated printing process(es).

16.1a Concept Proofs: *Early in the package development process, the designer creates several concept proofs incorporating input from all partners in the supply chain in an attempt to meet the customer's objectives.*

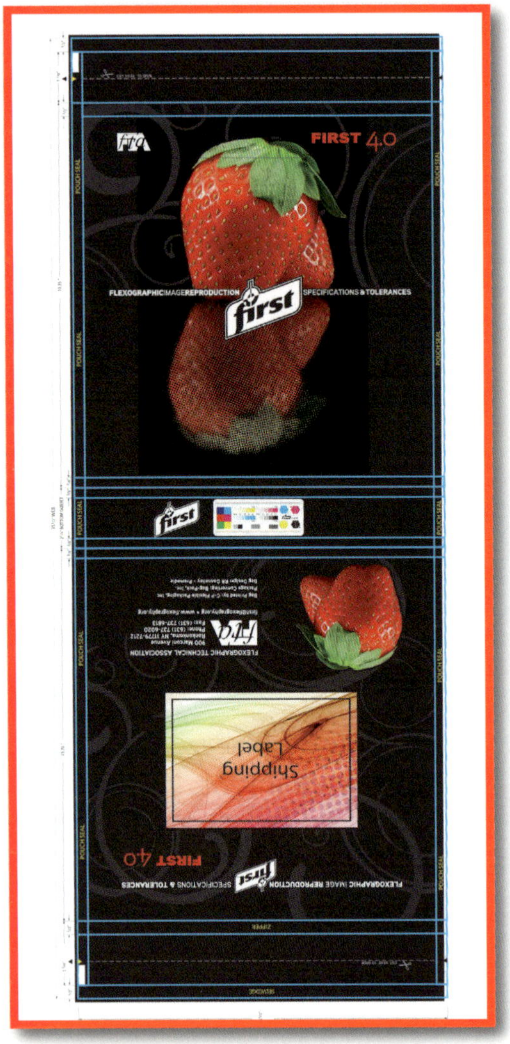

16.1b Profiled Contract Proof: *The profiled contract proof represents the customer's complete content and color expectations for the final printed product and is the basis for negotiations on project performance.*

16.1c Soft Proofing: *In order for soft proofing to be successful, the monitor must be color-calibrated.*

Comprehensive Proof (Comp)/Mock Up

This proof is formed to the shape of the final product and should indicate whether or not it is color accurate.

Profiled Contract Proof

The profiled contact proof represents the customer's expectation of the printed product.

The profiled contract proof represents the customer's complete content and color expectations for the final printed product and is the basis for negotiations on project performance. It illustrates how the printed image is expected to look when reproduced on press and is an important quality control tool and communication device. It is profiled using a color management system (CMS) and is prepared based on profiles provided by the specific printer or prepress provider and is produced according to *FIRST* specifications. The contract proof does not have to be a dot-for-dot reproduction, but it must exhibit a common visual appearance to the press or characterized reference printing condition (CRPC) dataset. Therefore, it must simulate the dot gain/TVI, color attributes, detail and contrast of the printed image. It must also contain a control target that is processed and imaged as part of the proof, which will be used to verify accuracy and consistency throughout the design, proofing and printing process. The control target must contain specific screen values, which should be determined with the printer, for any colors printing dots, including vignettes. Although most digital proofing devices may not reproduce a conventional dot pattern, the tone scales should be measured using a densitometer (or spectrodensitometer) in the dot area function. Each one of the tone scales must equal the weight (dot area) identified by the press profile. Before a contract proof can be accurately used, the entire reproduction system must be characterized so that the proofing system is calibrated to match the printed result. Afterward, both press and proofing systems must be maintained for consistency and repeatability. Refer to Section 14.4 CIELAB Color Management for additional information on profiling.

Soft Proof

This proof is viewed on a color-calibrated monitor. The soft proofing method can be used at any stage from concept proof to contract proof, depending on how well the system is calibrated. To use the soft proofing method, each party must have a color consistent monitor and a color management system (CMS).

16.2 Proofing Methods

Standard/Analog Proofs

Standard, or analog, proofs have been almost entirely replaced by digital proofs. They are made from halftone films and may utilize overlay, single layer, or surprint technologies. Although analog proofs are made from films, they do not usually simulate the dot gain/TVI associated with flexographic printing. Overlay proofs are not acceptable as a contract proof for approving process color separations. However, they are useful for verifying the content of film negatives, checking register, verifying traps between colors and proofreading copy for accuracy.

The most widely used contract analog proof for verifying the accuracy of color separations today is the surprint proof. Some proofing systems offer a limited number of process shades and density variations to match the color attributes of the process colors used in the printing process. Proofing films may be modified to allow the proof to simulate the dot gain/TVI of a particular flexographic printing system. Some prepress providers provide proofs on a wide variety of substrates to simulate the cast and brightness of the material being printed.

To correctly simulate the final printed result, the proofing films may be modified to allow the proof to simulate the dot gain/TVI that occurs on press. Therefore, the films used to create the proof are typically different from the films used for platemaking. The adjustments for dot gain/TVI should be developed in conjunction with the printer. Analog proofs must be exposed and processed per the manufacturer's recommendations and must contain the manufacturer's recommended control target. The proofing manufacturer's target is used to monitor the consistency of the proofing system and does not have a profile applied to it. In addition to the target used to monitor the consistency of the analog proofing system, the printer-specified control target should also be incorporated. The printer's control target must remain an integral part of the contract proof. Failure to provide a proof with a control target for verification is cause for rejection. All proofing densities must conform to the solid ink densities achieved during the press characterization. If the printer has not characterized the printing process, the proof should adhere to the general guidelines outlined in Section 16.4.1.1. Refer to Sections 12.7, 12.8 and 12.9 for control target test element information.

Direct Digital Color Proofs (DDCP)

Direct digital color proofs are produced directly from digital files without halftone films. There are many different technologies used to create digital proofs including office copiers, laser

16.2 Direct Digital Proofer

printers, thermal transfer and inkjet. The majority of office and desktop color printers are capable of accurately producing proofs to specifications and are consistent enough to be used as contract proofing devices provided they are used with the correct paper and an appropriate color management system (CMS) up front. Some DDCPs simulate a halftone dot that is used for the printing process, while others produce more of a continuous-tone reproduction. There is no advantage to one over the other because the dots are not the same dots as those produced on the imagesetter/platesetter. Correctly calibrated DDCP systems can produce an accurate proof of the expected print result.

Digital proofs must be produced according to the manufacturer's recommendations and must contain the control target recommended by the equipment manufacturer. In addition to the target used to monitor the consistency of the digital proofing system, the printer-specified control target must also be incorporated. The manufacturer's target does not have a profile applied to it and is used to monitor the consistency of the proofing system. The control target must remain an integral part of the contract proof. Failure to provide a proof with a control target for verification is cause for rejection. All proofing densities must conform to the solid ink densities achieved during the press characterization. If the printer has not characterized the printing process, the proof should adhere to the general guidelines outlined in Section 16.4.1.1. Refer to Sections 12.7, 12.8 and 12.9 for control target test element information.

Press Proofs

Press proofs have been almost entirely replaced by digital proofs; they require plates and possibly film (unless using direct-imaged plates). The same imaging device must be used to make both the production plates and the proofing plates. The proof must reflect the solid ink densities and dot gain/TVI experienced during the press characterization. If the dot gain/TVI experienced on the press proof is different from the results of the characterization, two things must be done:

1. Establish what the production press dot gain/TVI is and match it on the proof press controlling it within the same tolerance established for printing.
2. Develop a curve for the production film/file to compensate for the tonal difference between the press characterization and the press proof.

All proofs must incorporate the printer-specified control target. The control target must remain an integral part of the press proof. Failure to provide a proof with a control target for verification is cause for rejection. Refer to Sections 12.7, 12.8 and 12.9 for control target test element information.

16.3 Proofing Sequence & Colorants (Pigments/Dyes)

Off-press proofing material manufacturers (analog and digital proofs) often specify a particular sequence of colorant application. This is not necessarily the same as the production printing sequence. When a press proof is supplied, it is recommended that the ink lay-down sequence used for the production run be used when proofing. It is the responsibility of the press proof provider to obtain this information from the printer.

FIRST has endorsed the identification of inks using color index (CI) numbers to define the actual pigments. It is important that the proofing colors used be spectrally matched to *FIRST* recommended pigments. Specifying the CI number of the pigments used improves the ability to match color across platforms (proof-to-press) as well as press-to-press, run-to- run, plant-to-plant and even printer-to-printer. Standardizing pigments between proof and press, if possible, (or closest match if the same pigments are not available) greatly improves the printer's ability to actually match the proof on press. By standardizing pigments, the occurrence of metameric matches is also reduced. The potential for every package on the shelf (or newspaper in the news stand) to appear the same increases significantly when pigments are standardized.

16.4 Measurement of Contract Proofs

Measurement of printed materials consistent with ANSI/CGATS.4 2006 (Graphic Technology – Graphic Arts Reflection Densitometry Measurements – Terminology, Equations, Image Elements, and Procedures) and ANSI/CGATS.5 2003 (Graphic Technology – Spectral Measurement and Colorimetric Computation for Graphic Arts Images) is supported by *FIRST.* ANSI/CGATS.4 2006 is the primary reference for evaluation of graphic arts materials using reflection densitometry. ANSI/CGATS.5 2003 is the primary reference for evaluation of graphic arts materials using spectrophotometry. There are numerous definitions, equations and guidelines in both standards. Some of these may be found in the Glossary of this document. Measurement conditions and characteristics measured should be reported in a certificate of analysis (CoA) to be included with the contract proof.

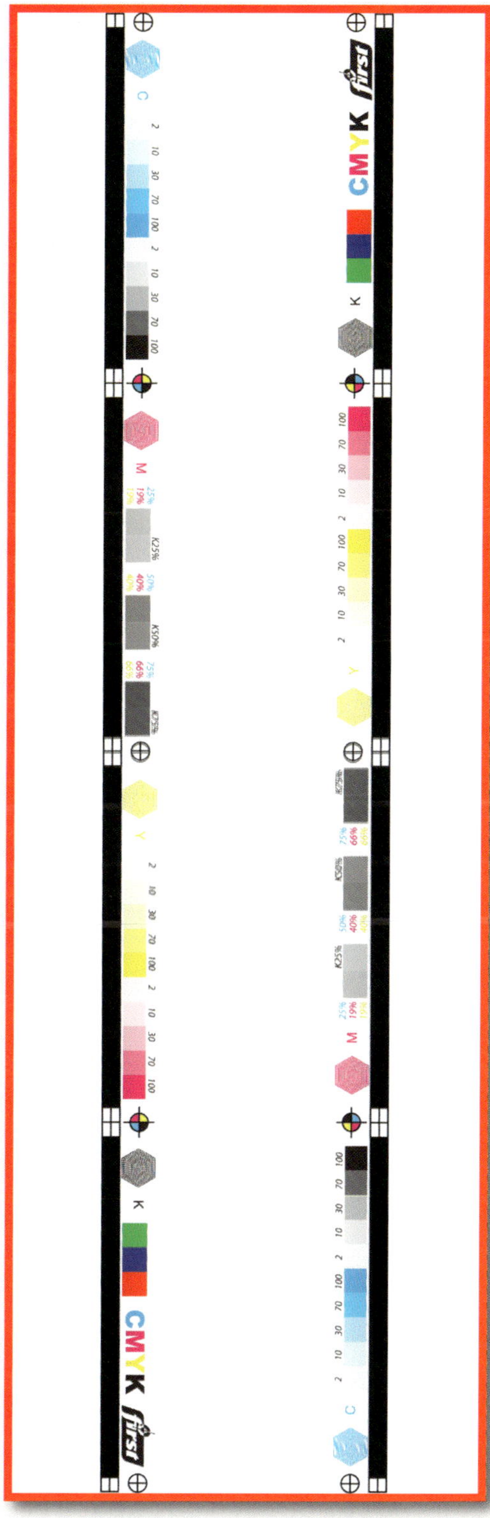

16.4 Measurement of Contract Proof: *A control target must be included on the contract proof for verification of proofing and printing accuracy.*

16.4.1a Spectrodensitometer

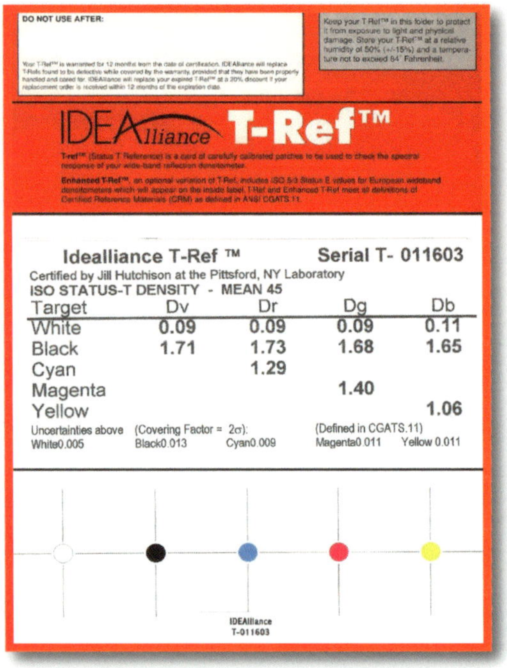

16.4.1b IDEAlliance T-Ref™

16.4.1 Densitometer Guidelines

Application: A reflection densitometer, or spectrodensitometer, is used to measure key print characteristics for process color images such as: density, dot area/dot gain/TVI, trap, print contrast and gray balance. These densitometric functions are critical for measuring and controlling the reproduction of continuous tone images. Using a densitometer improves the proofer's ability to consistently and accurately reproduce a continuous tone image. A densitometer is designed to be used only with the four traditional process inks (yellow, magenta, cyan and black). When using non-traditional colors for expanded gamut continuous tone images, the usefulness of the densitometric functions is limited because the filters and equations utilized by the instrument are specific to the traditional process colors.

Industry Standard: ANSI/CGATS.4 2006 (Graphic Technology – Graphic Arts Reflection Densitometry Measurements – Terminology, Equations, Image Elements, and Procedures) is the standard used for the measurement of printed materials using a reflection densitometer. For additional information on densitometry, refer to "Introduction to Densitometry – Users Guide to Print Production Measurement Using Densitometry," published by IDEAlliance. IDEAlliance contact information is listed in Appendix A.

Measurement Variables: Key instrument variables must be properly set to produce meaningful measurements. These variables should be documented and communicated with all densitometric data.

Densitometer/Spectrodensitometer: Manufacturer and model.

Spectral Response: The density value obtained is a function of the spectral characteristics of both the ink being measured and the instrument spectral response setting. The values obtained with different response functions may be similar or significantly different, depending on the particular material being measured. Status T is the preferred spectral response in North America. It is defined to closely match the characteristics of graphic arts materials normally used in the United States, such as ink-on-paper printed materials, off-press proofs and original art to be color separated. Status E is defined to closely match the characteristics of graphic arts materials normally used in Europe, such as ink-on-paper printed materials, off-press proofs and original art.

Sample Backing: Many flexographic substrates are translucent; consequently, the choice of backing material will greatly

Densitometric Instrument Variables	
Parameter	*FIRST* Recommendation
Standard	CGATS .4-2006
Spectral Response	Status T wide-band
Sample Backing	White with an: L* > 92, C* < 3 Non-Fluorescing
Aperture	3.4mm std. (2.0 - 6.0mm range)
Instrument Calibration Target	The IDEAlliance T-Ref™ target calibrated to +/- 0.02
Polarizing Filter	Yes or No (typically, no)

Table 16.4.1a

Applications and Use of M0, M1, M2, M3 *(Instead of Polarizing)*	
M0	Most legacy devices in use today, typically Tungston illuminant
M1	D50 illuminant onto the substrate to be measured
M2	UV filter regardless of the illuminant
M3	Polorized filter regardless of illuminant

Table 16.4.1b

influence any color measurements. The best choice is to use a white backing material that is spectrally non-selective, diffuse-reflecting and has a "L" value greater than 92 as specified in ISO 13655:2009 Annex A Sample Backing.

Sampling Aperture: While aperture sizes typically range from 2.0-6.0mm; 3.4mm is the standard aperture size. It is critical for each test element to be slightly larger than the instrument aperture for accurate measurement.

Calibration: *FIRST* supports assigning one person responsible for daily calibration of all densitometers/spectrodensitometers in accordance with the manufacturer's recommended procedures. A reflective calibration standard that is traceable to a standard reference (such as the T-Ref™ standard from IDEAlliance) should be used. Calibration results should be recorded. The T-Ref™ is a calibrated Status T standard reference on a laminated sheet with white, black, cyan, magenta and yellow circles with the associated readings for each color. Calibration to the T-Ref™ standard reference should be within +/-0.02 for each color. The T-Ref™

Densitometric Measurement Data		
Measurement	Information to Report	*FIRST* Recommendation
Density	Absolute or Relative	Absolute
	Color Channel (filter) Used	
Dot Area (Tone Value)	Solid Density	Murray-Davies
	Color Channel (filter) Used	
	Equation: Murray-Davies or Yule-Nielsen (include “n” factor for Yule-Neilsen)	
Dot Gain (Tone Value Increase)	Dot Area on Plate	Murray-Davies
	Equation: Murray-Davies or Yule-Nielsen (include “n” factor for Yule-Neilsen)	
Trap	Overprint(s) Measured	Preucil (apparent)
	Print Sequence	
	Equation: Brunner, Hamilton (newsprint), Preucil (apparent)	
Print Contrast	Dot Area on Plate	Solid Patch & 70% Tint Patch
	Paper (included or excluded)	Paper (included)

Table 16.4.1c

standard reference should be replaced at the manufacturer's recommended frequency to avoid the risk of fade or damage (creases, dirt, etc.), which could compromise the calibration. Refer to Appendix A for T-Ref™ ordering information from IDEAlliance.

Polarizing Filter: The polarizing filter reduces the substrate specular reflections measured by the instrument, thus increasing the reported density values. In situations where the substrate characteristics change during the manufacturing process resulting in differences in visual appearance and reported density values, the polarizing filter may reduce these differences in reported density value. Generally, the polarizing filter in not used for most flexographic print applications.

The vast majority of the world's population of spectrophotometers and densitometers used in graphic arts have incandescent lamps with spectra close to Commission Internationale de l'Eclariage [CIE] Standard Illuminant A, with a color temperature of 2856 K, ± 100 K. New illumination sources, including light emitting diodes [LED], allow hand-held color measurement instruments to measure with well-defined and controlled UV illumination components. To ensure consistency, new illumination sources and new substrates require new instrument and measurement standards for defining and measuring the relative UV content, and thus the degree of fluorescence of substrates containing optical brighter additives (OBAs). Defining and controlling the emitted UV component of the measuring device's illumination is essential to defining standard ways to measure and manage color printed on OBA-enhanced substrates. As part of ISO 13655-2009 (Spectral Measurement and Colorimetric Computation for Graphic Arts Images) a new measurements standards "M" series of measurement illumination conditions has been defined by the International Standards Organization (ISO) to standardize illumination conditions appropriate for different applications when substrates contain brightening agents. The new "M" series allows color management of OBA-enhanced substrates to be further refined to a very high degree.

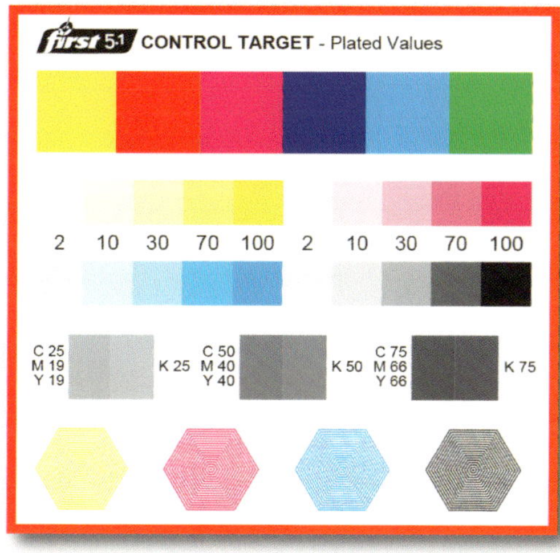

16.4.1c Example Control Target

Measurement Data: The instrument must be properly programmed to accurately capture the primary measurements used in process printing. The conditions under which these measurements are taken must be documented and reported along with the data.

Density: Absolute or relative. Absolute density includes the reflectance of the substrate. Relative density factors out the reflectance of the substrate; it is generally referred to as "density minus paper". Absolute density is the typical density value reported since the substrate is integral to viewing the printed image. Refer to Section 19.4.4 for more information on solid ink density.

Grayness: The grayness function measures the relative achromatic content of a printed area. It can be used to monitor gray balance of the yellow, magenta and cyan overprint. It is expressed as a percentage. Refer to Sections 12.9.1 and 19.4.2 for more information on gray balance.

Dot Area or Dot Gain/TVI: Also referred to as Tone Value and Tone Value Increase. The Murray-Davies equation is the preferred method for calculating both dot area and dot gain/TVI. If the Yule-Nielsen equation is used, the "n" factor must be empirically determined for each ink-substrate-screening-press configuration and reported with the data. Refer to Section 19.4.3 for more information on dot area and dot gain/TVI.

Print Contrast: The print contrast measurement indicates the printing systems ability to hold image detail in the shadow tones. The 70% tint is used to calculate print contrast for flexographic applications. Report the shadow tint used in the measurement along with the data. Refer to Section 19.4.5 for more information on print contrast.

Trap Equation: Hamilton (newsprint) or Preucil (apparent). Both the Hamilton and Preucil equations account for "missing" density arising from additivity failure. Additivity failure occurs when the density of the overprint is less than the sum of the solid densities of the two inks printed and measured separately. The Preucil (apparent trap) equation is typically used. Regardless of the equation used, it must be reported along with the trap data. Refer to Section 19.4.6 for more information on ink trap.

Measurement Accuracy: When the solid colors of a proof and press sheet are compared using both CIELAB spectral data and solid ink density data, it is possible for the proof and press sheet to match with one measurement method but not with the other. When this type of discrepancy occurs, the CIELAB spectral data takes precedence. The CIELAB data trumps the density data because spectrophotometric functions use the D50 light source and densitometric functions use the Status T spectral response function. Because Status T does not conform to the D50 standard, metameric color matches can result with the same density. This is especially problematic when using an inkjet proof because the dyes used in the proof contain different colorants compared to the pigments used on press, increasing the risk for a metameric match.

16.4.1.1 Solid Ink Density of Contract Proofs

The goal of any specification is to enable the production process to consistently produce the desired outcome. When the graphic is a continuous tone image, gray balance is the primary characteristic that determines whether or not the printed image will match the original. The purpose of measuring solid ink density and dot gain/TVI of the contract proof and printed product is to control gray balance across the tonal range: highlight, ¼-tone, midtone, ¾-tone and shadow. The *FIRST* "Solid Ink Density Specification" listed in Table 16.4.1.1, represents a "reasonable" target for both the proofer and the printer. If the printer is unable to achieve *FIRST* density targets, while minimizing dot gain/TVI with a given substrate, a compromise must be made. It is less objectionable to compromise on density than it is to compromise on dot gain/TVI. Increasing dot gain/TVI in order to achieve higher densities will result in "dirty" print with a loss of detail and undesirable color shifts when compared to the original. Invariably, increased make-ready time and production waste will occur. Refer to Section 19.4.4 for a more detailed discussion on Solid Ink Density.

Solid Ink Density: Starting Point
Density is print system dependent; determine optimum density during press optimization and fingerprint trials (ref 1.3.1 & 1.3.2)

Ink	Cyan			Magenta			Yellow			Black		
	Min	**Target**	Max	Min	**Target**	Max	Min	**Target**	Max	Min	**Target**	Max
Paper Products	1.30	**1.35**	1.40	1.20	**1.25**	1.30	0.95	**1.00**	1.05	1.45	**1.50**	1.55
Film Products	1.25	**1.30**	1.35	1.15	**1.20**	1.25	0.95	**1.00**	1.05	1.35	**1.40**	1.45
Newsprint	0.95	**0.97**	0.99	0.93	**0.95**	0.97	0.77	**0.79**	0.81	1.03	**1.05**	1.07
Printer Specific	-0.05 (target)		+0.05 (target)	-0.05 (target)		+0.05 (target)	-0.05 (target)		+0.05 (target)	-0.05 (target)		+0.05 (target)

Process for Achieving Color Balance (How to determine printer specific density targets)

Work with the Ink Supplier to:		
	1st Priority = Specify Each Ink Hue Value	2nd Priority = Check Gray Balance
	3rd Priority = Confirm Dot Gain	4th Priority = Obtain Solid Ink Density (no pinholes)

Table 16.4.1.1

The objective of the contract proof is to match the densities and dot gains/TVI established during the press fingerprint and characterization regardless of whether or not these densities match *FIRST* target densities. By proofing to the press established ICC profile, the contract proof represents what the customer can expect to see on the printed product with greater accuracy. Solid ink density values on dry proofs should fall within ±0.05 of the desired density value for each color as determined by the press fingerprint. As an example, if the magenta proof density is +0.05 the target, the remaining colors must be proofed at or above their respective target density to maintain balance. The contract proof must achieve gray balance.

FIRST has identified process color pigments by color index number. Proofing pigments need to match the inks being used. Using the recommended pigment (referenced by the color index number) will prevent metamerism. If the proofing method does not utilize pigments, the CI number of the pigments being used by the printer will facilitate the identification of the best match available for the given proofing colorant. Refer to Section 20.2.2 for *FIRST* recommended pigments.

16.4.1.2 Dot Gain/Tonal Value Increase of Contract Proofs

Failure to monitor and maintain accuracy of the contract proof will result in unpredictable, inconsistent proofs, rendering them unusable for the printer and customer to approve color. Tone scales must be imaged and plated (when providing press proofs) using the same devices and materials as the live job. Variation in dot shape and material can significantly alter print results.

Press Proofs

In order to ensure properly balanced tone reproduction, the total dot gain/TVI should be based on the press fingerprint produced by the printer. The allowable variation on the 30% and 70% original film value is ±5% of the dot area achieved during the fingerprint. For example, if the magenta proof dot gain/TVI is +5% the target, the remaining colors must be proofed at or above their respective target dot gains/TVI to maintain balance.

Prepress Proofs

All proofs must be made to the manufacturer's specifications using the recommended target for verification. The target should be outside the live image area of the graphics and must remain on the finished proof. The dot area on each of the colors containing a screen tint must be measured using the Murray-Davies equation for dot area measurement outlined in CGATS.4. The value of apparent dot area must be consistent with the supplier's recommendations for the respective proofing system. Table 16.4.1.2a is provided as an example. Actual values must be based on the printer's press fingerprint.

Example Press/Prepress Proof Deviation				
Ink	Cyan	Magenta	Yellow	Black
Film Value	50%	50%	50%	50%
Predicted Dot Gain	72%	72%	72%	72%
Actual Dot Area	72%	70%	68%	73%
Deviation from Target	0%	-2%	-4%	-1%

Table 16.4.1.2a

16.4.2 Spectrophotometer Guidelines

Description: A spectrophotometer measures the light reflectance of a color every 10-20 nanometers across the visible light spectrum (400-700 nanometers); this is referred to as "spectral data". This is the raw data to which mathematical formulas are applied to describe the given color. A "fingerprint", or reflectance curve, is created from the measurements, providing a precise definition of the color. The color can be plotted in

3-dimensional, CIELAB, color space and described in terms of lightness, chroma and hue.

Application: A spectrophotometer is used to measure a solid color and compare it to an established color target or standard. Using a spectrophotometer improves the proofer's ability to consistently and reliably match custom colors. A proofer may also use a spectrophotometer to measure and control proofing inks or dyes to ensure batch-to-batch consistency.

Industry Standard: ANSI/CGATS.5 2003 (Graphic Technology – Spectral Measurement and Colorimetric Computation for Graphic Arts Images) provides the equations used to calculate the colorimetric parameters and defines standard spectrophotometric instrument settings.

16.4.2a Spectrophotometer

Spectrophotometer Instrument Settings	
Standard	ANSI/CGATS .5-2003
Illuminant	D50 (5000 Kelvin)
Observer	2° Standard Observer
Measurement	Absolute (including substrate)
Geometry	0°/45° or 45°/0°
Tolerance Method	DE CMC or DE CIE94 or DE CIE2000
Aperture	3.4mm std. (2.0 - 6.0mm range)

Table 16.4.2a

Common CIE Illuminants		
CIE Illuminant	Common Name	Color Temperature (degrees Kelvin)
D50	D50 (horizon light)	5000K
D65	D65 (noon daylight)	6500K
F8	Fluorescent D50 (horizon light)	5000K
F7	Fluorescent D65 (noon daylight)	6500K
F2	Cool White Fluorescent	4230K
Illum. A	Incandescent Light	2856K
Illum. C	Average Daylight (northern sky)	6774K

Table 16.4.2b

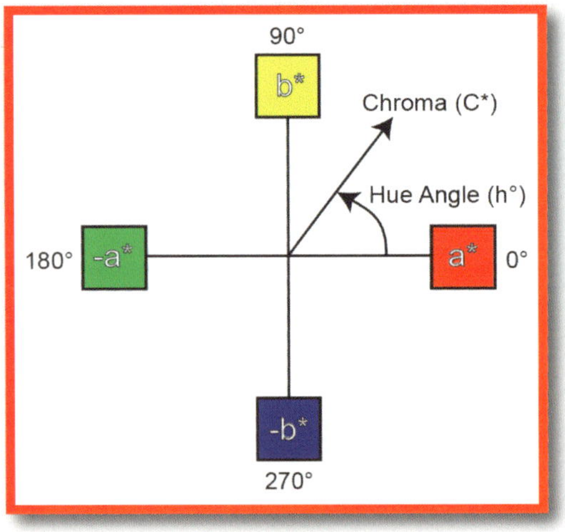

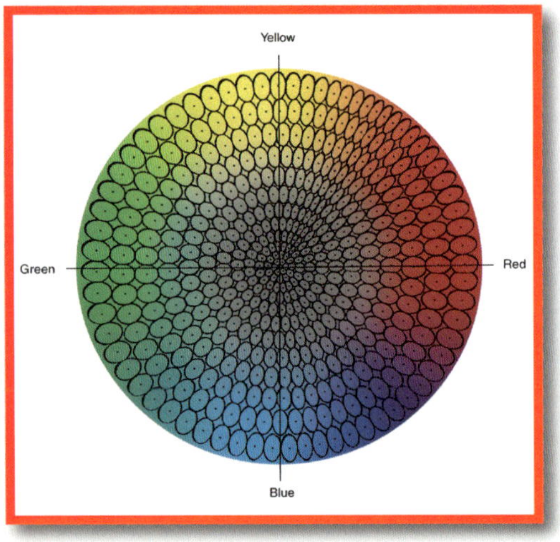

16.4.2b Determining Acceptable Color Difference: *The weighted color difference equations use slightly different mathematical formulas to calculate the shape of the acceptable color difference ellipse.*

Instrument Settings: When using a spectrophotometer to evaluate color, two equipment settings must be carefully chosen: the Illuminant and Standard Observer. Color measurements taken with different equipment settings (different illuminants and observers) are not comparable. The same spectral data (color fingerprint) will yield different colorimetric values (L*a*b*C*h° and DE) based on the selected illuminant and standard observer. Therefore, whenever colorimetric values are communicated, the illuminant and standard observer used to obtain the data must be included with the data.

The illuminant and standard observer must be standardized to accurately communicate colorimetric data. In the interest of a common communication standard, *FIRST* recommends ANSI/CGATS.5 2003 (Graphic Technology – Spectral Measurement and Colorimetric Computation for Graphic Arts Images) which specifies D50 as the illuminant and 2 degrees as the standard observer (D50/2°).

There may be situations where different equipment settings are specified and used, such as: customer requirements, the need to compare colorimetric data with other industries, or comparison to historical data. When settings other than D50/2° are used, *FIRST* recommends including the spectral data (light reflectance values every 10-20 nanometers across the visible light spectrum (400-700nm) for the measurements along with the colorimetric values (L*a*b*C*h° and DE). The spectral data allows the receiver to calculate the colorimetric values for any combination of illuminant and standard observer. This is important because colorimetric values derived using different equipment settings cannot be compared. Refer to Section 16.4.1 for more information.

Colorimetric Parameters: CIELAB L*a*b* and L*C*h° are the most common colorimetric parameters. L*a*b* describes the location of a color in CIELAB color space using the red/green and blue/yellow dimensions. L*C*h° is another method describing the location of a color in CIELAB color space, using the more native, or intuitive, language of lightness, chroma and hue. Chroma and hue are mathematically derived from the *a and *b values.

L = The lightness axis indicates how light or dark a color is. It moves from white (at the top 100) to black (at the bottom 0).

a* = Describes the red/green dimension of a color. The more positive the a* value, the more red the color. The more negative the a* value, the more green the color.

Comparison of Color Difference Equations					
Parameter	CIE76	CMC		CIE94	CIE2000
FIRST Recommended?	No	Accepted		Accepted	Preferred
Year Developed	1976	1984		1994	2000
Equation Developed By:	The International Commission on Illumination	The UK Society of Dyers & Colorist Color Measurement Committee		The International Commission on Illumination	The International Commission on Illumination
Pass/Fail Ellipsoid around the Standard with a set DE?		Yes		Yes	No
Is the DE the same if the "Standard" and "Sample" are reversed?		No		No	Yes
Weighted?	No	Yes		Yes	Yes
Typical Weighting Factors (for print applications)	N/A	(2:1)	(1:1)	K1 = 0.045 & K2 = 0.015	(1:1:1)
Example - DE Comparison: "Standard" L= 45.25, a= -40.35, b= -35.75 "Sample" L= 43.75, a= -39.50, b= -37.50	2.46	1.22	1.75	1.83	1.65

Table 16.4.2c

b* = Describes the yellow/blue dimension of a color. The more positive the b* value, the more yellow the color. The more negative the b* value, the more blue the color.

C = Chroma describes the color saturation; how strong or weak a color is. The closer to the center of the circle, the more neutral the color. The closer to the edge of the circle, the more saturated the color.

h° = The hue angle refers to the name of the color (red vs. blue) and identifies its position on the color wheel. Most hues have the greatest possible saturation at the mid-point of the lightness axis.

Color Tolerancing Method – Calculating DE: Color tolerancing is used to determine if a "sample" color, compared to a "reference" color, is acceptable. The accuracy of the color match is typically expressed as Delta E (ΔE or DE). The DE value represents the overall color difference and is derived from one of several equations. All color difference equations compare the location of a "reference" color to the "sample" color in CIELAB color space; however, CIELAB color space is not uniform. For example, shifts in hue are typically more easily perceived than shifts in lightness or chroma. Color difference equations are either weighted or unweighted.

16.4.3 Color Viewing Conditions: *Accurate, consistent, visual perception of color requires viewing of proofs and printed samples in a standard, chromatically neutral, controlled environment.*

Unweighted color difference equations, such as CIE 1976 (DE_{ab}), weight hue, chroma and lightness equally. Equal weighting does not correspond well with human perception of color differences; therefore, a sample may visually match the reference color but produce an unacceptable DE value or vice versa.

Weighted color difference equations, such as DE_{CMC}, $DECIE_{94}$ and DE_{2000}, achieve better agreement with human perception by weighting hue, chroma and lightness differently. The formula used to weight the three-color axis is slightly different for each weighted color difference equation.

The CMC tolerancing method includes three weighting factors set by the user, expressed as (l:c:cf), where: l=lightness, c=chroma and cf=commercial factor (or DE). The CMC ratio l:c (lightness:chroma) determines the shape of the ellipsoid, which is typically set at 2:1 for most applications; however, an l:c ratio of 1:1 is becoming more common. The cf (commercial factor) determines the overall size of the ellipsoid and the threshold, or tolerance, of acceptable color difference. The cf determines the DE limit, for example, if the cf = 1.5, then the acceptable DE = 1.5 as well.

FIRST recommends using one of the weighted color difference equations and, where possible, the latest applicable equation. Regardless of the color tolerancing equation used, it is critical to communicate both the equation and any weighting factors used to all parties receiving color data.

All parties measuring color must use the same tolerancing equation in order to have meaningful discussions of color differences. Indicate the color tolerancing equation used on all proofs or print samples containing color difference (DE) values. As instrumentation is purchased, *FIRST* recommends buying equipment that supports the latest methods in color tolerancing and has a robust upgrade path. Refer to Appendix A for The International Commission on Illumination (CIE) and The UK Society of Dyers & Colorist Color Measurement Committee (CMC) contact information.

Calibration: A spectrophotometer must be calibrated in accordance with the manufacturer's recommended procedure before use. The reflective calibration standard must be traceable to a standard reference. A documentation system should include all calibration steps, outcome and corrective actions. This documentation should be used for reference and historical evaluation. Measuring equipment should be re-certified as recommended by the manufacturer or by internal requirements of the user, such as ISO certification.

Communication: Spectral data should be accompanied by the following information:

- Originator of the data
- Date created
- Description of the purpose or contents of the data being exchanged
- Description of the instrument used, including the brand and model number
- Instrument parameters: illuminant, observer, measurement, filter, geometry, and aperture
- Wavelength interval used
- Tolerance Method: CIE_{2000}, CIE_{1994}, CMC
- When density data is reported using a spectrodensitometer, the spectral product's weighting function (status or type response) used should be identified

16.4.3 Viewing Artwork, Proofs & Printed Material

Application: A color-viewing booth is used to view printed images, proofs, and transparencies, under a controlled and standard light source. Accurate, consistent, visual perception of color requires the image to be viewed in a standard, chromatically neutral, controlled environment. When the printer, prepress provider and customer standardize viewing conditions, color discrepancies are minimized.

Industry Standard: *FIRST* supports the viewing standards defined in ISO 3664:2009. However, *FIRST* recognizes the D50 light source may not be optimal for all print segments.

ANSI 2.30 - 1989: Color Viewing Conditions	
Parameter	Specification
Lighting	5000 Kelvin Blubs
Light Source	Color Rendering Index >/= 90
Viewing Surface	Luminance = 204 +/- 44 Foot Candles
Viewing Area Color	Neutral Gray: Munsell N8 or Equivalent
Angles of Illumination	Minimize Glare

Table 16.4.3

16.6a Proofing for Expanded Gamut

Instrument Agreement: The illuminant used in the light booth should be the same as the equipment illuminant setting. For example, if the measurement equipment (spectrophotometer) is using D65 instead of D50, the light booth should use 6500 Kelvin bulbs (D65) instead of 5000 Kelvin bulbs (D50).

Communication: Regardless of the settings used, it is important to communicate the settings to all parties receiving proofs, printed samples and/or data measurements.

16.5 Proof Compliance Cover Sheet/Label

A "Proof Compliance Cover Sheet" or label should accompany the contract proof submitted for color match at press and approved by the customer. The cover sheet or label should identify the proofing product or system used and the company supplying the proof: contact name, telephone/fax numbers, as well as operator, date, job number and customer. The cover sheet must also contain information required to verify the proof's compliance to the technical attributes required for that proofing type. Refer to Section 16.2 for additional information on types of proofs.

It is a best practice approach for all proofs to include a "Certificate of Result". It should include all pertinent measurements: density, dot area, Delta E (@ 100% and 50%), trap, print contrast, bar code scan analysis, etc. Proof densities should be within the printer's on-press density specifications. The Proof Compliance Cover Sheet and Certificate of Result can be combined into one document. Refer to Section 19.4.4 for *FIRST* guidelines on solid ink density by print segment.

16.6 Proofing for Expanded Gamut Printing

Proofing for expanded gamut printing can be as diverse as the systems used to expand the gamut itself. At the very least, any proofing device used must be able to match the expanded gamut of the color separation being reproduced on press, with the chosen ink set. Expanded gamut refers to any process that expands the color availability beyond that typically available with standard four color printing (CMYK). It refers not only to the outer edges of the color gamut with the most saturated colors, but also to the purest pastels and semi-saturated colors that are often missing using traditional process color printing (CMYK).

Proofing Devices for Expanded Gamut

A range of devices are used with success, from analog devices with special ink colors, to digital proofing with CMYK and expanded CMYK inkjets to digital halftone devices using laser

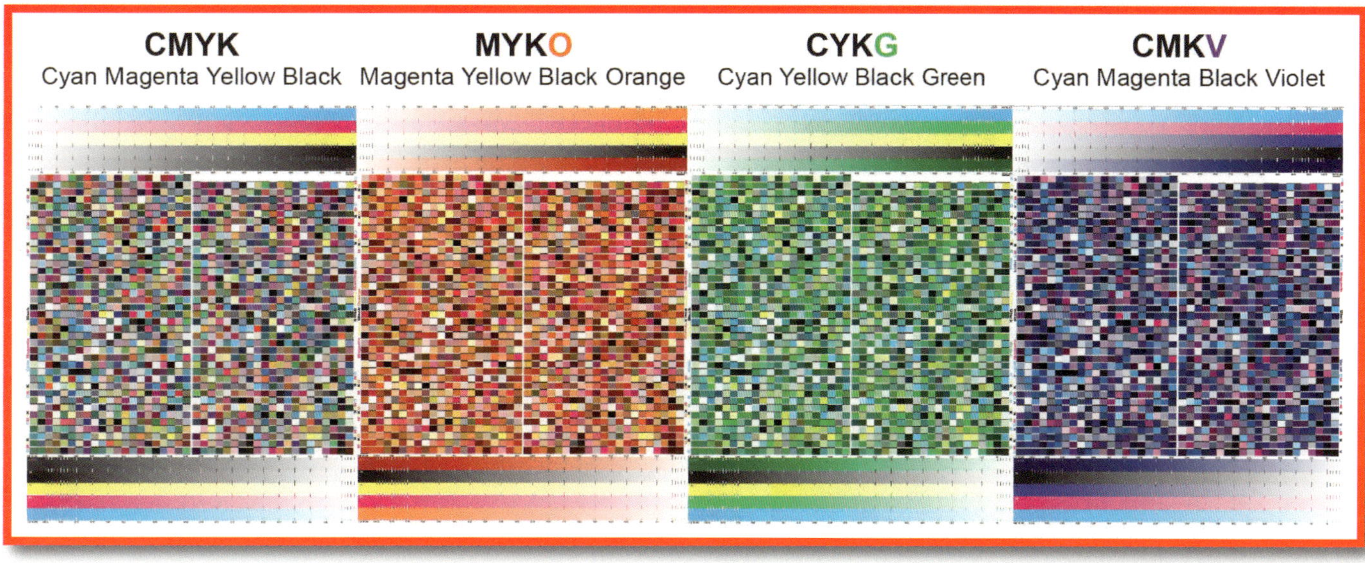

16.6b Characterization Targets: *This is one example of expanded gamut characterization targets. Due to the printing constraints of this publication, this graphic is a CMYK simulation.*

ablatable or thermal transferable ink sheets. Digital inkjet proofers using only CMYK use very pure single-pigmented inks. These pigments are more pure than the typical CMYK inks that are used commercially in flexographic printing. Because of the purity and saturation of these inks, and often the addition of a light cyan (fifth) and a light magenta (sixth) ink, the proofing device is capable of reproducing a much wider color gamut. Additionally, certain proofing devices utilize Orange, Green, and Violet inks. These devices may be acceptable for some expanded gamut proofing applications.

Expanded Gamut Color Space and Color Matching

There are numerous approaches to increasing the printed color gamut. Regardless of the chosen ink set, a characterization target must be printed. Targets are available from specific vendors. Each target contains patches for CMYK as well as for the additional inks. Once the data is analyzed with a spectrophotometer, a color gamut range is mapped. The same target is used to map the proofing device and a correlation of the color space between the press and proofer is determined. This correlation provides the information necessary to match the proof to the predicted print outcome. Refer to Section 1.3.4 for more information on press characterization and Section 20.2.5 for Printing with an Expanded Gamut.

Expanded gamut systems contain software that enable the printer to determine the specific colors capable of being reproduced on

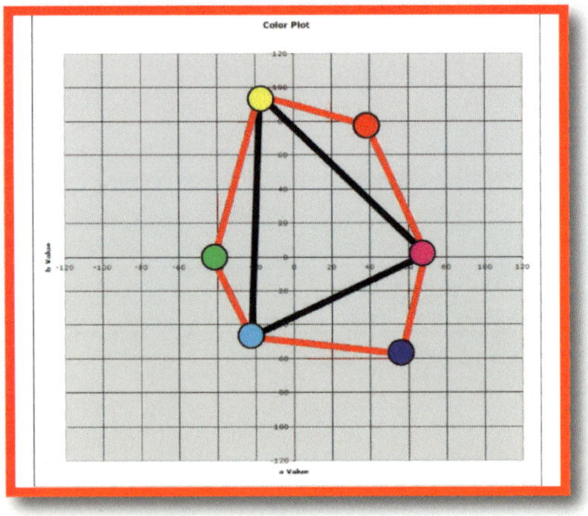

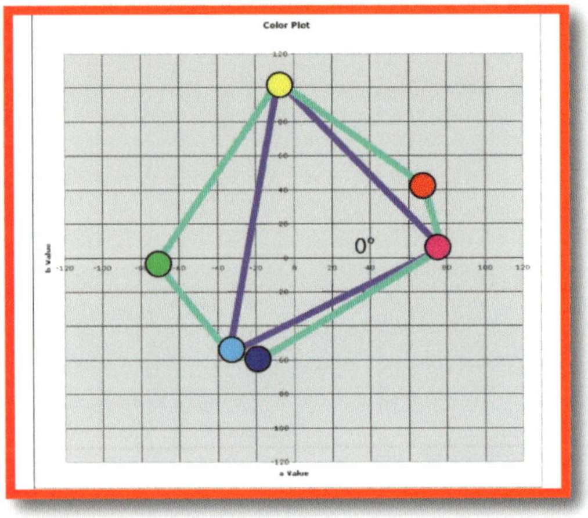

16.6c Expanded Gamut: *These graphs illustrate two different expanded gamuts compared to the traditional CMYK color gamut.*

press given the chosen ink set and print capabilities. Once the proofing device is correlated to the press, a proof can illustrate the special color matches that are achievable on press, based on L*a*b*/L*C*h° value comparisons.

Selecting Expanded Gamut Inks

The printable color gamut can be expanded by increasing ink density, increasing the whiteness of the substrate and/or by adding extra ink colors. The gamut of an ink set will only expand so far by increasing ink density, because the hue angle of the ink tends to curve away from pure with increased ink film thickness. When adding extra ink sets, *FIRST* recommends using single-pigment inks. An expanded gamut can be achieved using commercial systems with predefined ink sets or custom ink sets chosen by the user. The logical inks to add to Cyan, Magenta, and Yellow are pigments that are halfway between those values. Based on the closest hue values to the halfway point, the best ink choices are Violet 3, Red 2 and Green 7. Other colors used to expand the color gamut include:

- Red, Green, Blue
- Orange, Green, Violet
- Orange and Green
- Other colors chosen for a specific pictorial enhancement
- Using brand colors for pictorial enhancement ("Al's Cola Red")

Screen Angles and Expanded Gamut Printing

When printing with more than four colors, the choice of screen angles is an important consideration. One solution is to place each extra ink color on the same angle as the opposing complimentary color. For example, when adding Violet/Red/Green to the traditional CMYK, the green should be placed on the magenta angle, the red should be placed on the cyan angle, and the violet should be placed on either the yellow or black angle. Stochastic screening technology offers another solution for avoiding moiré problems by eliminating screen angles altogether.

Color Management Systems and Expanded Gamut Printing

The most problematic issue for conducting an expanded gamut printing trial is the implementation of color management. It is necessary to select software that can use custom colors and custom scripts. When utilizing expanded gamut printing, it is important to confirm that the CMS software used supports multi-color profiles. While the ICC format supports up to ten colors, few RIPs will support more than four colors. Therefore, *FIRST* recommends testing the process before implementing on a live job. Current color profiling software solutions do not feature industry approved standardized targets with color chips or swatches of pure additional inks and/or overprints of the additional inks with each other and the normal process hues. The evaluation process of expanded gamut printed material is not fully developed. Although expanded gamut printing can reproduce a more desirable product, the ability to control color consistency is challenging. Refer to Section 20.2.5 for printing considerations of expanded gamut printing.

Additional information regarding process control for expanded gamut can be found in the Appendix F Expanded Gamut: Reasonable Measurement for Process Control.

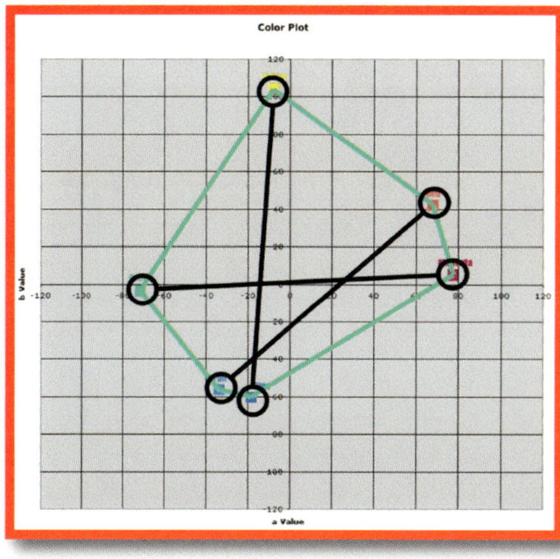

16.6d Complimentary Screen Angles: *The graph illustrates placing each extra ink set on the same angle as the opposing complimentary angle. A separation would never have red where there is cyan in the image; therefore, if red and cyan are on the same angle there will be no moiré.*

17.0 PRINTING PLATES

This section provides information on the processing, specifications, tolerances, and measurement methods for the various types of flexographic printing plates. This information will assist the print production team in selecting the most appropriate printing plate to best produce the design and ensure consistent performance.

Flexographic plates are supplied by the manufacturer in a variety of forms, which determines the platemaking process required to produce a press-ready printing plate.

Sheet Photopolymer: Photopolymer plates require ultraviolet light to initiate a chemical reaction, which creates the image. Photopolymer plates are supplied either as analog plates (require a film negative during the exposure step) or digital plates (do not use film negatives). Subsequent processing steps create the relief image.

Continuous Photopolymer on Sleeves: Round digital plates that require exposure and processing equipment designed to accommodate sleeves.

Liquid Photopolymer: Photo-reactive resins, supplied in a liquid form, that are converted into plates through an analog exposure and subsequent processing steps.

Laser Engraved Rubber/Cured Polymer Plates & Sleeves: Direct laser engraving creates the relief image on the plate or sleeve material.

Molded Rubber Plates: Molded rubber plates are duplicated from a mold or matrix, which is made from an original pattern plate.

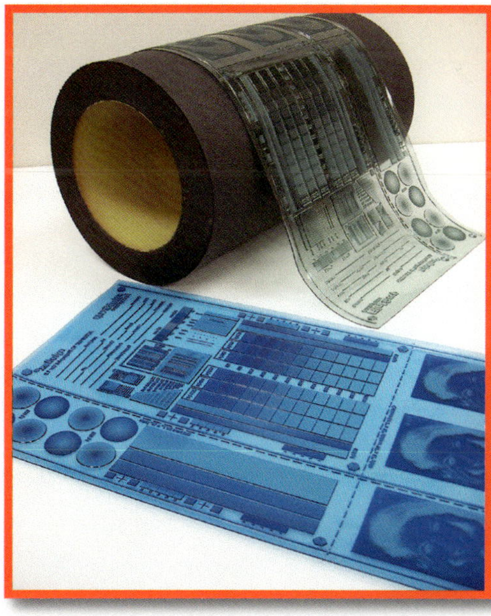

17.0 Printing Plates: *A flexographic photopolymer printing plate.*

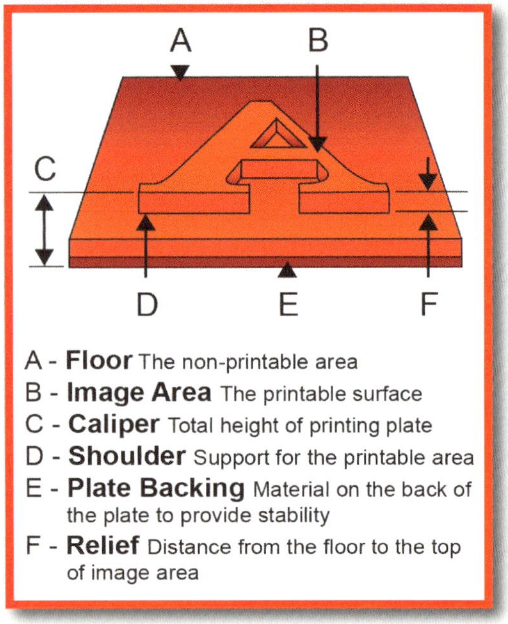

17.1a Printing Plate: *The illustration identifies plate attributes discussed in this section.*

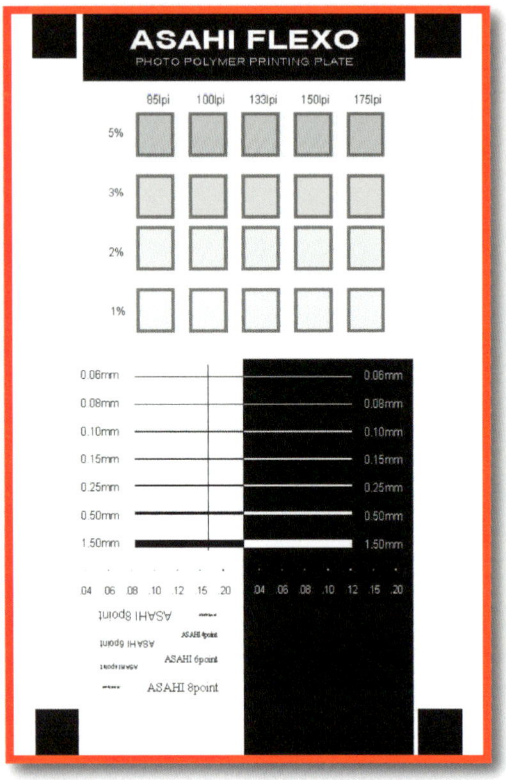

17.1b Example Plate Exposure Target

17.1 General Plate Specifications

Film/File Specifications

A review of incoming files/films is crucial to successful platemaking. Refer to Section 15.0 Final Films/Files/Mask Specifications for a comprehensive listing of film properties, handling, and critical considerations in file/film generation.

Plate Material Specifications

1. All plates for one print job should be made from the same batch of raw material.
2. Caliper tolerances are to nominal thickness from the raw material manufacturer.
3. During plate processing, main exposures should be made using the exposure target. This is to control and monitor accuracy of line widths and halftone dot integrity.

Plate Thickness Specification

When a plate thickness is specified, the nominal thickness, such as 0.067", represents the desired finished plate thickness. Plate manufacturers actually make the raw material thicker than the nominal thickness to allow for surface removal during plate washout. How much material is removed depends on the washout method employed. Therefore, it is likely that finished plates are actually thicker than the nominal thickness. Plates slightly thicker than the nominal thickness do not represent a printing problem. Whereas plates below the nominal thickness do represent a problem because the thinner plate will print at the bottom of the gear when the appropriate cylinder undercut is used. It is important for all of the plates for a given job to be within +/- 0.001" thickness of each other.

Plate Exposure Target

Micro-fine rules determine proper photopolymer face exposures. This target is available in five resolutions depending on the line screen used. Selecting and incorporating the appropriate resolution control target is important. After exposing and processing the plate, the microline must be straight to the touch. Failure to hold this line straight is an indication of under exposure. Under exposed plates can result in dirty print and dot bridging.

17.2 File Preparation for Digitally-Imaged & Laser-Engraved Plates

1-BIT Files (TIFF or LEN)

The use of 1-BIT files is common in digital platemaking applications. These rastered files are digitally stable and cannot be altered unless opened in a raster imaging software program.

Plate and Sleeve Relief Specifications - General Guidelines									
English Measurements									
Plate Thickness	.030”	.045”	.067”	.107”	.112”	.125”	.155”	.185”	.250”
Target Relief Plates	.023”	.018”-.022”	.018”-.022”	.018”-.022”	.020”-.025”	.035”-.060”	.050”-.080”	.050”-.080”	.075”-.125”
Target Relief Sleeves	.020”	.018”-.022”	.018”-.022”	.018”-.022”	.018”-.022”	N/A	N/A	N/A	N/A
Metric Measurements									
Plate Thickness	.762mm	1.14mm	1.70mm	2.72mm	2.85mm	3.18mm	3.94mm	4.70mm	6.35mm
Target Relief Plates	.584mm	.457mm-.559mm	.457mm-.559mm	.457mm-.559mm	.508mm-.635mm	.508mm-.635mm	.762mm-.889mm	.762mm-.889mm	1.27mm-1.78mm
Target Relief Sleeves	.508mm	.457mm-.559mm	.457mm-.559mm	.457mm-.559mm	.508mm-.635mm	N/A	N/A	N/A	N/A

Table 17.1a

Plate Exposure Target Micro Line Widths	
Line Screen	Micro Line Width
55 lpi (22 lpcm)	3.0 mils (0.08mm)
65 - 85 lpi (26 - 33 lpcm)	2.5 mils (0.06mm)
100 - 120 lpi (39 - 47 lpcm)	2.0 mils (0.05mm)
133 - 175 lpi (52 - 69 lpcm)	1.5 mils (0.04mm)
200 lpi (79 lpcm)	1.3 mils (0.03mm)

Table 17.1b

A 1-BIT file is essentially a digital negative created by a similar type of RIP (Raster Image Processor) that creates film negatives for an imagesetter/platesetter. 1-BIT files contain only black and white (positive or negative) information. 1-BIT files are written for individual colors, just as film negatives represent individual colors. All screening information is contained within the file. This information is used directly by the platemaking device to image a digital plate.

Variables to consider when transferring 1-BIT files include:

- **Compression:** File compression and FTP transfer are used to efficiently move data between facilities. The different formats represent different levels of compression. Uncompressed files should not be transmitted. Refer to Section 11.1.1 for more information.
- **Proofing:** *FIRST* recommends the producer of the 1-BIT file proof the content of the file in the same manner and for the same reasons traditional negatives are proofed. Several manufacturers produce software that allows the composite viewing of screened files in a soft-proof environment. Screen ruling, screen angle and other measurement tools are built into these soft-proof systems.
- **Inkjet RIPs:** Several inkjet manufacturers' RIPs also allow the compositing of screened 1-BIT files for output to a drop-on-demand (DOD) inkjet device. However, accuracy of these proofs is compromised when pushed beyond the capability of the inkjet device. Out-of-gamut color and spot-color overprints are two potential problems. Using 1-BIT files with inkjet proofers limits the control of proofing to only nine colors. A custom file must be generated for the proof and the file cannot be proofed on multiple proofing systems.
- **Required Parameters:** All job parameters, such as screen ruling, screen angles, image trapping, distortion and dot gain/TVI compensation, must be known and incorporated by the producer of the 1-BIT file. The 1-BIT file is typically the final plating file; therefore, all stepping parameters, bearer bars, density blocks, registration marks, etc. must be included. In a digital platemaking workflow, specific criteria, such as calibration for minimum dot (bump curve) and resolution of output device, should be agreed on by the producer of the 1-BIT file and the platemaker.
- **Resolution:** Digital plate imaging resolutions will commonly be found between 2400 and 5080dpi. Some specialty screening may require resolutions at 4000dpi and above. Engraving systems using higher resolution files may take longer to engrave. Consult with the image engraving company and RIP manufacturer to optimize resolution for output and screening.

Pre-RIP Settings Prior to Imaging

The following attributes of a job require proper data input prior to RIPing the electronic file to the digital-imager or laser-engraver. Specifications may vary by equipment manufacturer.

- **Circumference:** Confirms the job size will be correct in both dimensions on the imaged plate. This setting also ensures screens image correctly around the circumference of the cylinder. A setting for circumference is not necessary for all imagers.
- **Resolution:** Specifies the image resolution of the file on the laser device. It is specified as either: dpi (dots per inch) or ppi (pixels per inch). Resolution will vary depending on the RIP and the screen ruling. For high-quality direct laser engraving of fine linework and screens, a resolution of 635dpi is generally recommended. Higher resolutions only benefit special applications, such as when engraving pin register holes where positioning is critical. Higher resolution files typically take longer to engrave. Consult with the image engraving company to optimize resolution.
- **Distortion:**
  1. For photopolymer plates processed flat and subsequently mounted on a cylinder or sleeve, a distortion factor must be applied to compensate for the change in image size when the plate is mounted around the print cylinder. Refer to Section 15.6 for additional information on calculating distortion factors.
  2. For photopolymer plates or sleeves that are imaged and processed on the same cylinder size as used to print, a distortion factor is not applied; the distortion automatically occurs based on the circumference.
- **Dot-Gain Compensation:** Determined by identifying the total dot gain/TVI expected in the printing process as measured by the press fingerprint and applied to the final electronic file. Dots, as well as positive and reverse line elements, are influenced by dot gain/TVI and should be compensated accordingly. For some laser-engraving systems, the dot gain/TVI curve can be entered as one parameter of the system's internal dot generation and applied to a 100% file during plate engraving.
- **Bump (Highlight Compensation Curve):** Used for digitally- imaged photopolymer plates only. A bump curve does not apply to laser-engraving; the bump prevents imaging dots in the mask that are too small for full dot formation on the plate during UV exposure. Apply a bump to increase the size of the minimum dot. If the minimum dot in the file is large enough, a bump is not necessary. The bump curve will vary with screen ruling; the finer the screen ruling, the greater the bump required to image the minimum dot.

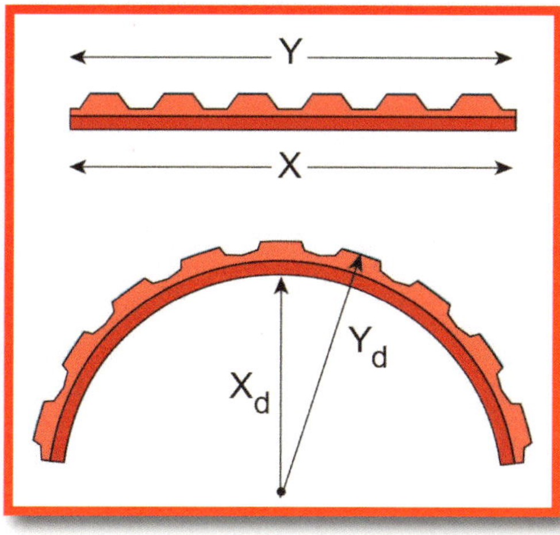

17.2 Calculating Distortion: *When a photopolymer plate is laying flat, the top of the plate and the bottom of the plate is the same length (X=Y). However, when the plate is wrapped around a printing cylinder, the surface of the plate becomes stretched because the distance around the top of the plate is greater than the distance around the bottom of the plate ($Y_d > X_d$).*

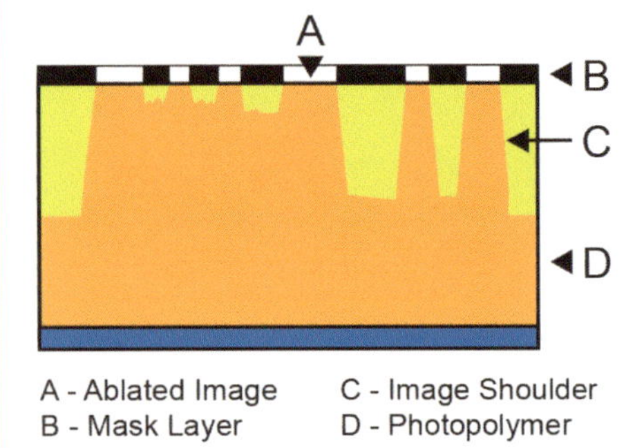

DIRECT-TO-PLATE IMAGING

17.3 Digitally-Imaged Photopolymer Plates: *The digitally imaged photopolymer plate is processed without the use of photographic film. A laser ablative mask acts as a mask, blocking the UV light, during the plate exposure.*

- **Line Screen:** Refer to Section 15.3 for print segment specifications. Consult the printer to confirm printer specific requirements.
- **Seamless Sleeves:** Because a seamless image requires a perfect "pixel-to-pixel" match all the way around the cylinder, it may be necessary to adjust the line screen prior to imaging. Some RIPs can automatically calculate and adjust output to maintain whole dots across the seam. Circumference and file resolution must be considered.

17.3 Digitally-Imaged Photopolymer Plates

The digitally-imaged photopolymer plate is processed without the use of photographic film. The sheet photopolymer material is supplied to the platemaker with a laser ablative mask on the surface that will be imaged from an electronic file using a laser. The laser removes, or ablates, the laser ablative layer in the image areas of the design. After ablation, the carbon black layer acts as a mask (or negative), blocking the ultraviolet light, during the platemaking exposure process. The plate is back exposed, main exposed, (no vacuum needed) processed (solvent, thermal or water-wash), dried (solvent or water-wash only), detacked and post exposed just as with a conventional sheet photopolymer plate.

In addition to standard digital photopolymer printing plates, plates may be digitally imaged, and then exposed using a procedure to create "flat top dots". Flat top dot printing plates are characterized by the following:

1. The image produced is defined by the mask used to create them. The two-dimensional mask opening and resulting two-dimensional dot surface are almost identical. This is often referred to as 1:1 reproduction.
2. The printing surface has a substantially planar top surface and a sharp transition between the top and side (shoulder) of the dot.

17.3.1 Mask Specifications

Laser Ablative Mask Density Specifications

The laser ablative mask should have a density greater than 3.0, a uniform consistency of coating and be free of visual defects (such as pinholes, scratches, abrasions and smudges) prior to ablation. After ablation, the mask opening "stain level" should have a density of less than 0.07 on a transmission densitometer.

Mask Ablation on Carriers or Sleeves

Guidelines for preparing to image on carriers or sleeves:

1. Back expose the digital photopolymer plate prior to

Digitally-Imaged Photopolymer Plate and Sleeve Specifications *Raw Plate Specifications: Finished plate uniformity is dependent on platemaking process*	
Halftone Dot Range on Plate	1% to 98% @ 175 lpi (69 lpcm) line screen and lower
Minimum Rule Width	0.004" (0.102mm)
Minimum Positive/Reverse Type	2 point
Thickness of Uniformity:	
Within a Plate	+/- 0.0005" (0.0127mm)
Within a Job Plate Set	+/- 0.001" (0.025mm)
Sleeve/Roll Uniformity:	
Total Indicated Runout (TIR)	+/- 0.001" (0.025mm)
Microfinish	< 0.000025" Ra
Plate Sizes: *(As availble from supplier of plate or roll stock)*	
Gauges </= 0.125" (3.2mm)	From roll stock up to 50" x 360" (127cm x 914cm)
Gauges > 0.125" (3.2mm)	From roll stock up to 50" x 180" (127cm x 457cm)
Plate Thickness	0.030" - 0.280" (0.076mm - 7.11mm)
Repeat Circumference	Capabilities to 62" (1575mm)

Table 17.3

17.3.1 Mask Imaging Inspection: *The mask should be inspected prior to imaging, after imaging, and after UV face exposure.*

Density Specifications	
Prior to Ablation: Mask: D-Max	> 3.0
Post Ablation: Opening: Stain Level: D-Min	< 0.07

Table 17.3.1

placement on the sleeve or cylinder. This exposure determines the relief of the image.

2. Verify relief height with the printer. In general, there is less relief when imaging on a carrier or sleeve.
3. Laser ablative mask layer should be uniform and consistent. Variation in the stain creates problems with process printing, such as inconsistent dot size and shape.
4. As sleeve diameters vary, a consistent main exposure is important for highlight dot reproduction.

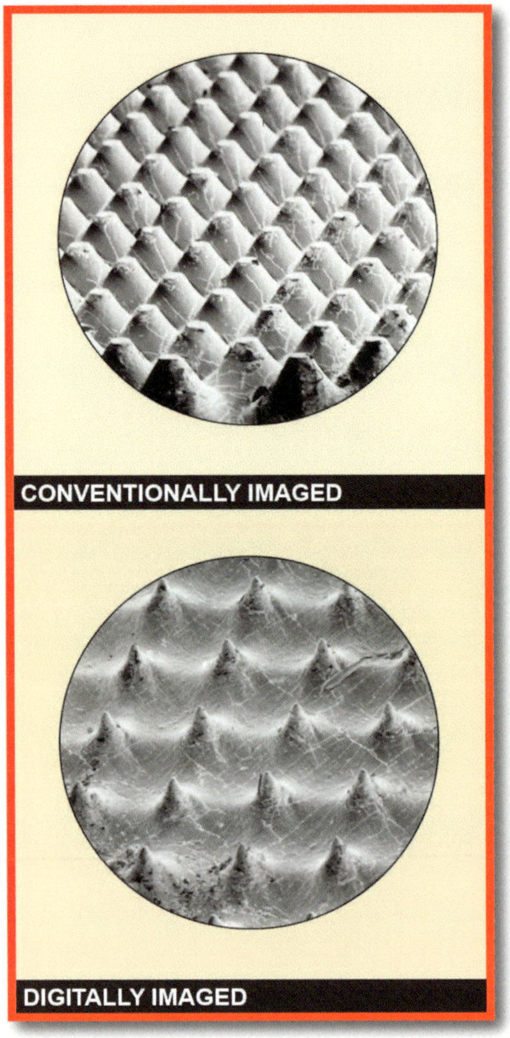

17.3.2a Dot Formation: *Dots imaged direct-to-plate have steeper shoulders than conventionally imaged dots.*

Mask Imaging Inspection

1. Prior to imaging, check the mask layer for any visible scratches, abrasions, or marks. Plate material should be clean and flat. The laser ablative mask can be very sensitive to abrasions; therefore plates should be handled carefully throughout the imaging and exposure process.
2. After imaging and prior to UV face exposure, re-inspect for damage to the mask layer. Flaws in the non-image area of the mask can be corrected with tools, such as high-density opaque or red litho tape. Be sure to remove tape prior to plate processing.
3. After UV face exposure, measure the device calibration target using a transmission densitometer. A video imaging system may also be used for stain and dot area measurements of the image on the plate. Check the 100% area to ensure complete clearing. Check the 50% dot area using a densitometer. Should the density be outside that range, the imaging process should be checked. Process checks should include the imaging equipment, setup parameters and raw material.

The device calibration target is a tone scale without a bump curve. It identifies where dots begin to fully form at printing height in the digital imaging process. The scale should include, but not be limited to, values between 0-10%, 25%, 50%, 75%, 90%, and solid. A device calibration target should be placed in a void, or non-print area, for each plate being imaged.

17.3.2 Plate Evaluation

While plate processing is generally a stable process, it is important that plate makers employ a process for evaluating plate quality and ensuring consistency from plate to plate. The following are guidelines for developing an evaluation process. However, it is advised that measurement device variation be considered before determining final specifications and tolerances for print height, relief, and dot measurement.

Finished Plate Characteristics

Print Height: All imaged areas must be at full printing height. Recessed dots on a digital photopolymer plate are not recommended by *FIRST*. An inadequate bump curve can result in fine details, such as highlight dots, not achieving full printing height. Measure the height of the minimum dot patch with a micrometer and compare to a solid area. Or, if that is not possible, carefully inspect the highlight dots to ensure they are fully formed and not recessed.

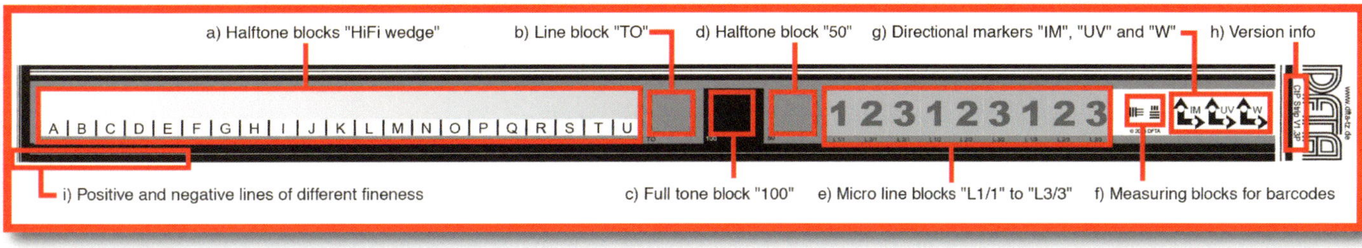

17.3.2b DFTA Control Strip for Digitally-Imaged Photopolymer Plates

Relief Specifications: Digital photopolymer plates image with a unique dot profile and structure. Digital dots have a straighter shoulder than conventional dots; therefore, they require less relief to achieve maximum dot support. Variances in cylinder build-up, press tolerances, or press operation may influence relief requirements for a particular application. Confirm target relief specifications with the printer.

Caliper: Refer to Section 17.9 for measurement method and Table 17.3 for specifications.

Dot Accuracy: Test elements for verifying dot accuracy are discussed in Section 12.9.2; these elements should be used to verify dot accuracy after platemaking. Use a calibrated "Flexo Plate Analyzer" (Section 17.9) to measure screen values and confirm dot accuracy.

Dot Tolerances for Digital Platemaking	
Dot Percentage	Tolerance
2 - 9%	+/- 0.50%
10 - 24%	+/- 1.00%
25 - 39%	+/- 1.50%
40 - 49%	+/- 1.75%
50 - 98%	+/- 2.00%

Table 17.3.2

Control Strip for Digitally Imaged Photopolymer Plates
In order to control the digital platemaking process and ensure each photopolymer plate is properly imaged and processed, a control strip should be included on each plate. Each control strip should include as a minimum:

- A minimum dot patch is used to ensure the minimum gray level in an electronic file is not lost. This patch is

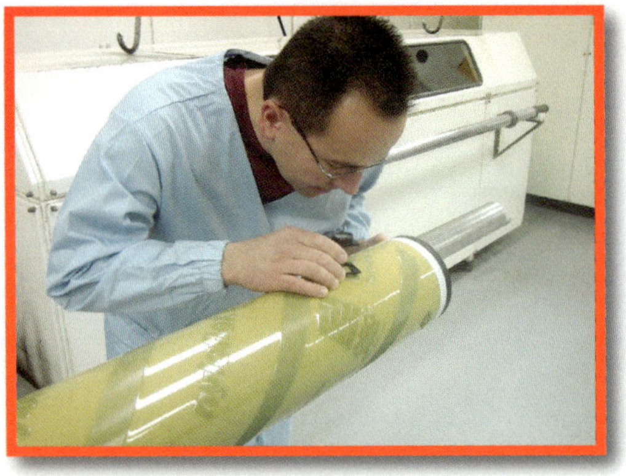

17.4a Laser-Engraved Rubber/Cured-Polymer Plates and Sleeves: *The digitally-engraved rubber, or cured-polymer, platemaking process is a direct-to-plate system that does not use photographic film, etched molding masters, conventional rubber vulcanization, or subsequent processing steps.*

not measured; it is visually evaluated to confirm proper formation of the dots.

- A dot percentage of a known value should be included so that it can be measured using a flexo plate analyzer. Refer to Section 17.9 for more information on plate measurement devices.

One example of a control strip for digitally imaged photopolymer plates is the DFTA Control Strip. The DFTA CtP strip V1.3 consists of two versions: the vector version and the pixel version.

Both versions are available at: www.flexography.org/FIRST_extras. The two versions are very similar; however, they differ slightly in the way they provide information.

1. **Vector Version:** The DFTA CtP strip V1.3 vector version is used to characterize and control the entire imaging system including the RIP, the platesetter and the printing plate material. The vector version of the control strip must be processed like a normal image file by the RIP used with the platesetter. It is available in both Postscript (PS) and PDF file formats.
2. **Pixel Version:** The DFTA CtP strip V1.3 pixel version is used to characterize and control only the actual output system including the platesetter and printing plate material. The pixel version of the control strip must be introduced into the workflow behind the RIP. It is available in both TIFF and LEN file formats.

The screen ruling is the same for both versions, 105.83lpi (41.66 L/cm). It is important to note that the vector version screen ruling can be changed by the RIP if not monitored and controlled by the operator. The resolution of both versions is 2116.67dpi (833.33 L/cm). It is important that the resolution of the control strip match the resolution of the plate setter. Other resolutions can be ordered. Scaling the control target when RIPing can generate different resolutions of the vector control strip. Refer to Appendix A for DFTA contact information.

17.4 Laser-Engraved Rubber/Cured-Polymer Plates & Sleeves

The digitally-engraved rubber, or cured-polymer, platemaking process is a direct-to-plate system that does not use photographic film, etched molding masters, conventional rubber vulcanization, or subsequent processing steps. Rubber or polymer sheets, or seamless covered sleeves, are supplied to the platemaker fully "cured" and ready for the one-step production of the digitally-imaged plate or sleeve. The image files are sent to a laser; it

Laser-Engraved Rubber/Cured-Polymer Plate and Sleeve Specifications	
Halftone Minimum Plate Dot	3% @ 133 lpi (52 lpcm) line screen (~38micron dot)
Minimum Rule Width	0.004" (0.102mm)
Minimum Positive/Reverse Type	3 point (0.042" = 1.067mm)
Thickness of Uniformity:	
Within a Plate	+/- 0.00075" (0.019mm)
Within a Job Plate Set	+/- 0.001" (0.025mm)
Sleeve/Roll Uniformity:	
Total Indicated Runout (TIR)	+/- 0.001" (0.025mm)
Microfinish	< 0.000025" Ra
Plate Sizes: *(As availble from supplier of plate or roll stock)*	
Gauges </= 0.125" (3.2mm)	From roll stock up to 50" x 360" (127cm x 914cm)
Gauges > 0.125" (3.2mm)	From roll stock up to 50" x 180" (127cm x 457cm)
Plate Thickness	0.030" - 0.280" (0.076mm - 7.11mm)
Sleeve/Roll Sizes Face Length	Capabilities >150" (3810mm)
Repeat Circumference	Capabilities to 62" (1575mm)

Table 17.4a

Dot Tolerances for Laser-Engraved Platemaking	
Dot Percentage	Tolerance
2 - 9%	+/- 0.50%
10 - 24%	+/- 1.00%
25 - 39%	+/- 1.50%
40 - 49%	+/- 1.75%
50 - 98%	+/- 2.00%

Table 17.4b

ablates (removes) the rubber or cured polymer from the non-image areas of the design. As a final step, after laser engraving, remove any engraved debris from the plate material.

The laser-engraved rubber process also produces seamless rolls or sleeves. In this process, a rubber compound adheres to a

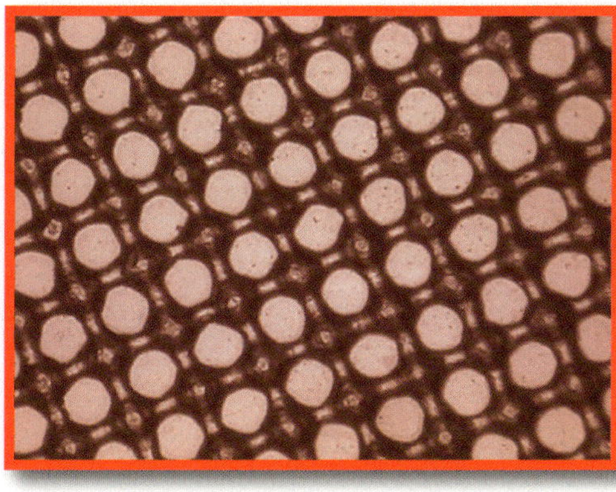

17.4b Finished Polymer Plate Surface

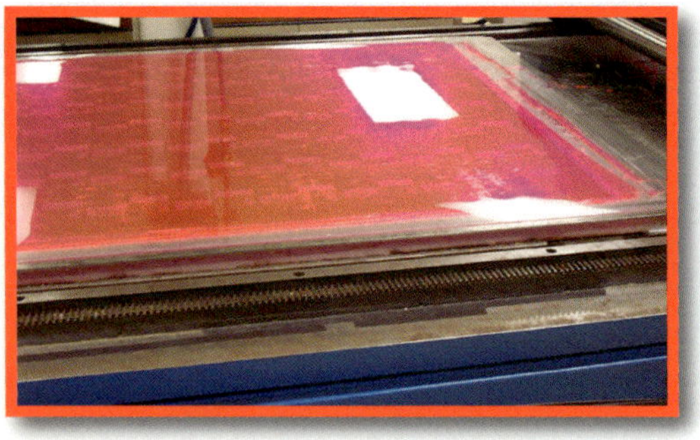

17.5 Liquid Photopolymer Plate Making

removable sleeve, or sometimes directly to a plate cylinder. The laser engraves the image into the rubber roll. Caution is advised when using demountable sleeves; the process of adhering rubber to the sleeve uses heat that may distort some sleeve material. Not all sleeve types can be used in the seamless process.

Finished Plate/Sleeve Characteristics

Relief Specifications: Laser-engraving results in finer highlight and quarter-tone dot profiles than plates produced through conventional engraving to molded matrix or molded rubber platemaking. Digital dots have a straighter shoulder than conventional dots; therefore, they require less relief to achieve maximum dot support and lengthened plate life. Variances in cylinder build-up, press tolerances, or press operation may influence relief requirements for a particular application. Confirm target relief specifications with the printer.

Print Height: Confirm all imaged areas are at full printing height. Although recessed dots are not recommended by *FIRST*, some software packages allow discretionary engraving of controlled "below-the-surface" dots in fine highlight areas. If used, verify imaged areas above 10% dot area are at full printing height.

Caliper: Refer to Section 17.9 for measurement method and Table 17.4a for caliper specifications.

Dot Accuracy: Tone scales (Section 12.9.2) should be used to verify dot accuracy after laser engraving. Use a calibrated Flexographic Plate Analyzer (Section 17.9) to measure the screen values and confirm dot accuracy.

17.5 Liquid Photopolymer Printing Plates

In the liquid photopolymer platemaking process, place the photographic negative on the plate exposure unit with the emulsion side of the film up. Place a cover film over the negative and remove air with the vacuum. Apply a layer of liquid photopolymer resin and a polyester-backing sheet on top of the cover film and negative. Close the exposure unit and expose the material to UV light. The photopolymer is hardened by the UV exposure from two banks of lamps. The top lights create the floor (back exposure) and the bottom lights set the image from the negative into the plate material (main exposure). Photopolymer not exposed to light remains a liquid and should be recycled for future use. Wash out the exposed plate in a detergent and water solution removing unexposed polymer in the non-image areas. Dry the printing plate in a hot air dryer and post

Liquid Photopolymer Printing Plate Specifications	
Halftone Minimum Plate Dot:	
Plates </= 0.125" (0.025" - 0.030" relief)	2% @ 120 lpi (47 lpcm)
Plates = 0.250" (0.110" relief)	2% @ 100 lpi (39 lpcm)
Minimum Rule Width:	
Plates </= 0.125" (0.025" - 0.030" relief)	0.003" (0.08mm)
Plates = 0.250" (0.110" relief)	0.005" (0.127mm)
Durometer	25 Shore A - 50 Shore A
Plate Size	52" x 80" (132cm x 203cm)
Plate Thickness	0.020" - 0.300" (0.5mm - 7.62mm)
Thickness of Uniformity:	
Within a Plate	
</= 0.125" (3.2mm)	+/- 0.0005" (0.013mm)
> 0.125" (3.2mm)	+/- 0.001" (0.025mm)
Within a Job Plate Set	
</= 0.125" (3.2mm)	+/- 0.0015" (0.038mm)
> 0.125" (3.2mm)	+/- 0.002" (0.05mm)

Table 17.5

expose. An optional germicidal exposure finishes the plate to a tack-free surface. Cap liquid photopolymer plates to extend the tonal range of the printed image.

17.6 Conventional Sheet Photopolymer Printing Plates

Sheet photopolymer printing plates are derived from pre-cut sheets of photopolymer material manufactured to a specified thickness. Because raw photopolymer sheets are pre-manufactured, the specifications and tolerances of the finished plate are determined by the raw material specifications of the manufacturer. The platemaker controls the plate making process so that the finished plate remains within the manufacturer's specifications for the raw material.

Make an overall exposure through the back of the plate (back exposure) to determine the relief of the image. After removing the cover sheet, expose the front of the plate, under vacuum,

17.7 Continuous Photopolymer Covered Sleeve

Sheet Photopolymer Printing Plate Specifications	
Halftone Minimum Plate Dot	1% @ 150 lpi (60 lpcm) line screen and lower
Minimum Rule Width:	
Plates </= 0.125" (3.2mm)	0.008" (0.20mm)
Plates > 0.125" (3.2mm)	0.015" (0.38mm)
Plate Sizes (raw material)	Up to 52" x 80" (132cm x 203cm)
Plate Thickness	0.045" (1.143mm)
	0.067" (1.70mm)
	0.107" (2.72mm)
	0.112" (2.84mm)
	0.125" (3.18mm)
	0.250" (6.35mm)
Caliper Uniformity:	
Within a Plate	
</= 0.125" (3.2mm)	+/- 0.0005" (0.013mm)
Within a Job Plate Set	
</= 0.125" (3.2mm)	+/- 0.001" (0.025mm)
> 0.125" (3.2mm)	+/- 0.0015" (0.038mm)

Table 17.6

through a matte-finish negative (main exposure). The non-exposed material is washed away in a solvent, or melted onto a fabric and removed, leaving a relief image in the plate material. Dry the plate in the dryer to remove solvents. Complete the platemaking process with the post-exposure and finishing steps. Run an exposure test to determine the exposure time required to achieve optimum plate relief. Use the control target during the main exposure to control the integrity of the image in the material. The control target contains exposure rules to verify proper exposure of the photopolymer. The width of the rule is approximately equal to the diameter of a 3% dot; therefore, five different targets are provided for different screen rulings. After exposure, the lines should be straight both visually and to the touch.

17.7 Continuous Photopolymer-Covered Printing Sleeves

Bonding specially prepared sheet photopolymer materials to a sleeve produces the continuous photopolymer-covered printing sleeve. The sleeve material must be stable and within the caliper and runout tolerances of the final product specifications. Once the photopolymer material is bonded to the sleeve, a seamless image can be made across and around the sleeve.

After bonding the sheet photopolymer to the sleeve, the plate height is not correct in caliper for the finished product. The diameter of the sleeve is oversized and exhibits some imperfections in the surface of the material. The photopolymer surface is finished. The flexibility of the process allows the finished diameter to vary by several thousandths of an inch above, at, or below the pitch diameter of the cylinder based on the requirements and specifications of the customer.

Continuous Photopolymer Covered Printing Sleeve Specifications	
Halftone Minimum Plate Dot	2% @ 150 lpi (59 lpcm) line screen and lower
Minimum Rule Width	0.008" (0.18mm)
Total Indicated Runout (TIR)	+/- 0.0005" (0.013mm) within a sleeve

Table 17.7

17.8 Molded Rubber Printing Plates

In the molded rubber plating process, a film negative is made of the image. The film negative is exposed onto magnesium or copper that has a light sensitive coating. The metal is placed in a washout solution that removes the metal in the non-image areas, resulting in a relief image. The metal engraving is used to make an impression into a thermal setting matrix material. A charge of non-vulcanized rubber is placed into the matrix, and under heat and pressure a rubber plate is made from the image in the matrix. The total plate thickness and the relief of the image is controlled by spacers or bearers that are built up between upper and lower heated platens in the plate molding press. The control of bearer thickness, heat, pressure and processing time determine the quality of the finished plate.

If rubber plates are to be ground, the molded thickness should be ±0.002" (0.05mm) within the plate. Plate thickness variations beyond 0.004" (0.10mm) cannot be corrected by grinding due to image distortion. It is not recommended to grind process plates due to distortion of process dots.

17.9a **Plate Durometer:** *A Shore A gauge is used to measure plate hardness, or durometer.*

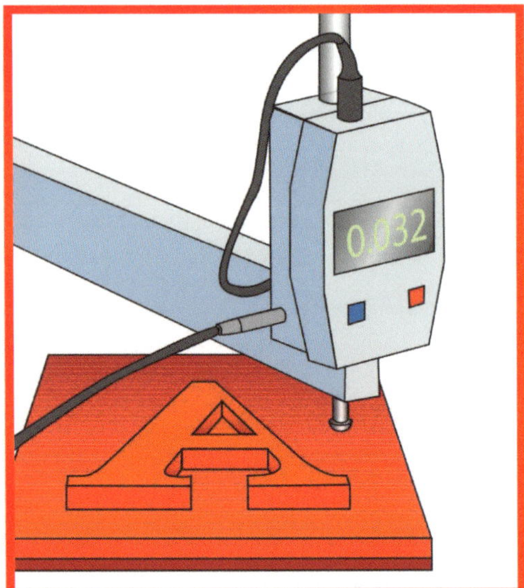

17.9b **Plate Micrometer:** *Plates should be inspected and measured with a micrometer. Gauge readings should be marked on the plates for use during plate mounting.*

Molded Rubber Printing Plate Specifications	
Halftone Minimum Plate Dot	3% @ 120 lpi (47 lpcm) line screen and lower
Minimum Rule Width	0.010" (0.25mm)
Durometer	25 Shore A - 50 Shore A
Thickness of Uniformity:	
Within a Plate	
</= 0.125" (3.2mm)	+/- 0.0005" (0.013mm)
> 0.125" (3.2mm)	+/- 0.0015" (0.038mm)
Within a Job Plate Set	
</= 0.125" (3.2mm)	+/- 0.001" (0.025mm)
> 0.125" (3.2mm)	+/- 0.002" (0.05mm)

Table 17.8

17.9 Printing Plate Measurement and Control

Shore A Gauge

Measure the durometer, or hardness, of polymer materials, such as flexographic printing plates with a Shore A gauge. Test the calibration of the Shore A gauge using the test block supplied with the instrument. The test block is matched with the serial number of each instrument and is stamped with a durometer reading for calibration. When properly calibrated, the instrument should read within plus or minus one point of the number stamped on the test block.

When using the Shore A gauge, a minimum area of 2" x 2" (50mm x 50mm) is required to get an accurate reading. Multiple

Printing Plate Measurement & Control	
Properties to Measure	Measurement Device/Tool
Durometer	Shore A Gauge For material < 0.250" (6.35mm)
Thickness Uniformity	Plate Micrometer
Plate Relief	Plate Micrometer
Halftone Dot Size	Flexo Plate Analyzer; Microscope
Bulb Output	Radiometer
Line Width	Microscope

Table 17.9a

measurements should be taken. The ASTM standard for measuring Shore A requires a plate caliper of 0.250" (6.4mm). The durometer for thinner plates may still be measured; be sure to indicate the plate thickness actually measured. Readings are only taken in the solid area of the plate. Do not take readings on process areas of the plate due to distortion of the plate material.

17.9c Flexo Plate Analyzer: *Flexo Plate Analyzers measures halftone dot size on the finished plate.*

Plate Micrometer

Measure the thickness uniformity and relief of flexographic printing plates with a plate micrometer (either analog or digital readings). Calibrate the instrument using a calibrated gage block inserted between the surfaces. Accuracy of calibration should be within ±0.0005" (0.013mm) of the calibration block.

When measuring caliper and relief with the plate micrometer, the test area should be at least 1" x 1" (25mm x 25mm). Take multiple measurements across the plate to determine the uniformity of plate thickness. Digital plate micrometers equipped with a printer can output statistical data of the plate measurements. To confirm the plates comply with certification requirements, the printer should verify the data.

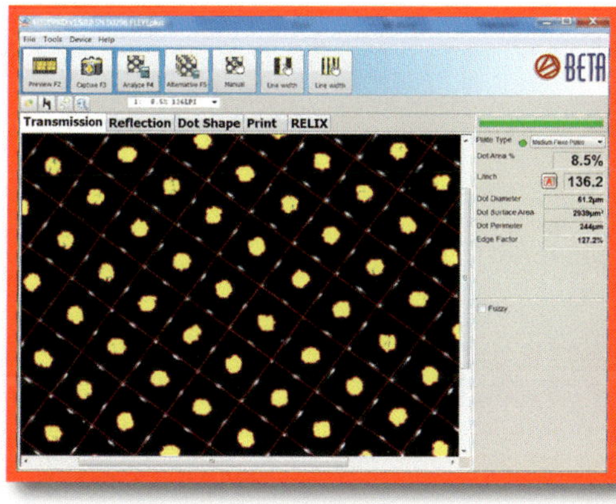

17.9d Flexo Plate Analyzer

Flexographic Plate Analyzer

The flexographic plate analyzer measures halftone dot size in the two-dimensional mode. Three-dimensional imaging is used for analysis of extremely small, non-printing dots used in advanced screening techniques. A high-resolution video camera allows for precise measurement; it can read halftone plates regardless of the contrast, color, or graining of the image. Excellent correlation is possible with conventional densitometers when measuring film- or mask-to-plate. Stochastic (FM) screen images can also be verified for quality. Multiple measurements should be taken to determine the repeatability of the measurements, especially with highlight dots.

Microscope

A microscope, or 100x magnifier with a scale, can be used to inspect images. Halftone dot size, rule widths and plate relief can be measured.

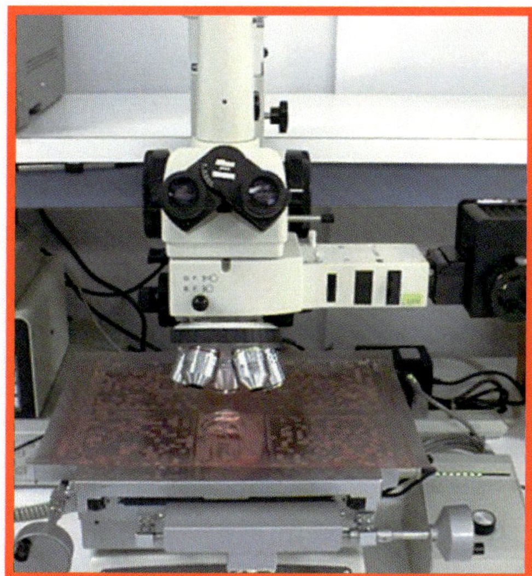

17.9e Microscope: *Dot size, rule widths and plate relief can be measured on the finished plate using a microscope.*

Exposure Guide

Radiometers: Exposure is the most important step in platemaking because it is the image formation step. Exposure bulbs degrade over time and should be monitored using a radiometer to ensure consistent image formation. Radiometers are available which can measure UVA bulbs (exposure and post exposure) as well as UVC bulbs (light finishing). When exposures

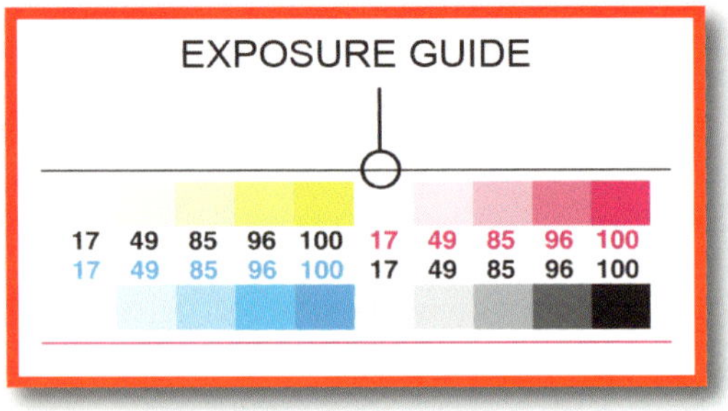

17.9f Exposure Guide: *The imaged exposure guide line should feel straight to the touch. If it is wavy, the plate is under-exposed. The highlight dot patch on the tone scale should be fully imaged. If it is not fully imaged, the plate is under-exposed.*

are established, bulb output (irradiance) should be measured and documented. As bulb output declines, exposure times must be increased to ensure consistent energy is delivered to the plate. The adjustment to exposure can be calculated using the equation:

Energy (millijoules/cm^2) =
Irradiance (milliwatts/cm^2) x Time (seconds)

For example: Main exposure was originally established to be 600 seconds at a bulb irradiance of 20 mW/cm^2. So the total energy required is 20 x 600 = 12,000mJ/cm^2. One month later, the bulb irradiance was measured at 18 mW/cm^2. To ensure the same energy is delivered to the plate, the exposure time needs to be increased to 667 seconds (12,000mJ/cm^2 ÷ 18mW/cm^2).

Control Targets: During the main exposure process, the control target is imaged within each plate. The control target is used as the exposure guide; it contains positive line and halftone scales. The line screen value of the control target should correspond with the line screen value of the image. The imaged exposure guide should be measured to confirm the printing plate is produced within specifications.

For example: If 133lpi is used, the exposure rules will be 0.0015" (0.11pts) thick. This line weight corresponds to the diameter of a 3% dot at 133lpi.

Reasons for Various Control Targets

There are five control targets for two reasons: The exposure rule varies in width with screen ruling and to provide a coarse and fine slur target for different screen rulings.

Plate Exposure Control Targets		
Line Screen	Exposure Line Weight Thickness	Star Target
55 lpi (22 lpcm)	0.22 point (0.078mm)	Coarse
65 - 85 lpi (26 - 33 lpcm)	0.18 point (0.064mm)	Coarse
100 - 120 lpi (39 - 47 lpcm)	0.14 point (0.049mm)	Fine
133 - 175 lpi (52 - 69 lpcm)	0.11 point (0.039mm)	Fine
200 lpi (79 lpcm)	0.09 point (0.032mm)	Fine

Table 17.9b

DOWNLOAD FIRST 5.1 Extras Referenced in this Section at:
http://www.flexography.org/FIRST_extras

18.0 PRINT INTRODUCTION

18.1 Overview

FIRST is created to facilitate communication among all participants involved in the design, preparation and printing of flexographic materials. A more detailed review of the Package Development Process and the roles and responsibilities of each party are provided in Section 1.4.1.1. The printer is responsible for reproducing the design accurately and consistently throughout the press run and between press runs. The ability of the printer to measure and control the printing and converting process directly impacts the quality of the printed piece, the efficiency of the press run, and the overall time-to-market for the new product. The approach outlined in the Print Section of *FIRST* assists the printer in understanding the flexographic print variables, achieving print consistency and maximizing process capability; thereby, enhancing profitability.

18.2 Responsibility

As graphics continue to increase in complexity and production timelines continue to compress, the clear assignment of responsibilities is necessary to ensure a quality printed product in a timely manner. The assignment of responsibilities requires planning and collaboration among all involved parties.

Consumer Product Company: Ultimately, the customer defines expectations and therefore, must drive the collaboration process. The customer determines the effort expended to reach satisfaction. The Consumer Product Company (CPC) must facilitate communication between the supply chain parties: designer, prepress provider and printer.

Designer: The designer must work with both the prepress provider and the printer to understand the capability of the printing/converting process being utilized. Based upon the print capability, the designer must provide a design concept that will enable the printer to meet the expectations of the customer (CPC). The earlier in the design development process the prepress provider and printer are involved, the better able the team is to determine specific capabilities and ensure the final product meets the customers design objectives.

Prepress Provider: The prepress provider must work with the printer to understand the capability of the printing/converting process being utilized. The prepress provider is responsible for providing the designer with accurate and timely information regarding print capabilities at the beginning of the design phase

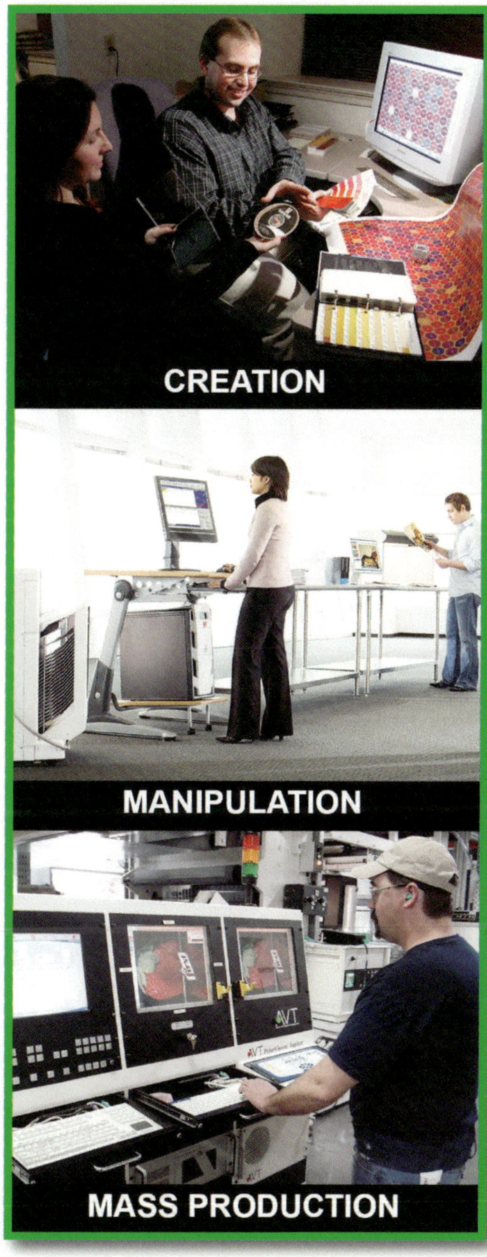

18.2 Package Development Responsibilities: *In short, the designer creates the image, the prepress provider manipulates the image, and the printer mass produces the image.*

to facilitate the creation of a printable design. Based upon the print capability, prepress produces appropriate films/files/plates that will enable the printer to meet the expectations of the customer (CPC). They must document the controls that ensure the consistency and accuracy of the supplied media (films/files/plates). The prepress provider is responsible for producing a contract proof calibrated to accurately predict the printed result. Additionally, they must provide the printer the ability to objectively confirm the accuracy of the prepress work and the printing process. This can be accomplished through the use of agreed-upon control targets.

Printer: The printer is responsible for consistently reproducing the graphic design to the satisfaction of the customer (CPC). The printer must utilize and document the process controls necessary to ensure that accuracy and consistency are achieved. They must work with the other parties and suppliers to define the capability of the printing process. The printer is responsible for providing the designer with accurate and timely information regarding process capabilities at the beginning of the design phase to facilitate the creation of a printable design.

Ink Supplier: The ink supplier is responsible for consistently providing ink that meets the functional and color requirements of the job. They must utilize and document the process controls necessary to ensure the accuracy and consistency of the ink. They must work with the printer to understand the capability of the printing process and the requirements of the particular job.

Substrate Supplier: The substrate supplier must provide the requested substrate consistent to defined specifications. They must utilize and document the process controls necessary to ensure the accuracy and consistency of the substrate. They must work with the printer to understand the capability of the printing process and the requirements of the particular job. For paper substrates, graphic image reproduction variables include: colorimetric values, absorption, hold out characteristics and brightness. For the specific variables and characteristics of the substrates refer to Section 20.1.

19.0 PRINT EVALUATION

In order to print a job accurately, consistently and cost efficiently, the printer must measure and control key print characteristics. Objective measurement, using standard measurement conditions, enables the printer to produce the image consistently throughout the production run every time. A thorough understanding of the key print characteristics, the targets used to monitor them and the press variables that influence them is necessary by all press crews if process control is to be successfully implemented.

In this Section, *FIRST* outlines the key print characteristics for both line and process work, explains the control target used to measure each characteristic and identifies the variables that influence each print characteristic. Prior to explaining the primary print characteristics, this section provides a detailed explanation of the equipment used to measure each print characteristic.

19.1 Measurement Tools

19.1.1 Spectrophotometer

Purpose
Measures the light reflectance of a color every 10-20 nanometers across the visible light spectrum (400-700nm). A precise definition of the color is derived from reflectance curves created from these measurements. The color can be plotted in 3-dimensional color space and described in terms of lightness, chroma and hue.

Application
Used to measure printed line/spot colors and compare them to established color targets/standards, allowing for the consistent matching of custom colors throughout and between production runs, as well as measuring and controlling incoming inks to ensure consistency from batch to batch.

Industry Standard
ANSI/CGATS.5 2003 (Graphic Technology – Spectral Measurement and Colorimetric Computation for Graphic Arts Images) provides the equations used to calculate colorimetric parameters and defines standard spectrophotometric instrument settings.

Settings
Two equipment settings on the spectrophotometer must be chosen: the illuminant and standard observer. The same color, measured with different illuminants and standard observers, will yield different measurements, and are thus incompatible. Indicate the illuminant and standard observer used whenever colorimetric data is communicated.

FIRST recommends ANSI/CGATS.5 2003 (Graphic Technology – Spectral Measurement and Colorimetric Computation for Graphic Arts Images) which specifies D50 as the illuminant and 2-degrees as the standard observer (D50/2°). In situations where

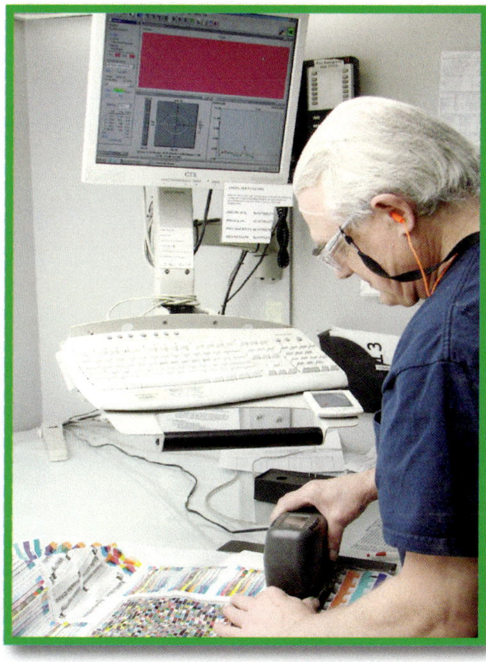

19.1.1a Press Side Spectrophotometer: *Using a spectrophotometer improves the printer's ability to consistently match custom colors throughout a production run and between production runs.*

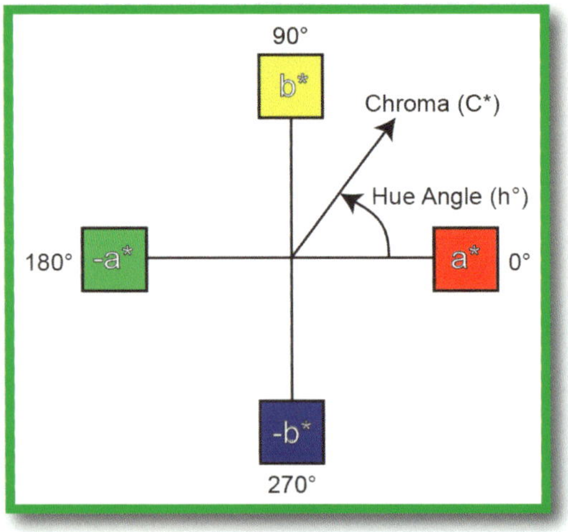

19.1.1b Color Tolerancing: *Expresses, as a numerical value, the accuracy of the color match.*

other settings need to be used, *FIRST* recommends including the spectral data for the measurements, to allow the receiver to calculate colorimetric values for any combination of illuminant and standard observer.

Spectrophotometer Instrument Settings	
Standard	ANSI/CGATS .5-2003
Illuminant	D50 (5000 Kelvin)
Observer	2° Standard Observer
Measurement	Absolute (including substrate)
Geometry	0°/45° or 45°/0°
Tolerance Method	DE CMC or DE CIE94 or DE CIE2000
Aperture	3.4mm std. (2.0 - 6.0mm range)

Table 19.1.1a

Calibration

A spectrophotometer must be routinely calibrated in accordance with the manufacturer's recommended procedure using a calibration standard that is traceable to a standard reference. A documentation system should include all calibration steps, outcomes and corrective actions.

Communication

Colorimetric data should be accompanied by the following information:

- Originator of the data
- Date created
- Description of the purpose or contents of the data being exchanged
- Description of the instrument used, including the brand and model number
- Instrument parameters: illuminant, observer, measurement filter, geometry, aperture
- Wavelength interval used
- Tolerance method: CIE2000, CIE1994, CMC
- When density data is reported using a spectrodensitometer, the spectral product's weighting function (status or type response) used should be identified.

Colorimetric Parameters

CIE L*a*b* and CIE L*C*h° are the most common colorimetric parameters. L*a*b* describes the location of a color in CIE color

Common CIE Illuminants		
CIE Illuminant	Common Name	Color Temperature (degrees Kelvin)
D50	D50 (horizon light)	5000K
D65	D65 (noon daylight)	6500K
F8	Fluorescent D50 (horizon light)	5000K
F7	Fluorescent D65 (noon daylight)	6500K
F2	Cool White Fluorescent	4230K
Illum. A	Incandescent Light	2856K
Illum. C	Average Daylight (northern sky)	6774K

Table 19.1.1b

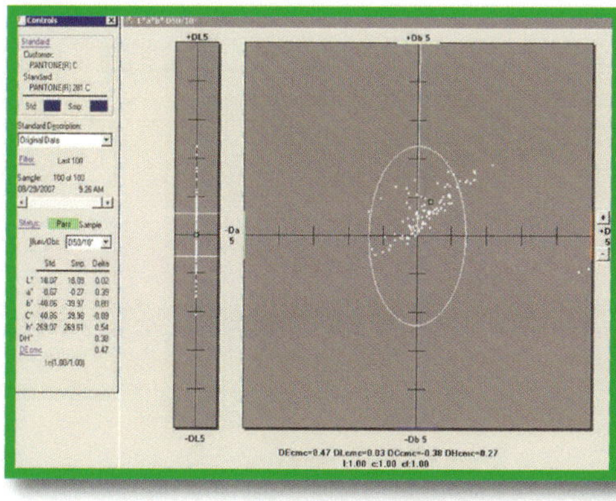

space according to lightness (L*), the red/green (a*) and blue/yellow (b*) dimensions. L*C*h° color in terms of lightness (L*), chroma (C* - the neutrality of the color) and hue (h°), which are mathematically derived from the *a and *b values. As a general reference, hue angles for red, yellow, green and blue begin as follows: 0°=Red, 90°=Yellow, 180°=Green, 270°=Blue.

Color Tolerancing Method – Calculating DE

Color tolerancing is used to determine if a "sample" color, compared to a "reference" color, is acceptable. The accuracy of the color match is typically expressed as Delta E (ΔE or DE). The DE value represents the overall color difference between two colors, and is derived from one of several equations. Since differences in color are not uniform, DE equations can be either weighted or unweighted.

Unweighted color difference equations, such as CIE 1976 (DEab), weight hue, chroma and lightness equally. This does not correspond well with human perception of color differences. Therefore, a sample may visually match the reference color, but produce an unacceptable DE value or vice versa.

Weighted color difference equations, such as DECMC, DECIE94 and DE2000, achieve better agreement with human perception by weighting hue, chroma and lightness differently. The formula used to weigh the three-color axis is slightly different for each weighted color difference equation.

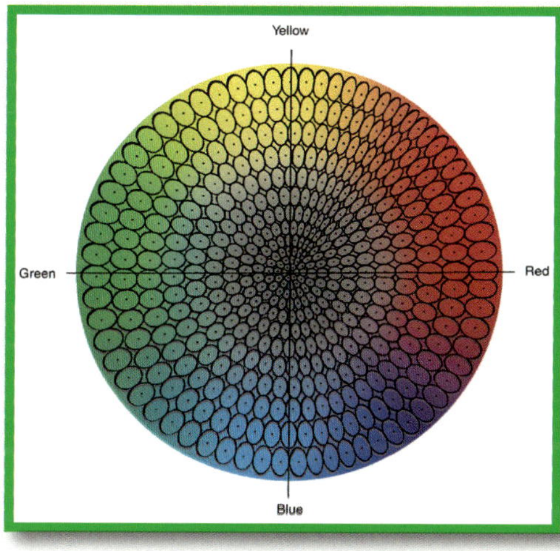

19.1.1c Acceptable Color Difference: *The weighted color difference equations use slightly different mathematical formulas to calculate the shape of the acceptable color difference ellipse.*

Comparison of Color Difference Equations

Parameter	CIE76	CMC		CIE94	CIE2000
FIRST Recommended?	No	Accepted		Accepted	Preferred
Year Developed	1976	1984		1994	2000
Equation Developed By:	The International Commission on Illumination	The UK Society of Dyers & Colorist Color Measurement Committee		The International Commission on Illumination	The International Commission on Illumination
Pass/Fail Ellipsoid around the Standard with a set DE?		Yes		Yes	No
Is the DE the same if the "Standard" and "Sample" are reversed?		No		No	Yes
Weighted?	No	Yes		Yes	Yes
Typical Weighting Factors (for print applications)	N/A	(2:1)	(1:1)	K1 = 0.045 & K2 = 0.015	(1:1:1)
Example - DE Comparison: "Standard" L= 45.25, a= -40.35, b= -35.75 "Sample" L= 43.75, a= -39.50, b= -37.50	2.46	1.22	1.75	1.83	1.65

Table 19.1.1c

The CMC tolerancing method includes three weighting factors set by the user, expressed as (l:c:cf), where l=lightness, c=chroma and cf=commercial factor, or DE. The CMC ratio l:c (lightness: chroma) determines the shape of an ellipsoid, while the cf (commercial factor) determines the overall size of the ellipsoid and the threshold, or tolerance, of acceptable color difference. The cf determines the acceptable DE limit. For example, if the cf=1.5, then the acceptable DE value is 1.5 as well.

FIRST recommends using one of the weighted color difference equations and, where possible, the latest applicable equation. Regardless of the color tolerancing equation used, it is critical to communicate both the equation and any weighting factors used to all parties receiving color data.

All parties measuring color must use the same tolerancing equation in order to have meaningful discussions of color differences. Indicate the color tolerancing equation used on all proofs or print samples containing color difference (DE) values. As instrumentation is purchased, *FIRST* recommends buying equipment that supports the latest methods in color tolerancing and has a robust upgrade path. Refer to Appendix A for The International Commission on Illumination (CIE) and the UK Society of Dyers & Colorist Color Measurement Committee (CMC) contact information.

19.1.2 Densitometer/Spectrodensitometer

Application

A reflection densitometer (or spectrodensitometer) is used to measure important print characteristics for process color images such as: density, dot area or dot gain/TVI, trap, print contrast and gray balance. These densitometric functions are critical for measuring and controlling the reproduction of continuous tone images. Using a densitometer/spectrodensitometer improves the printer's ability to consistently and accurately reproduce a continuous tone image throughout a production run and across production runs. Densitometers/spectrodensitometers should be located press-side for quick and accurate press set-up and run using process control methodology as discussed in Section 1.3.3. The densitometric functions are designed for use with the four traditional process inks (yellow, magenta, cyan and black). When using non-traditional colors for expanded gamut continuous tone images, the usefulness of the densitometric functions is limited due to a lack of filters and equations for those colors. Refer to Section 20.2.5 for additional information on printing with expanded gamut technology.

19.1.2a Spectrophotometer

Standard

ANSI/CGATS.4 2006 (Graphic Technology – Graphic Arts Reflection Densitometry Measurements – Terminology, Equations, Image Elements, and Procedures) is the standard for the measurement of printed materials using a reflection densitometer. For additional information on densitometry, refer to "Introduction to Densitometry – Users guide to Print Production Management Using Densitometry," published by IDEAlliance. Their contact information is listed in Appendix A.

Variables

Key instrument variables must be properly set to produce meaningful measurements. These variables should be documented and communicated with all densitometric data:

Densitometer: Manufacturer and Model.

Spectral Response: The density value obtained is a function of the spectral characteristics of both the ink being measured and the spectral response. The values obtained with different response functions may be similar or significantly different, depending on the particular material being measured. Status T is the preferred spectral response in North America. Status E is defined to closely match the characteristics of graphic arts materials normally used in Europe.

Sample Backing: Many flexographic substrates are translucent; consequently, the choice of backing material will significantly impact the color measurements. The best choice is to use a

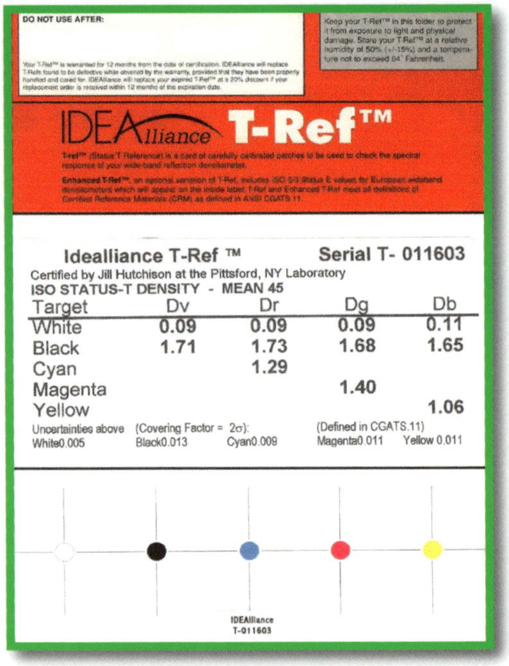

19.1.2b T-Ref™ Calibration Standard: *The T-Ref™ calibration standard should be routinely replaced to avoid the risk of fade or damage (creases, dirt, etc.) which could compromise the instrument calibration.*

white backing material that is spectrally non-selective, diffuse-reflecting, non-fluorescing and has a L* value greater than 92 and C* value of less than 3. It is important to use the same backing for all measurements.

Sampling Aperture: While aperture sizes range from 2.0-6.0mm, 3.4mm is the standard aperture size. It is critical for each test element to be slightly larger than the aperture for accurate measurement.

Calibration: *FIRST* supports a daily calibration of all densitometers/spectrodensitometers in accordance with the manufacturer's recommended procedures using a reflective calibration standard that is traceable to a standard. The calibration standard should be replaced at the manufacturer's recommended frequency to avoid the risk of fade or damage, which could compromise the calibration.

Polarizing Filter: This filter reduces the substrate specular reflections picked up by the instrument, thus increasing the reported density values. While the polarizing filter is not typically used, it is useful in situations where the substrate characteristics change during the manufacturing process, resulting in differences in visual appearance and reported density values. The polarizing filter may reduce these differences in reported density value.

Densitometric Instrument Variables	
Parameter	*FIRST* Recommendation
Standard	CGATS .4-2006
Spectral Response	Status T wide-band
Sample Backing	White with an: L* > 92, C* < 3 Non-Fluorescing
Aperture	3.4mm std. (2.0 - 6.0mm range)
Instrument Calibration Target	The IDEAlliance T-Ref™ target calibrated to +/- 0.02
Polarizing Filter	Yes or No (typically, no)

Table 19.1.2a

Measurement Data

The densitometer/spectrodensitometer must be properly set to accurately capture the desired data. The conditions under which these measurements are taken must be documented and reported with all densitometric data.

Density: Absolute or relative. Absolute density includes the reflectance of the substrate. Relative density factors out the reflectance of the substrate; it is generally referred to as "density minus paper." Absolute density is the typical density value reported since the substrate is integral to viewing the printed image. Refer to Section 19.4.4 for more information on solid ink density.

Grayness: The grayness function measures the relative achromatic content of a printed area. It can be used to monitor gray balance of the yellow, magenta and cyan overprint. It is expressed as a percentage. Refer to Section 19.4.2 for more information on gray balance.

Dot Area or Dot Gain/TVI: Also referred to as Tone Value and Tone Value Increase. The Murray-Davies equation is the preferred method for calculating both dot area and dot gain/TVI. If the Yule-Nielsen equation is used, the "n" factor must be empirically determined for each ink-substrate-screening-press configuration and reported with the data. Refer to Section 19.4.3 for more information on dot area or dot gain/TVI.

Print Contrast: The print contrast measurement indicates the printing system's capability to hold image detail in the shadow tones. The 70% tint is typically used to calculate print contrast for flexographic applications. Report the shadow tint used in

Densitometric Measurement Data		
Measurement	Information to Report	*FIRST* Recommendation
Density	Absolute or Relative	Absolute
	Color Channel (filter) Used	
Dot Area (Tone Value)	Solid Density	Murray-Davies
	Color Channel (filter) Used	
	Equation: Murray-Davies or Yule-Nielsen (include "n" factor for Yule-Neilsen)	
Dot Gain (Tone Value Increase)	Dot Area on Plate	Murray-Davies
	Equation: Murray-Davies or Yule-Nielsen (include "n" factor for Yule-Neilsen)	
Trap	Overprint(s) Measured	Preucil (apparent)
	Print Sequence	
	Equation: Brunner, Hamilton (newsprint), Preucil (apparent)	
Print Contrast	Dot Area on Plate	Solid Patch & 70% Tint Patch
	Paper (included or excluded)	Paper (included)

Table 19.1.2b

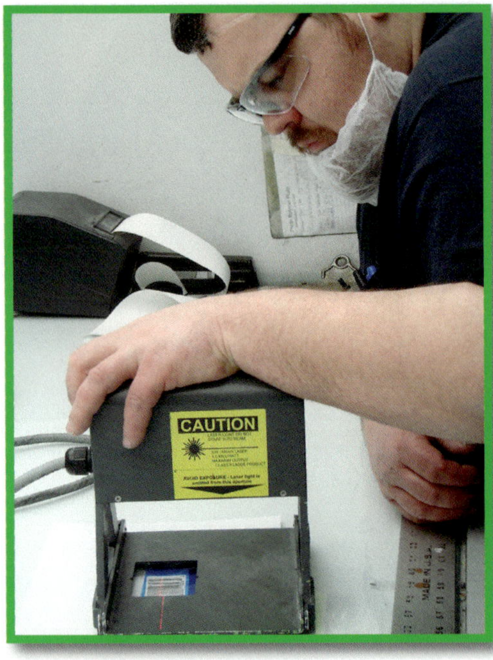

19.1.3 Bar Code Verifier: *It is critical to verify the scannability of the bar code as it is being printed in order to prevent correctable problems and minimize waste.*

the measurement along with the data. Refer to Section 19.4.5 for more information on print contrast.

Trap Equation: Hamilton (newsprint) or Preucil (apparent). Both the Hamilton and Preucil equations account for "missing" density arising from additivity failure. Additivity failure occurs when the density of the overprint is less than the sum of the solid densities of the two inks printed and measured separately. The Preucil (apparent trap) equation is typically used. Regardless of the equation used, it must be reported along with the trap data. Refer to Section 19.4.6 for more information on ink trap.

19.1.3 Bar Code Verifier

Application

Retailers impose large fines and penalties for bar codes that do not scan; therefore, most CPCs require ANSI grading of bar codes. Printers use bar code verifiers to confirm the bar code is printing within ANSI specifications at press setup and throughout the production run. The bar code parameters measured include symbol contrast, edge contrast, minimum reflectance, edge determination, modulation, defects, decodability and quiet zones. An ANSI grade ("A" through "F") is given based on these parameters, with the overall grade corresponding to the lowest scoring parameter. For most applications, a 2.5 ("C") grade is the minimum acceptable grade. Refer to Sections 19.3.4 and 22.0 for detailed information on bar codes.

Industry Standard

ANSI X3.182 was the original bar code quality specification in the USA; however the current international specifications are ISO 15416:2000 and ISO/IEC 15415:2004. The minimum symbol grade required, as well as the verifier's measuring aperture and peak wavelength of light, are determined by industry application standards (symbol specifications) depending on the symbol being analyzed and where it will ultimately be scanned. Minimum bar code print quality specifications outside those determined by the accredited standards organizations are not supported by *FIRST*.

Calibration

Bar code verifiers require periodic calibration, either by using an accompanying calibration target/patch, or through maintenance by the manufacturer. *FIRST* recommends following the manufacturer's recommended calibration procedures and frequency. Additionally, ANSI/ISO-based bar code verifiers should be validated by determining how well they agree with a universally accepted and traceable conformance standard (such as ISO 15426-1:2006 verifier conformance specifications).

Variables Reducing ANSI/ISO Grades for Bar Code Quality	
ANSI/ISO Parameter	Potential Problem Areas
Symbol Contrast	Specify adequate colors for bar code and monitor color throughout press run.
	Monitor ink "application"; density, quality, ink receptivity.
	Monitor the substrate qualities that relate to overall reflectance of the bars and spaces (reflectance uniformity and properties of printed substrates).
	Insufficient reflectance of substrate (translucency).
	To simulate the best and worst case conditions for a filled package, make measurements against an opaque white and opaque black background.
Edge Contrast	Monitor anything that would reduce the symbol's edge acuity or "print sharpness".
	Uneven impression or over impression (halos).
	Ink density and application quality (feathering or improper adhesion).
	Plate problems (cylinders out of round, low and high spots, or nicks in the plate).
Minimum Reflectance	Refer to symbol contrast.
Edge Determination	Monitor anything that could cause and additional bar to appear or space to disappear.
	Verify match between bar code size and the specified imaging resolution (bars or spaces may be "rounded" in or out of a symbol).
	Monitor conformance to "predicted" bar width growth in prepress (allocated bar width reduction) and on press (running within press characterization range).
	Significant designer manipulation of the symbol (resizing, ungrouping).
	Damaged plate or impression cylinder.
	Substantial ink bridging/feathering.
Modulation	Monitor bar width growth.
	Insufficient BWR (bar width reduction) for print conditions.
	Excessive impression (narrow spaces).
	Ink spread (ink bridging/ink feathering into substrate).
Defects	Monitor any variable that can cause "spots" in spaces or "voids" in bars.
	Process: bar code design, bar code negatives, plate cylinder, plate wear.
	Ink: metering (pooling, absorption, splatter, "ghosting").
	Substrate: surface texture, dirt, recycled content, absorption.
Decodability	Monitor conditions that cause element width variations within individual elements.
	Verify match between bar code size and the specified imaging output resolution to avoid individual element (bar or space) widths being "rounded" up or down.
	Monitor conditions that can result in incorrect positioning of an individual element (defective plate cylinder, printing plate defects, or improper mounting).
	Distortion for plate circumference (if bars are printed in the transverse directions).
Quiet Zones	Not enough room reserved for the quiet zones.
	Debris (ink spots) in the quiet zones.
	Symbol is partially located underneath a fold, flap, die-cut, emboss, or laminate in the final filled package.

Table 19.1.3

19.1.4 Color Viewing Booth: *Accurate, consistent, visual perception of color requires viewing of printed material in a standard, chromatically neutral, controlled environment.*

19.1.5 Magnifier: *A magnifier enables the press operator to visually examine the detail of the printed image.*

19.1.4 Color Viewing Booth

Application

Use a color-viewing booth to view printed images, proofs, or transparencies, under a controlled and standard light source. Accurate and consistent visual perception of color requires the image to be viewed in a standard, chromatically neutral, controlled environment. When the printer, separator and customer standardize viewing conditions, color discrepancies are minimized.

Industry Standard

FIRST supports the standards established in ANSI 2.30 1989. However, *FIRST* recognizes the lighting specified may not be optimal for all print segments.

ANSI 2.30 - 1989: Color Viewing Conditions	
Parameter	Specification
Lighting	5000 Kelvin Blubs
Light Source	Color Rendering Index >/= 90
Viewing Surface	Luminance = 204 +/- 44 Foot Candles
Viewing Area Color	Neutral Gray: Munsell N8 or Equivalent
Angles of Illumination	Minimize Glare

Table 19.1.2a

Instrument Agreement

The light source used in the light booth should be the same as the equipment illuminant setting. For example, if the spectrophotometer is using D65 instead of D50, the light booth should use 6500-Kelvin bulbs (D65) rather than 5000-Kelvin bulbs (D50).

Communication

Regardless of the viewing conditions used, it is important to communicate the viewing conditions to all parties receiving data measurements.

19.1.5 Magnifier & Tape Measure

Magnifier

Locate 10X and 30X magnifiers press-side for visual examination of printed image detail. Characteristics evaluated with a magnifier include: dots (for roundness), halo(s), donuts, slurring, color-to-color register, ink lay smoothness and sharpness.

Tape Measure

Use a tape measure to verify image repeat, cut-off, web or sheet width and critical package elements such as glue zones. Use a tape measure, in the scale appropriate for the press, to determine the amount of misregistration in order to correct the press more precisely.

19.2 Using Process Control Test Elements

Application

To ensure repeatability and consistency, space must be allocated on the sheet, web, or printed product for appropriate process control test elements, such as those found in Sections 19.3 (line work) and 19.4 (process color work). These elements should be used during print optimization, fingerprint and characterization trials, as well as on "live" jobs. These elements should be measured at set-up and throughout the production run to facilitate process control. By using similar test elements on press trials and production jobs, the printer can verify current print conditions and flag any changes since the press was last fingerprinted. Additional information regarding process control for expanded gamut can be found in the Appendix F Expanded Gamut: Reasonable Measurement for Process Control.

19.2a Example of Control Targets on Live Packaging

Placement

Some packaging lends itself to placing test elements under flaps, in glue zones, or on the waste matrix. Other packaging requires test elements to remain visible on the finished package. Therefore, each print application should determine where to place the individual test elements.

Format

Sections 19.3 and 19.4 describe the key print characteristics for both line and process work, and the test element used to measure each characteristic. All test elements discussed in these sections will be supplied for construction into a suitable control target, optimization or fingerprint test design for each print application. These elements are available to download at: http://www.flexography.org/FIRST_extras

Process Control Test Elements

FIRST recognizes that certain press configurations and product types may not have large enough trim areas or glue zones to maintain all recommended process control elements. At minimum, include a test element to verify density and at least one dot area for each color to maintain consistency.

19.2b Placement of Control Targets: *Control targets can be placed in folds, glue flaps, the waste matrix and, when necessary, on the "live" print area of the package.*

Ideally, these five test elements should be on all process color jobs:

1. Registration: color-to-color and print-to-cut
2. SID/Trap
3. Tone scales
4. Impression: anilox-to-plate and plate-to-substrate
5. Gray balance

Test Element Configuration

Size: In general, for measurements using spectrophotometers or densitometers, the largest aperture that the test element can support is the most ideal, as larger apertures result in less measurement variability. While *FIRST* does not recommend using smaller test elements and apertures, smaller apertures are available, and are a better option than not measuring at all. See Table 19.2 for minimum and recommended aperture sizes specified by line screen, according to ANSI/CGATS.5. In addition to these guidelines, print application must also be considered, as the minimum acceptable aperture may be larger in some cases. The designer and prepress provider should confirm test element size with the printer.

Imaging: Test elements should be imaged at the same time and with the same specifications (line screen, angle, dot shape, resolution, etc.) as the actual image. Surprinting, plate slugs, and plate build-up of the control target are not accurate representations of the live image and are not supported by *FIRST*. Refer to Section 19.4.3 for a detailed explanation of the type of tone scales required on press trials and production runs.

19.3 Line Colors: Print Characteristics Measured

19.3.1 Positive & Reverse Type Elements

Description

Type specifications are normally expressed in "points," which is derived from the height of the characters (1 point = 1/72"). The specifications differentiate between Serif fonts (such as Times New Roman) and Sans Serif fonts (such as Helvetica). Because of "non-structural details" in Serif fonts, their minimum type size is typically larger than the minimum size for Sans Serif fonts, as these details are typically very small, and fill in faster in reverse print, or print dirty with positive print. Fonts with unique designs (Kanji characters, script fonts, etc.) may not relate to "Minimum Type" recommendations; in these cases, use "Minimum Rule" specifications instead, to ensure that the thinnest lines of the font are not thinner than the appropriate minimum rule width specification. Refer to Section 12.2 for more information on text elements.

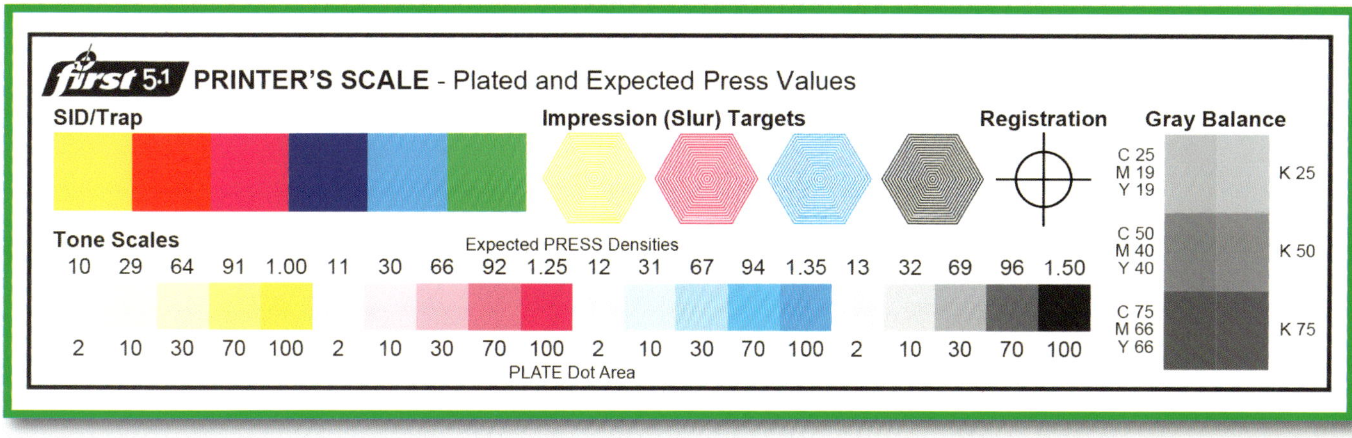

19.2c Printers Control Target

CGATS.5 Densitometer & Spectrophotometer Aperture Size					
Screen Frequency		Round Sampling Aperture (mm)		Non-Round Sampling Aperture (mm)	
Lines per inch	Lines per cm	Minimum	Recommended	Minimum	Recommended
65.0	26.0	3.8	5.7	11.3	25.5
85.0	33.0	3.0	4.5	7.1	15.9
100.0	39.0	2.6	3.9	5.3	11.9
120.0	47.0	2.1	3.2	3.5	7.8
133.0	52.0	1.9	2.9	2.8	6.4
150.0	59.0	1.7	2.6	2.3	5.1
175.0	69.0	1.4	2.1	1.5	3.5
200.0	79.0	1.3	2.0	1.3	3.0

Table 19.2

Print Challenges

When printing reverse type, the biggest challenge is ink filling in the type element, rendering it illegible. For positive type, plugging or "dirty" type is the primary challenge. This occurs when ink spreads beyond the individual characters, eventually connecting multiple type elements together, or when the inner portions of characters such as "o," "p," "b," etc. fill in. Other challenging type elements include web addresses (http://...), @, ® and ©. The smaller the type element, the less ink can spread before words become illegible. Excessive pinholing can also make type difficult to read, particularly at smaller sizes.

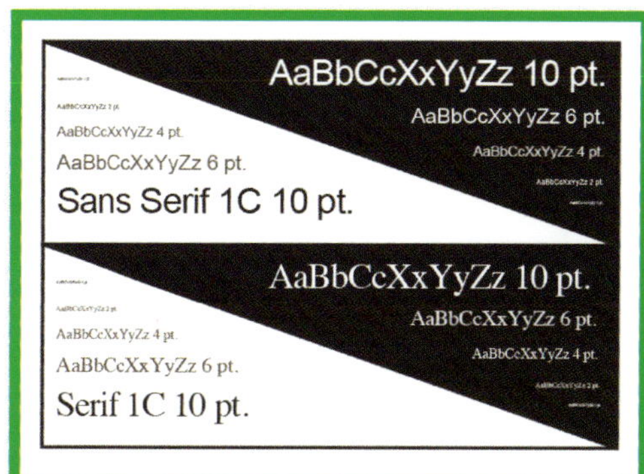

19.3.1a Type Elements: *A test element for type should include both positive and reverse type in varying point sizes using both Serif and Sans Serif fonts.*

19.3.1b Minimum Rule Width: *Determine the minimum positive and reverse rule width for the specific print conditions during the press fingerprint.*

Print Variables

Print variables that influence ink spread (substrate absorbency, anilox volume, ink metering system, mounting tape compression, ink viscosity, etc.) will determine the minimum achievable type size for each condition. Variables that influence ink pinholing (ink dry rate, between-deck dryer balance, mounting tape compression and plate cleanliness) will influence small type legibility as well.

Test Element

Minimum positive and reverse type size and rule width are determined during the fingerprint trial. Test various rule widths and type sizes, for both Serif and Sans Serif typefaces, in positive and reverse format to determine minimum specifications for each print condition.

19.3.2 Custom/Spot/Line Colors

Description

Non-process colors, specified by the customer (often brand colors) may be specified as a PMS or GCMI match (such as PMS 186c red), or the customer may provide a sample to match. The customer may approve the color on-press; the approved press

Minimum Type Size: General Guidelines
Minimum type size is print system dependent; determine minimum type size with press fingerprint (ref. 1.3.2)

Print Segment		Substrate	Positive		Reverse		Printer Specific			
							Positive		Reverse	
			Serif	Sans Serif	Serif	Sans Serif	Serif	Sans Serif	Serif	Sans Serif
Wide Web	Preprint Linerboard	All	8 pt	6 pt	10 pt	8 pt				
	Combined Corrugated	White Top	8 pt	6 pt	10 pt	8 pt				
		Coated Paper	6 pt	4 pt	8 pt	6 pt				
	Folding Carton	All	6 pt	4 pt	8 pt	6 pt				
	Multiwall Bag	Coated Paper	8 pt	6 pt	12 pt	10 pt				
		Uncoated Paper	10 pt	8 pt	18 pt	12 pt				
	Film Products	Polyester	8 pt	6 pt	12 pt	10 pt				
		Polypropylene, Polyethylene & Metallized	8 pt	6 pt	10 pt	8 pt				
	Newsprint	Uncoated Paper	10 pt	7 pt	11 pt	10 pt				
Narrow Web	Paper Products	All	6 pt	4 pt	8 pt	6 pt				
	Film Products	All	6 pt	4 pt	8 pt	6 pt				
	Envelope	All	6 pt	4 pt	8 pt	6 pt				

Table 19.3.1a

Minimum Rule Width: General Guidelines *Minimum rule width is print system dependent; determine minimum rule width with press fingerprint (ref. 1.3.2)*						
Segment		**Substrate**	**Positive Rule**	**Reverse Rule**	**Printer Specific**	
					Positive Rule	**Reverse Rule**
Wide Web	Preprint Linerboard	All	0.010"	0.015"		
			0.254mm	0.38mm		
	Combined Corrugated	White Top	0.013"	0.020"		
			0.33mm	0.51mm		
		Coated Paper	0.007"	0.010"		
			0.18mm	0.254mm		
	Folding Carton	All	0.006"	0.008"		
			0.15mm	0.20mm		
	Multiwall Bag	Coated Paper	0.007"	0.010"		
			0.18mm	0.254mm		
		Uncoated Paper	0.013"	0.020"		
			0.33mm	0.51mm		
	Film Products	All	0.007"	0.013"		
			0.18mm	0.33mm		
	Newsprint	All	0.007"	0.013"		
			0.18mm	0.33mm		
Narrow Web	Paper Products	All	0.005"	0.010"		
			0.13mm	0.245mm		
	Film Products	All	0.004"	0.008"		
			0.10mm	0.20mm		
	Envelope	All	0.007"	0.010"		
			0.18mm	0.254mm		

Table 19.3.1b

sample or ink drawdown becomes the new standard. In this case, it is important that the standard be renamed to avoid conflicts with existing PMS colors.

Print Challenge

Three of the most common challenges associated with printing line colors are mottle (the appearance of an uneven solid), pinholing (tiny holes in the solid where the substrate or underlying ink appears), and color match (achieving a visual match to the customer standard).

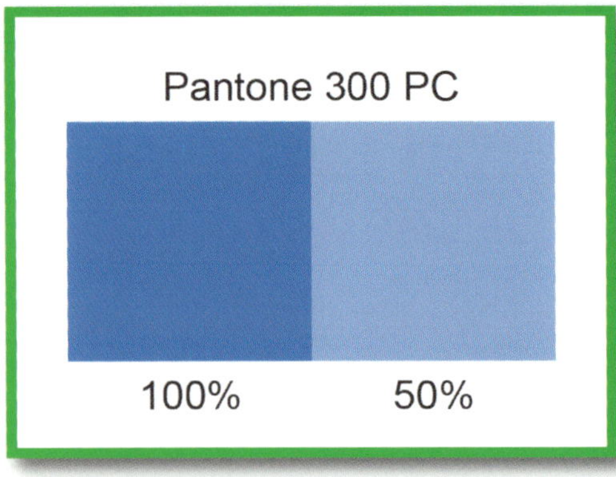

19.3.2 Example of a Spot Color Test Element

19.3.3a Blends/Vignettes/Gradations

Print Variables

Print variables that influence the delivery of ink to the substrate (anilox volume, ink metering method, mounting tape compression and plate surface modification) will influence how the ink lays (mottle and pinholing), as well as the appearance of the color. Controlling ink chemistry on press (viscosity, pH and dry rate) is critical to achieving and maintaining smooth ink lay and color match. Finally, variables such as the substrate (absorbency and color) and ink formulation (pigments and vehicle chemistry) will influence both visual color match and ink lay characteristics. Avoid matching custom colors using more than one combination of pigments; this may result in a "metameric" match, where colors match under one light source, but not another.

Test Element

FIRST recommends including a solid square of the custom color if the graphics do not provide a large enough solid area to measure with a spectrophotometer. Store the customer "standard" color in color matching software that supports the spectrophotometer. Compare printed samples to the customer "standard" at set-up and throughout every production run. Refer to Section 19.1.1 for information on spectrophotometry and color tolerancing. Visually compare print samples to the color standard in a color viewing booth under standardized viewing conditions. Refer to Section 19.1.4 for *FIRST* recommended color viewing conditions. If there are screens or vignettes of a custom color, include at least one tint patch to monitor dot gain/ TVI during the press run.

19.3.3 Blends/Vignettes/Gradations

Description

A halftone graphic (design element) that changes smoothly in tonal values from light to dark or vice-versa. It may or may not go all the way to a specular highlight (zero tone value). Also referred to as a gradient or blend.

Print Challenges and Variables

There are four primary challenges when printing vignettes: barring/banding, dirty print, hard edges and rainbowing. Print variables vary according to the challenges encountered:

1. **Barring/Banding:** Bands of light and dark color throughout the vignette, perpendicular to the web direction. This is typically attributed to mechanical problems in the print deck, such as gear wear or doctor blade chatter. Increasing press speed with exaggerate mechanical problems.

2. **Dirty Print:** Typically, the result of dot bridging due to excessive dot gain/TVI on press, or dirty printing plates. Variables influencing dirty print in a vignette are the same as the variables influencing dot gain/TVI: substrate absorbency, line screen, anilox volume, ink metering method, mounting tape compression, plate durometer, etc. Additionally, dirty print can be caused by paper dust on printing plates, or between-deck dryers blowing on plates, causing ink to dry on the plate.
3. **Hard Edge:** This occurs when a vignette fades to zero, rather than the printer's minimum dot. This is primarily a design/prepress problem that cannot be corrected on press.
4. **Rainbowing:** This occurs when vignettes are composed of multiple colors, with unbalanced dot gain/TVI or which end in different places throughout the vignette. Rainbowing can result from both prepress and print variables. From the prepress perspective, a vignette composed of more than one color must have all colors stop at the same point and maintain the printer's minimum dot in all colors. The appropriate dot gain/TVI compensation curve must be applied to each color to achieve color balance. On press, the printer must control dot gain/TVI in each color used to create the vignette and run to the targets established during the fingerprint trial.

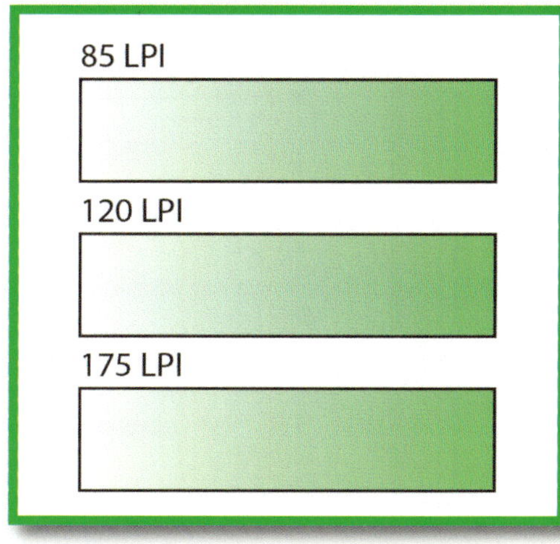

19.3.3b Vignette Test Element: *Include the line screens most appropriate for the print application being evaluated.*

Test Element

FIRST recommends printing vignettes in varying line screens on all print decks during the fingerprint trial to evaluate the mechanical capability of each deck. Multiple line screens are used to identify the optimum line screen for a vignette given the set-up of the print deck (ink metering system, anilox, mounting tape, plate durometer, ink chemistry, substrate, etc.). Typically vignettes print with a coarser line screen than process images in order to stay clean and print smoothly at production speeds. Evaluate vignettes visually under standard viewing conditions. All press variables should be documented. Refer to Section 19.1.4 for more information on standard viewing conditions.

19.3.4 Bar Codes: Minimum Size and Bar Width Reduction

Description

There are two major variables that the printer is responsible for specifying to the designer and prepress provider: the minimum size (or magnification) for a particular symbol, and the corresponding bar width reduction (BWR). These variables are unique to each print condition, and are therefore the printer's responsibility to identify and communicate. Additional information on bar code scanning ability and verification can be found in Sections 4.3, 12.4, 19.1.3 and 22.0. For more information regarding 2D Bar Codes reference Appendix G.

Industry Standard

GS1 (formerly the Unified Code Council) specifies bar code symbol sizes in an acceptable range for scanning applications. Printing a bar code below the minimum size specified by the applicable symbol specification is not supported by *FIRST*. Some symbols also have a fixed relationship between height and width, while others have a minimum height specification. Reduction in symbol height below the application standard or symbol specification is known as truncation and is not supported by *FIRST*.

For corrugated UPC symbols, GS1 specifies a minimum 150% magnification and maximum 200% magnification for all containers that will be scanned on automated conveyor lines during distribution.

For Code 25 (formerly ITF-14) symbols, GS1 has provided several specifications specific to usage and container size. The nominal size specified for Code 25 symbols carrying the SCC-14 number is based on an X-dimension of 0.040" (1.0mm) and a height of 1.25" (31.8mm). Generally, the 100% magnification is recommended; however truncation may be necessary depending on height restrictions. The current magnification range for Code 25 symbols is 70% to 120%. For containers to be scanned on automated conveyor lines, the minimum height is fixed at 1.25", with a maximum X-dimension of 0.040" (1.0mm), or 100% magnification.

Print Challenge

Each bar code symbology (UPC-A, Code 25, Code 128, etc.) has a unique bar-to-space width relationship. The width of the narrowest bar (the "X-dimension") varies based on the symbol's magnification. Similar to the relationship between dot gain/TVI and line screen, the necessary BWR will increase as bar code magnification decreases. On-press challenges include controlling bar growth and minimizing pinholing and dirty print.

Print Variables

Variables influencing the optimum BWR include substrate absorbency, anilox volume, ink metering system, mounting tape compression, plate durometer and ink chemistry (viscosity, pH, strength, etc.). These variables must be identified during a print optimization or fingerprint trial.

Test Element

Minimum symbol magnification and optimum BWR for each magnification can be determined with a print trial. When the range of magnifications and BWRs is very broad, it is best to test them during a print optimization trial. When the range of options is narrower (based on previous testing or process control data analysis), a fingerprint trial can be designed to fine-tune the optimum BWR for a given magnification under the required print conditions of the new graphic design.

1. Print symbols with varying magnification and BWR combinations; for example, an optimization trial for UPC codes may include four symbol magnifications. Each magnification might be produced five times at different BWRs for a total of 20 symbols on the test target (as demonstrated in Image 19.3.4).
2. Print symbols in the machine direction (ie. picket fence). *FIRST* does not support running bar codes in the cross direction, as distortion inherent in the printing process may alter bar width, producing unacceptable results.
3. Use a "live" symbol that can be evaluated using an ANSI-based verifier. Refer to Section 19.1.3 for detailed information on using bar code verifiers.
4. The file containing the symbols must be processed under standard conditions (software, output device, resolution, plate material, mounting tape, etc.). The press must also be set-up and run under standard operating conditions for the results to accurately predict the desired outcome on live jobs.
5. Pull multiple samples at both kiss (optimum) and maximum impression.
6. Scan and record the average percent decodability for each magnification/BWR combination at each impression.
7. For each symbol magnification, the BWR that produces the highest percent decodability will be the optimum BWR for that magnification under the given print conditions. Communicate this information to the designer and prepress provider.
8. If a particular magnification does not have an acceptable percent decodability, regardless of BWR, then the magnification is not suitable for the given print conditions.

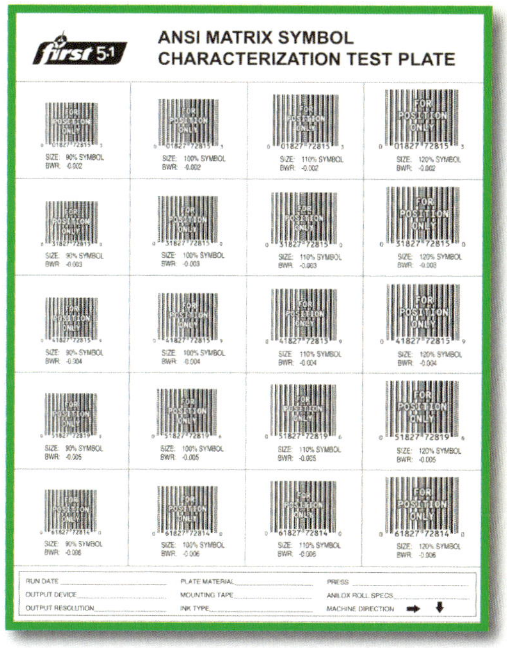

19.3.4 Sample Test Element: *Create a grid of the appropriate symbology. Vary the size of the symbol and the BWR in order to identify the optimum BWR for each magnification.*

Minimum Bar Code Magnification: General Guidelines *Bar code magnification is print system dependent; determine optimum magnification with press fingerprint (ref 1.3.2)*			
Print Segment		**Magnification** *(Machine Direction)*	**Printer Specific Magnification** (*Machine Direction)*
Wide Web	Preprint Linerboard	100%	
	Combined Corrugated *(flute dependent)*	UPC: 110% - 200% ITF -14: 100%	
	Folding Carton	100%	
	Multiwall Bag	115%	
	Film Products	100%	
Narrow Web	Paper Products	80%	
	Film Products	100%	

Table 19.3.4

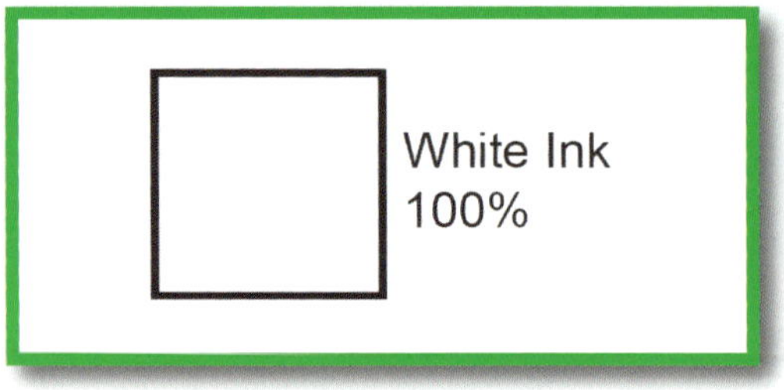

19.3.5 White Ink Test Element

19.3.5 Opacity of White Ink or Substrate

Description

Image quality is influenced by the way light reflects from the substrate, or white ink, under the printed image. When printing on clear or non-white substrates, a background of printed white ink is recommended.

Print Challenge

White ink opacity does not increase linearly with ink volume; small increases in opacity may require a relatively large increase in ink volume. In some cases, a second layer of white ink is required to produce sufficient opacity; this requires two printing decks and presents some significant drying challenges. Ink lay is critical; pinholing poses a particular challenge with white inks.

Print Variables

The primary variables influencing white ink opacity include substrate absorbency, ink strength (pigment load), ink chemistry (viscosity, pH, etc.), anilox volume, mounting tape compression, ink dry rate and plate durometer.

Test Element

A large solid white patch should be included on the press optimization and fingerprint trials as well as on each production run. Use a spectrophotometer to monitor ink contamination, and visually monitor pinholing. Use an opacity meter to measure percent opacity. Alternatively, use a densitometer/ spectrodensitometer to measure white opacity on a black backing using the black density setting; the lower the number, the higher the opacity. Refer to Section 23.3.2 for opacity/transparency testing.

19.4 Process Color: Print Characteristics Measured

When printing a continuous tone image, the primary print characteristics that must be measured and controlled are gray balance (near neutral density), dot gain/TVI, solid ink density and print contrast. Other variables for the printer to optimize and control include print sequence, ink trap, registration, image slur and impression. A detailed explanation of each of these variables is provided in this section.

19.4.1 Achieving Color Balance: *Gray balance is the most critical variable in achieving and maintaining color balance.*

19.4.1 How to Achieve Color Balance

The Relationship Between Gray Balance, Dot Gain/TVI & Solid Ink Density

The recommended method for achieving color balance can be found in FTA's FLEXOGRAPHY: Principles & Practices 6.0. In continuous tone images, gray balance (near neutral density) is the primary characteristic that determines whether or not the printed image will look like the original. Gray balance is controlled by monitoring dot gain/TVI and solid ink density of the printed image; a larger color gamut and gray balance can be achieved when the print process is optimized to run strong inks with thin ink film thicknesses.

A compromise must be made when the printer is unable to achieve target densities while minimizing dot gain/TVI on a given substrate; it is less objectionable to compromise on density than to compromise on dot gain/TVI. If a compromise is necessary, follow these guidelines:

1. Ensure that the hue angle of process inks conforms to the CIELAB values defined in Section 20.2.2. Hue angle can be determined by measuring a solid ink patch for each process color with a spectrophotometer or spectrodensitometer, and should be consistent with hue angles achieved during fingerprint and characterization trials. Refer to Section 19.1.1 for more information on color tolerancing.
2. Process colors must balance with each other for proper image reproduction and neutral gray balance. CMYK densities should remain within +/-0.05 of the target density. If one color has a higher density than the target, the other colors' densities should be increased to match.
3. Minimize dot gain/TVI for each process color and keep dot gain/TVI relatively consistent for all process colors. If dot gain/TVI varies radically, maintaining gray balance will be unlikely, resulting in a color shift in the image.
4. For process printing, the optimal set-up is to run the lowest anilox volume possible for a given substrate and ink formulation, while achieving target solid ink density

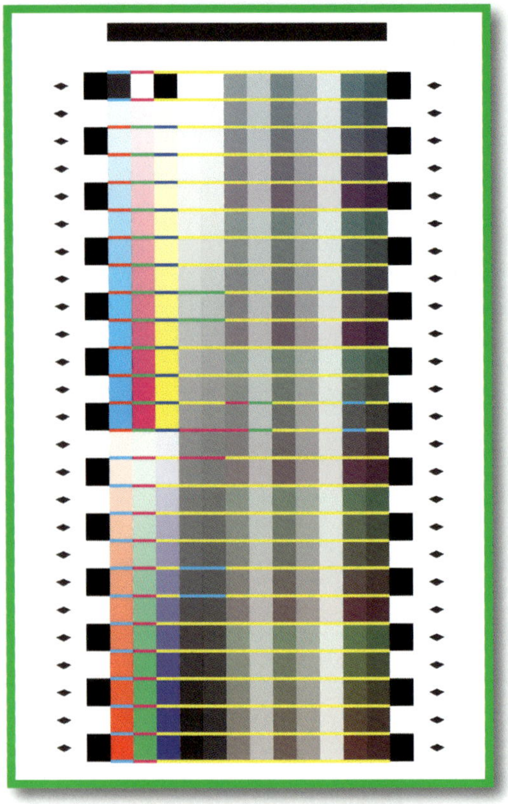

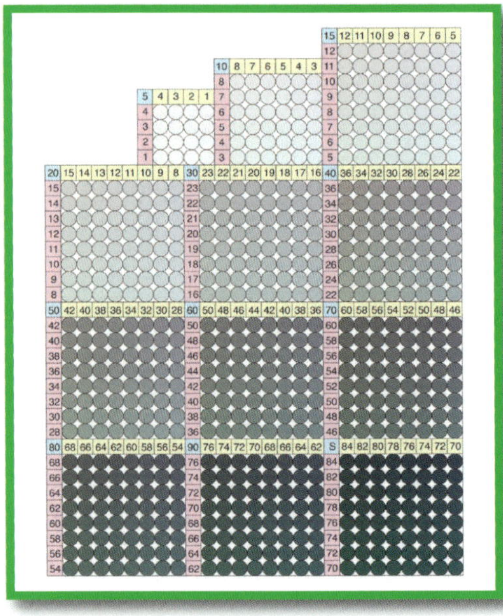

19.4.2 Gray Balance Test Elements: *The target on the top is the P2P25X target which is used in conjunction with the G7™ calibration technique. The target on the bottom is used to manually identify the combinations of CMY that produce a neutral gray throughout the tonal range.*

(SID) and minimizing dot gain/TVI. If target densities are unachievable, work with the ink supplier to re-formulate the ink for a higher pigment load (including adjusting pigment suspension, dry rate and other performance properties). Refer to Sections 20.2.1 and 20.4.2.2 for more information on ink components and anilox cell volume.

5. Once anilox volume and ink strength have been optimized, if SID is still too low, other variables to consider include mounting tape and plate material, as both variables will also impact SID and dot gain/TVI. Refer to Sections 17.0 and 20.5.2 for additional information on printing plates and mounting materials.
6. Once all above variables have been optimized, if SID is still below target, the printer may consider increasing anilox volume slightly; however if this adversely affects dot gain/TVI, it is better to revert to lower volume anilox roll, even if SID remains below target.

In general, larger color gamut on press is achieved with higher solid ink densities, resulting in a closer match to the original image. If target densities are not achievable, and lower target densities are established that maintain balance across all process colors, gray balance will be achieved, allowing the printed image to "match" the original in color balance. Images printed with lower densities will have a smaller gamut, and may appear "washed out," but not out-of-balance. However, this should be an acceptable printed outcome as opposed to running out-of-balance. Running out-of-balance will result in "muddy" print, loss of detail and color shifts when compared to the original.

19.4.2 Gray Balance/Near Neutral Density

Description

Gray balance is the proper combination of cyan, magenta and yellow ink to produce a near neutral gray as measured by a densitometer or spectrodensitometer. The dot percentages required to produce a near neutral gray are print system dependent and must be determined during the fingerprint and characterization trials.

Print Challenge

Gray balance affects all but the most saturated tones in print reproduction. Color shifts will occur when the three colors are not balanced. The optimal dot combination of CMY to produce neutral gray across the tonal range must be identified during the fingerprint and characterization trials, and must be applied to color separations by the prepress provider. The same conditions

Sample Gray Balance Densities & Dot Percentages					
Density	K Dot %	Black (K)	3-Color (CMY)	File C/M/Y Dot %	Density
0.25	25%			25/19/19%	0.22
0.54	50%			50/40/40%	0.50
0.90	75%			75/66/66%	0.90

Table 19.4.2

used during the fingerprint and characterization trials must be used during all production runs to maintain consistent gray balance. Density and dot gain/TVI must be carefully monitored and controlled throughout the run.

Print Variables

The variables that affect gray balance are those that affect density and dot gain/TVI; these variables must be documented during the fingerprint and characterization trials and must be repeated during production runs.

Test Element

During fingerprint and characterization trials, a test element is used to determine the optimal dot percentage of CMY to achieve gray balance across the tonal range. Use a densitometer or spectrodensitometer to determine the combination of CMY tint values that best achieves neutral gray. In production runs, use a smaller target, including ¼-tone, midtone and ¾-tone dot areas. One patch per area is black only, while the adjacent patch is a combination of CMY to match the weight and neutrality of the black patch as determined during the fingerprint and characterization trials. These patches are monitored by the printer visually and using a spectrodensitometer.

19.4.3 Dot Area/Dot Gain/Tonal Value Increase

Description

Accurate tone reproduction is dependent on correct compensation for dot gain/TVI based on the measurements taken during the fingerprint and characterization trials. Dot area refers to the physical area covered by dots of a color plus optical effects that cause the dots to appear larger in size. Tonal Value (TV) is an alternate term for dot area. Dot gain/TVI refers to the difference between the dot area on the plate and the printed dot area. Tonal value increase (TVI) is an alternate term for dot gain. The printer must achieve consistent dot reproduction in order

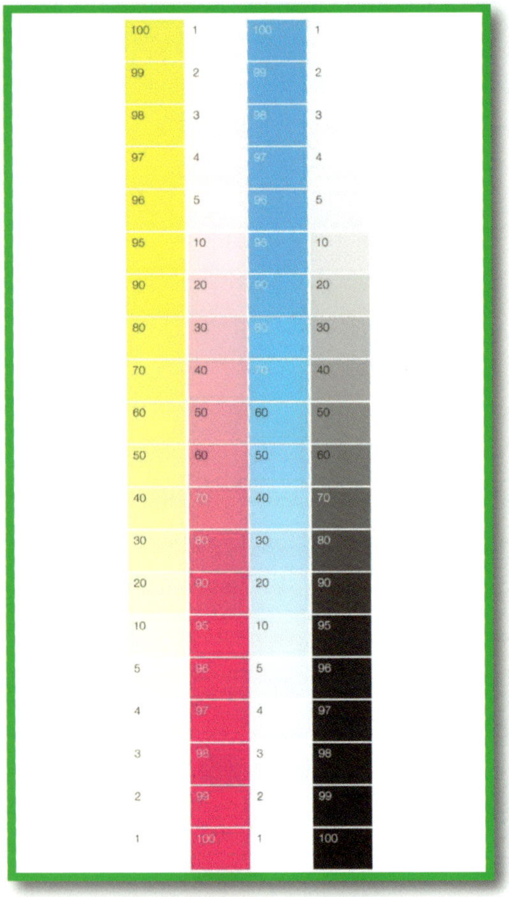

19.4.3a Dot Area Test Element: *Typically, multiple dot area scales, at varying line screens, are included on the fingerprint trial to identify the optimum line screen for the print condition.*

to match the contract proof. Measuring and controlling dot area and dot gain/TVI enables the printer to achieve similar results during production runs as established during fingerprint and characterization trials.

Print Variables

Many print variables influence dot gain/TVI, such as line screen, dot shape, substrate absorbency and smoothness, ink film thickness, ink viscosity, mounting tape compression, plate durometer, plate thickness and imaging method. To minimize dot gain/TVI, these variables must be optimized and controlled. For more information on line screen and dot shape refer to Sections 15.2 and 15.3. Refer to Section 17.0 for more information on printing plates. For more information on substrate refer to Section 20.1. Refer to Sections 20.2 and 20.4 for more information on ink, anilox rolls and metering systems. For more information on mounting tapes refer to Section 20.5.2.

Measurement

There are two basic approaches to measuring a tint patch. Both methods are outlined in ANSI/CGATS.4 2006 (Graphic Technology – Graphic Arts Reflection Densitometry Measurements – Terminology, Equation, Image Elements, and Procedures). Dot % is determined by measuring a tint patch using a densitometer/spectrodensitometer in the dot area or dot gain/TVI function, set to the Murray-Davies formula. Density measurement can also be used by measuring the tint patch with a densitometer/spectrodensitometer with the density function set to "absolute" using Status T wide-band spectral response.

Test Element

Dot area and dot gain/TVI are measured using tone scales, which contain patches of known tint values from highlights, midtones and shadows. During the optimization, fingerprint and characterization trials, use an extended tone scale containing patches for each process color. Typically, these tone patches include a larger number of tints throughout the tonal range, which are measured to determine appropriate dot gain/TVI curves.

During production runs, it is important to include tint patches to monitor dot area and dot gain/TVI, to allow for better control and monitoring of the printed image. At a minimum, for each color, include a solid patch and the tint patch that is most sensitive to pressure adjustments. Include additional patches if more space is available.

FIRST recommends using tone scales to control both the platemaking process as well as the printing process. To facilitate process control in platemaking, tone scales can be imaged two ways: "linear," where the desired values are achieved on the printed plate; and "adjusted," where the values are different from the input digital values due to a purposeful adjustment. Regardless of the approach, if a patch is labeled with a value indicating its dot area, it should match the final plated dot area.

For the printer, it is vital to have a control target imaged consistently with known values on every job, in order to print in a consistent repeatable manner. The values for the desired plate tint patches should be discussed and agreed upon with the printer.

Printer (Linear) Scale: The Printer Scale is used by the printer to monitor and control press settings, components and materials. The Printer Scale is labeled with known finished plate values, rather than input (mask or film) values. For example, a plated value labeled "30%" must measure ~30% (within acceptable tolerances) on the plate after exposure, processing and finishing. These "linear" values provide the printer with the necessary tool to monitor and control dot reproduction regardless of platemaking variables. The Printer Scale must be included on all trial runs to facilitate proper press set-up and run conditions.

Prepress (Adjusted) Scale: The Prepress Scale is created under the same conditions as the live image, including any curves and adjustments that are applied to the live image. Therefore, this scale represents the live image, and is used by the prepress provider, platemaker and printer to confirm the appropriate compensation has been applied to the live image. While printing this scale is optional, the Prepress Scale must be plated in order to be measured on the finished plate as part of the plate's CoA.

Labeling Tone Scales: Each type of scale must be clearly labeled in the file and on the plate to avoid confusion. The printer may choose to have values for each patch placed in the file above or below the patch for clarity. There are two approaches to labeling tint patches and either (or both) may be used:

1. Label each patch with the plated tint value. *FIRST* recommends this approach for the Prepress Scale because it is a QC tool for the prepress/platemaking process.
2. Label each patch with the expected printed tint value. For example, if the 30% patch is expected to print as a 55% on press, label the 30% plated tint patch as "55%." The expected printed tint value is determined during the fingerprint trial.

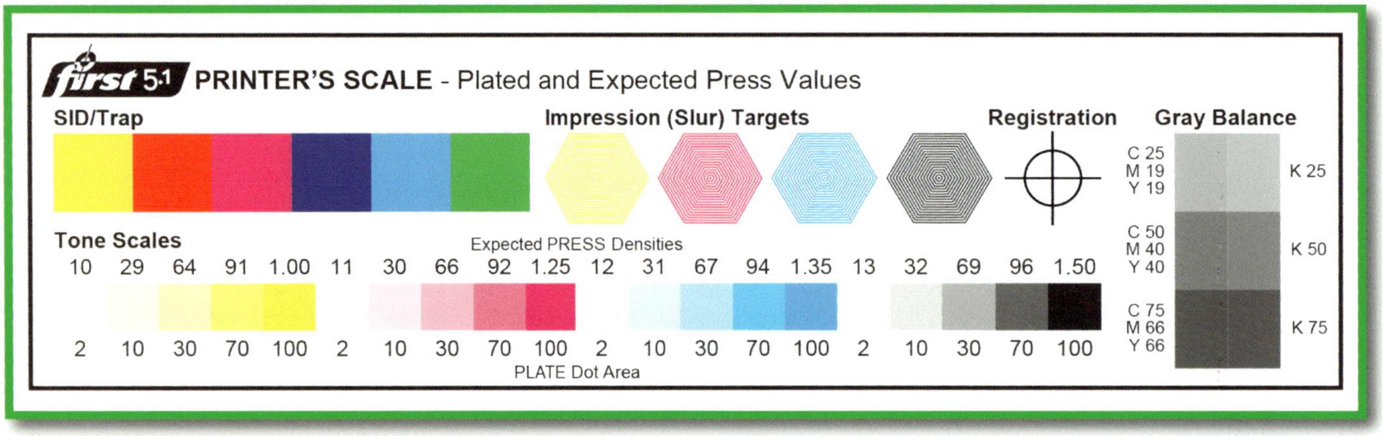

19.4.3b Printers Control Target: *The critical difference between the Printer and Prepress control target is the dot size placed in the file (and output onto the plate) for each dot area patch.*

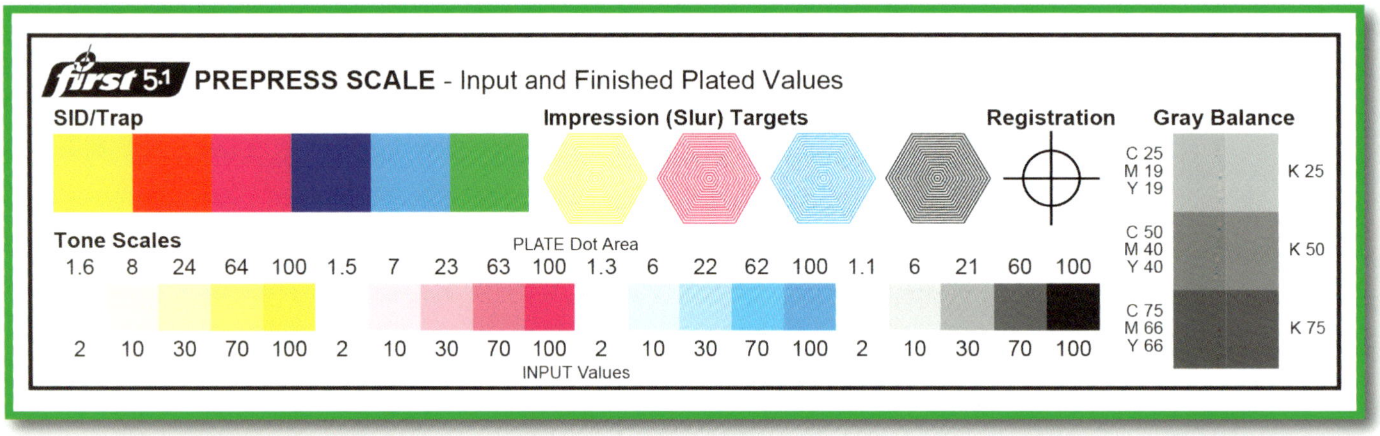

19.4.3c Prepress Control Target

Tone Scale Placement: Ideally, a Printer Scale should be placed on both sides of the web so the press operator can control pressure settings across the press. *FIRST* recognizes that this is not always possible. Typically, the Prepress Scale is located in a non-print area of the plate, and removed prior to printing; only one is required on each plate.

19.4.4 Solid Ink Density

Description

Solid Ink Density (SID) is a measure of the light absorbing property of a printed ink. It is the reflected value of a solid area of color measured with a densitometer or spectrodensitometer using a specific filter with a specific light source. A higher density indicates more light is absorbed; therefore, a darker surface. The *FIRST* suggested solid ink densities in Table19.4.4 represent a reasonable starting point and range for flexographic printed materials.

Solid Ink Density: Starting Point *Density is print system dependent; determine optimum density during press optimization and fingerprint trials (ref 1.3.1 & 1.3.2)*												
Ink	Cyan			Magenta			Yellow			Black		
	Min	**Target**	Max	Min	**Target**	Max	Min	**Target**	Max	Min	**Target**	Max
Paper Products	1.30	**1.35**	1.40	1.20	**1.25**	1.30	0.95	**1.00**	1.05	1.45	**1.50**	1.55
Film Products	1.25	**1.30**	1.35	1.15	**1.20**	1.25	0.95	**1.00**	1.05	1.35	**1.40**	1.45
Newsprint	0.95	**0.97**	0.99	0.93	**0.95**	0.97	0.77	**0.79**	0.81	1.03	**1.05**	1.07
Printer Specific	-0.05 (target)		+0.05 (target)	-0.05 (target)		+0.05 (target)	-0.05 (target)		+0.05 (target)	-0.05 (target)		+0.05 (target)

Process for Achieving Color Balance (How to determine printer specific density targets)		
Work with the Ink Supplier to:	1st Priority = Specify Each Ink Hue Value	2nd Priority = Check Gray Balance
	3rd Priority = Confirm Dot Gain	4th Priority = Obtain Solid Ink Density (no pinholes)

Table 19.4.4

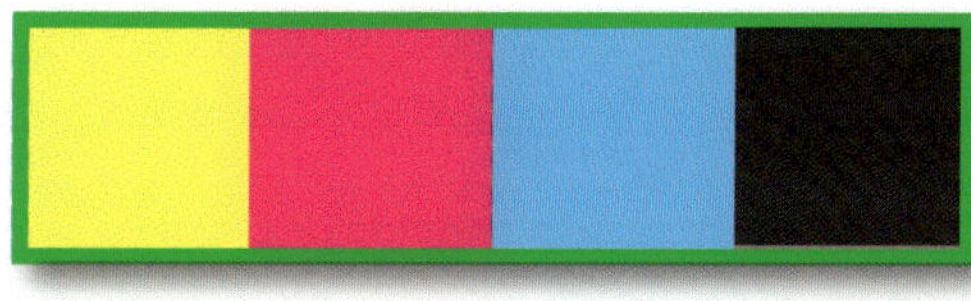

19.4.4 Solid Ink Density Test Element

Application

The key to consistency is to match the density and dot gain/TVI values established during the press fingerprint trial and achieved during the characterization trial. By running to these values, the printer is able to consistently match the contract proof, assuming the prepress provider applies the correct characterization data to the job.

Print Challenge

Process-only print decks should be set-up to minimize dot gain/TVI. However, this makes printing a smooth strong solid very challenging. Achieving *FIRST* suggested densities is secondary to minimizing dot gain/TVI. Combination decks should be set-up to balance dot gain/TVI with smooth strong solids. Refer to Section 19.4.1 for more information on the relationship between SID, dot gain/TVI and gray balance.

Print Variables

Most variables influencing SID also influence dot gain/TVI, including anilox volume, mounting tape compression, plate durometer, substrate absorbency, ink strength and viscosity. Remember: on process decks, minimal dot gain/TVI is more important than achieving *FIRST* recommended SIDs. On combination decks, a balance between low dot gain/TVI and high SID should be achieved. Press speed can also affect SID. Record press speeds during the optimization and fingerprint trials and print at that speed during production press runs.

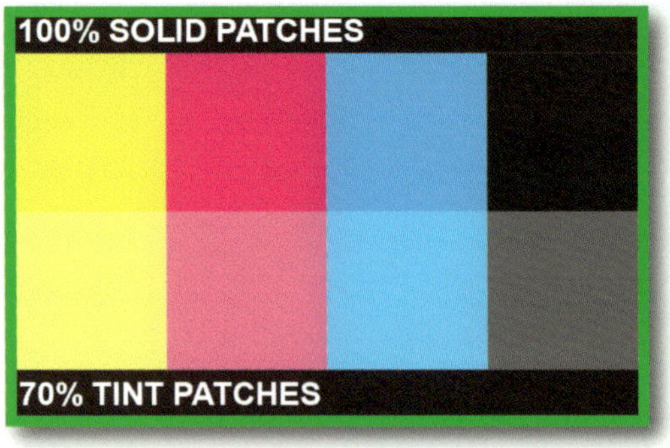

19.4.5 Print Contrast Test Element

Test Element

The SID patches should represent the value used for maximum solid ink coverage for each ink color in an image, including any plate surface modification. Typically this is 100%. The SID patch and at least one dot patch for each color are the minimum test elements required on every process job.

19.4.5 Print Contrast

Description

Print contrast indicates the ability to hold shadow detail, expressed as a percentage. The most desirable print contrast occurs when dot gain/TVI is minimized and high solid ink densities are achieved. It is calculated by the following equation:

$$\%\ PC = \frac{D_S - D_T}{D_S} \times 100$$

Where D_S is the printed solid density and D_T is the density of the 70% tint.

Application

Print with the highest print contrast possible, while balancing any variation in print contrast between process colors (within 5% of each other). Table 19.4.5 provides print contrast percentages as starting points for process printing.

Print Challenge

Keeping shadow detail open can be very challenging, because there is less room for dot gain/TVI before the dots bridge and begin to print solid. High SID and low dot gain/TVI is key to good print contrast. Low densities and high dot gain/TVI compromise print contrast because the difference between 70% tint and the solid is smaller.

Print Variables

Print contrast is affected by the same variables that affect dot area and density, such as substrate absorbency and smoothness, ink strength, anilox volume, metering system, mounting tape compression, plate durometer, plate thickness, imaging method, line screen, dot shape and general press condition. Refer to Sections 19.4.3 and 19.4.4 for detailed discussions on the variables influencing density and dot area.

Test Element

Include a solid patch (or the tint value used for maximum solid ink coverage, including plate surface modification) and a 70% tint patch for each process color to measure print contrast.

Print Contrast: Starting Points *Density is print system dependent; determine optimum density with press fingerprint trial (ref 1.3.2)*				
Ink	C Minimum	M Minimum	Y Minimum	K Minimum
Starting Point	>/= 20%	>/= 20%	>/= 15%	>/= 25%
Printer Specific				
Process for Achieving Color Balance (How to determine printer specific print contrast)				
Work with the Ink Supplier to:	1st Priority = Specify Each Ink Hue Value			
	2nd Priority = Achieve Gray Balance			
	3rd Priority = Minimize Dot Gain			
	4th Priority = Maximize Solid Ink Denisty			

Table 19.4.5

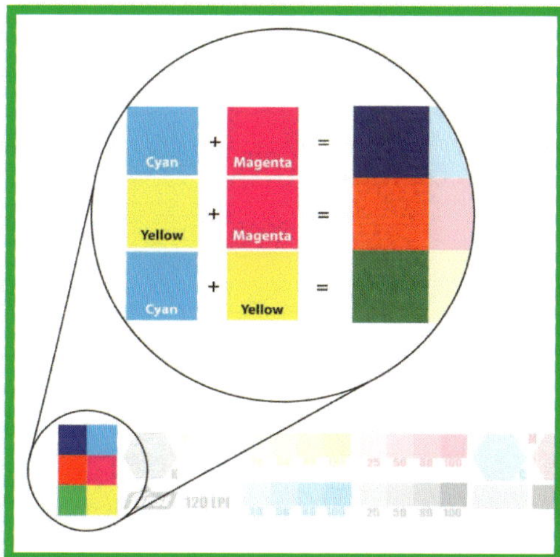

19.4.6a Ink Trap: *This graphic illustrates how process colors combine to produce the overprint colors red, green, and blue.*

19.4.6 Ink Trap

Description

Ink trap refers to the ability of one ink to lay smoothly over another. This is crucial in process printing, as the process requires each color to overprint the previous colors in order to produce secondary and tertiary colors. Measure ink trap with a densitometer in Status T mode using the Preucil formula:

$$\%\ \text{Apparent Trap} = 100 \times \frac{D_{OP} - D_1}{D_2 - D_O}$$

Where:

D_O = *Density of the substrate*
D_{OP} = *Density of the overprint*
D_1 = *Density of the first-down ink*
D_2 = *Density of the second-down ink*

Application

Generally, a trap value of 80% or better is considered desirable. However the achievable trap will vary based on press conditions. By evaluating trap, the printer can quickly identify ink transfer and substrate problems.

Print Variables

There are several press variables that influence ink trap, including press dryer configuration, substrate, anilox volume, ink metering system, print sequence and ink formulation. The key objective in ensuring high trap values is to ensure that the first-down ink is sufficiently dry before the second-down ink is printed on top.

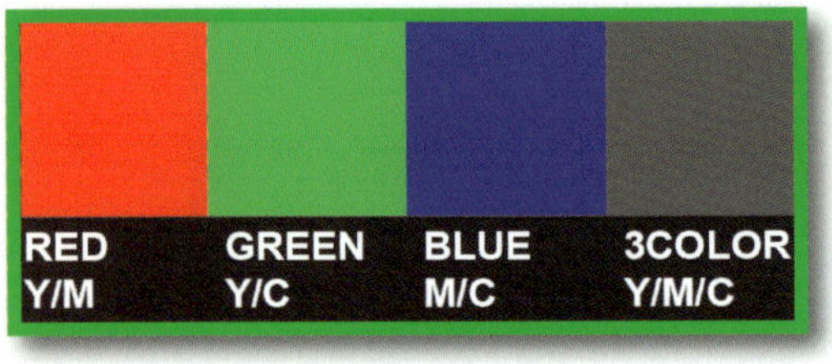

19.4.6b Ink Trap Test Element

19.4.7a Total Image Trap Tolerance: *An overlap of two abutting colors is required to account for inherent movements in press registration during the production run. The amount of overlap, or trap, is print system dependent.*

Press Dryers: Between-deck dryers are critical for good ink trap. Each layer of ink must be able to dry before the next layer is applied. Otherwise the second-down ink will remove the first-down ink from the printed material, resulting in a printed image with "picking." Refer to Section 21.1 for additional information on press dryer configuration.

Substrate: Generally, uncoated stock will produce higher trap values, as some of the ink solvent (including water and amine) will absorb into the substrate, reducing the amount of solvent required to evaporate. On coated stock, if the press dryers are functioning properly and the anilox volume and ink dry rate are optimized, high trap values are possible. Refer to Section 20.1 for more information on typical flexographic substrates.

Anilox Volume & Ink Metering System: With higher-volume anilox rolls, ink film thickness increases. The thicker the ink film applied, the more solvent that must be evaporated by the dryers. Therefore, thicker ink films generally result in slower dry rates and lower trap values. Refer to Section 20.4.2.2 for more information on anilox cell volume.

Ink Formulation: The optimal ink film thickness is the least amount required to print an even solid. Ensuring a proper balance between pigment load and solvent, as well as proper pigment transparency, will improve trap value. Refer to Section 20.2.2 for *FIRST* recommended ink pigments. Refer to Section 23.3 for ink functionality testing including drying, adhesion to nonporous substrates, blocking and rub resistance (all of which affect ink trap).

Print Sequence: Percent ink trap can be increased by skipping a print station between each process color, if the press configuration allows this luxury. This will allow each ink color to dry longer.

Test Element

Two- and three-color overprints in combinations of CM, MY, CY and CMY are used to evaluate ink trap. A solid ink patch for each process color is also required in order to calculate % ink trap. The print sequence must be recorded.

19.4.7 Registration & Total Image Trap Tolerance

Description

Registration is defined as the proper alignment of successive colors and/or images with each other and with the sheet or web. Image trapping is a tool utilized during the prepress process to minimize the objectionable appearance of misregister. It is accomplished through the use of chokes and spreads and should be used when two color are adjacent to each other, whether the graphics are line or screen.

Print Variables

Print system variables that influence image trap include: general press design (width, gear vs. gearless, etc.) and condition (ie. gear wear, print cylinder TIR, parallelism, tension controls and substrates). Printers must first optimize the press and print variables and then determine the specific trap tolerance during optimization trials.

Test Element

There are many different registration marks available for manual or automatic assessment. Register marks belong on the left and right side of the image area, and should be used for every color, both process and spot colors. When all marks are in register, the elements of the printed image should be in register as well. When space allows, *FIRST* supports the use of the RTT (Railroad Track

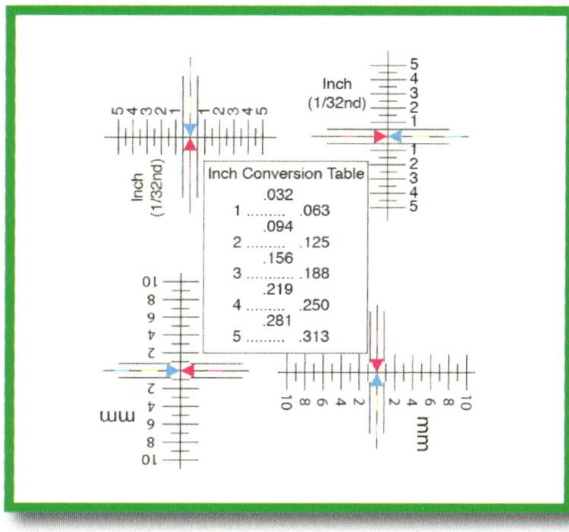

19.4.7b RTT Registration Target

Total Trap Tolerance: General Guidelines *Trap tolerance is print system dependent; determine minimum trap with print optimization and fingerprint trials (ref. 1.3.1 & 1.3.2)*				
Print Segment		**Color-to-Color**		**Printer Specific**
Wide Web	Preprint Linerboard	Total Trap	</= 0.0156" (1/64)	
			</= 0.3969mm	
	Combined Corrugated	Between Station	</= 0.0625" (1/16)	
			</= 1.5875mm	
		Through the Press	</= 0.125" (1/8)	
			</= 3.175mm	
	Folding Carton	Total Trap	</= 0.0156" (1/64)	
			</= 0.3969mm	
	Multiwall Bag	Total Trap	</= 0.0313" (1/32)	
			</= 0.7938mm	
	Film Products	Total Trap	</= 0.0156" (1/64)	
			</= 0.3969mm	
	Newsprint	Total Trap	</= 0.0156" (1/64)	
			</= 0.3969mm	
Narrow Web	Paper Products	Total Trap	</= 0.0156" (1/64)	
			</= 0.3969mm	
	Film Products	Total Trap	</= 0.0156" (1/64)	
			</= 0.3969mm	
	Envelope	Total Trap	</= 0.008" (1/125)	
			</= 0.2032mm	

Table 19.4.7

19.4.7c Traditional Registration Target

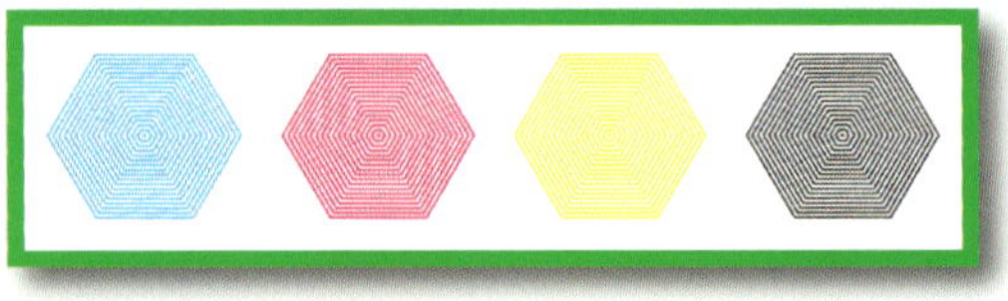

19.4.8a Hexagon Target: *The hourglass pattern indicates over-impression plate-to-substrate.*

19.4.8b Star/Flower Target: *The flower printing solid in the middle indicates over-impression anilox-to-plate.*

Target) registration mark because it quantifies the exact amount of misregister without the use of a tape measure. The size and tolerance should be determined during the fingerprint trial. The objective of the RTT target is to place each color's triangle on the centerline in both lateral and circumferential directions.

19.4.8 Image Slur & Impression

Description

Optimum (kiss) impression is the minimum necessary pressure required to transfer ink from the anilox to the printing plate and from the plate to the substrate. Slur is a blurred printed image, and is caused by the surface of the plate traveling at a different speed than the substrate or anilox roll. Slur and impression targets quickly and easily indicate when impression is not optimal, and when slur exists.

Print Variables

Slur targets indicate plate surface speed discrepancy. Using only the minimum necessary pressure to transfer ink from the anilox-to-plate and the plate-to-substrate is critical to successful flexographic printing. In addition to proper impression setting of the anilox roll and plate cylinder, proper tensions must be set throughout the press to prevent slur, particularly with film substrates.

Test Element

There are two types of impression targets: the hexagon and star/flower target. The hexagon target consists of concentric hexagonal shapes, and is produced in two resolutions. One is used for coarse screen rulings under 85lpi, and the other is used for 100lpi and finer screens. These scales must be used at the appropriate size; reducing/enlarging the scales will alter the line weights and compromise their function. When an hourglass pattern appears in the hexagon, there is too much plate-to-substrate impression. The hexagon target can also indicate ink balance and drying issues on press.

The star/flower target contains "petals" growing in width from the center to the outer edge of the target. This target is useful for monitoring anilox-to-plate impression. Over impression from anilox-to-plate results in the center of the flower filling in. Under impression results in the middle of the flower not printing. The size of the flower and the individual petals should be adjusted for anilox volume. Higher volume anilox rolls require a larger target. The appropriate size target should be determined during the optimization or fingerprint trials.

20.0 JOB-SPECIFIC PRINT VARIABLES

20.1 Substrates

The flexographic process is capable of printing on a multitude of substrates. Substrates can be divided into two major classes: wood-based products and petroleum/mineral based products. Wood products include categories such as: paper, paperboard, linerboard, tissue, newsprint and corrugated. Petroleum or mineral based products include categories such as: film, foil and laminated structures. Each substrate category contains many groups; for example, the film category includes polyethylene, polypropylene, polyester, polystyrene, etc. Each of these groups can be further divided into subsets (ie. polyethylene includes HDPE, LDPE, LLPE, etc.). The appropriate specifications are specific to each class, category, group and subset as well as end-use application.

20.1.1 Substrates: *There are many substrate variables that influence print quality and should be considered when selecting a substrate.*

The substrate chosen for a particular job is primarily determined by the finished product requirements, typically package requirements such as strength, flexibility and protection. However, there are many substrate variables that influence print quality and should be considered during substrate selection. The primary substrate characteristic influencing print quality is consistency. Regardless of the substrate property being evaluated, the consistency of that property throughout the production run and from run-to-run has the greatest impact on the printer's ability to achieve reproducible print results. Control of key performance properties by the substrate manufacturer is critical to the achievable print quality downstream.

20.1.1 Substrate Selection Process

In order to identify the optimum substrate for a given job, the printing and converting characteristics of each substrate must be considered along with the substrate's ability to achieve the required package and product function characteristics. Below, *FIRST* outlines a systematic approach to achieving this objective.

1. Identify the package and product function requirements.
2. Determine the optimum substrate category (ie. paper, corrugated, film, foil, etc.) to achieve the product requirements.
3. Identify the appropriate group and subset of the substrate category (ie. corrugated, bleached white and E-flute).
4. Identify potential suppliers of defined substrate.
5. Request samples, specifications and pricing from each supplier.
6. Identify important substrate attributes influencing printing/converting quality and compare them to

20.1.2 Substrate Properties: *While all of the properties reviewed in this section influence print quality, the primary substrate characteristic influencing print quality is consistency.*

the available products. Suppliers should provide specifications and tolerances for key printing/converting attributes. *FIRST* lists many substrate properties that influence print quality in Section 20.1.2 along with recommended procedures, specifications and tolerances. The specifications and tolerances listed in Section 20.1.2 should be viewed as generally acceptable limits. Tighter specifications and tolerances may be required for specific applications. If the appropriate specification and tolerance for a given application is unknown, it can be determined through optimization trials.

7. Shorten the list of potential suppliers by comparing supplier specifications and tolerances against customer and printer requirements.
8. Utilize optimization trials, sample testing and process variability data (provided by substrate suppliers) to identify the supplier(s) whose product meets the necessary print and end-use requirements with the least amount of variability. The printer may choose to request process variability data both from within a production run and between production runs (ie. sampling shipments of the substrate for several months) to assess both types of variability. If the printer is considering using more than one supplier (or multiple mills/extrusion plants from a single supplier), the printer may choose to evaluate product and data from all potential sites simultaneously to ensure product compatibility and to identify any problems due to site-to-site product variability.
9. Implement a Certificate of Analysis (CoA) or Certificate of Compliance (CoC) program. This is an important part of the process control system. A CoA/CoC report should be included with each substrate shipment. It should verify that the shipment conforms to defined specifications. If a substrate supplier has multiple manufacturing locations (mills, extrusion plants, etc.), determine which manufacturing locations are approved locations. Substrates produced by the same company at different manufacturing facilities must conform to the same specifications. This is critical if the variability introduced by the substrate, which influences print quality and product function, is to be minimized.

20.1.2 Substrate Properties Influencing Print Quality

The substrate properties influencing print quality can be divided into three categories: structural properties, surface properties, and chemical properties. The properties identified below have a significant impact on print quality.

1. **Structural Properties – Influencing Print Quality**
    - Clarity/Haze
    - Dimensional stability
    - Dirt/Gels
    - Flatness
    - Formation
    - Porosity
    - Thickness: Caliper/Gauge
2. **Surface Properties – Influencing Print Quality**
    - Brightness
    - Color
    - Coefficient of Friction
    - Gloss
    - Opacity
    - Smoothness
    - Surface strength/Pick resistance
    - Surface tension/Treat level
    - Wash boarding
3. **Chemical Properties – Influencing Print Quality**
    - Aging/Fade
    - Moisture
    - Sizing

The International Organization of Standardization (ISO) has developed many widely accepted substrate test methods. In addition, the Technical Association of the Pulp and Paper Industry (TAPPI) has developed some of the most widely used paper and paperboard test methods. Several of these are referenced in this section. Refer to Appendix A for ISO and TAPPI contact information. The customer and supplier should establish the specifications and tolerances based on print quality requirements, package function requirements, and manufacturing process capabilities. The specifications defined within *FIRST* should be viewed as generally acceptable limits; tighter specifications may be required for specific applications.

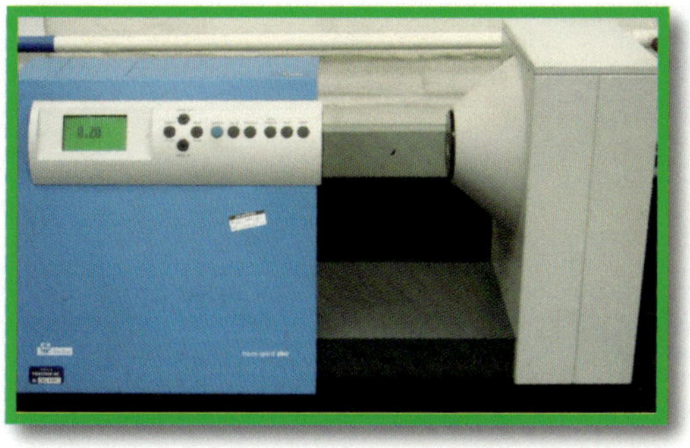

20.1.2.1a Haze Meter: *Haze creates a milky appearance which reduces the clarity of objects when viewed through the substrate.*

20.1.2.1 Structural Properties – Influencing Print Quality

1. **Clarity/Haze**

    Importance

    Haze is the scattering of light by a transparent or translucent substrate. It results in a milky appearance or reduced clarity of objects when viewed through the substrate. Variables influencing haze include: the crystallinity and molecular weight distribution of resins, using copolymers (increase haze), gauge (haze generally increases with film gauge), processing temperatures during extrusion and additives/coatings (can increase haze).

Procedure
ASTM D1003 and Visual Evaluation.

Specification
Established by the customer.

Tolerance
Film: +/- 10%

2. **Dimensional Stability**
   Importance
   Dimensional stability reflects the ability of a substrate to hold its original size, or constant dimension, in all directions when exposed to physical stress or variable moisture. Substrates with poor dimensional stability will not hold color-to-color register. Dimensional stability is especially important when water based inks (which add more moisture) and unit print stations are used (such as corrugated and some envelope). The proper setting of interstation dryers can minimize any influence the added moisture that ink may have on the substrate. Excessive heat levels can dry out the substrate and negatively impact dimensional stability. Paper dimension changes as a percentage of its size (% strain). Therefore, larger dimension papers grow and shrink more, than smaller dimension papers exposed to the same amount of moisture change. Dimensional stability of paper, or board, is also influenced by ambient moisture. Changes in the humidity level (increase or decrease) can cause paper to absorb or lose moisture. The ambient moisture level can influence paper, or board, even after printing if the finished product is not protected from the environment.

   Procedure
   Measure registration using a tape measure.

   Specification
   Established by the customer.

   Tolerance
   Paper: +/- 0.005" (0.127mm)
   Paperboard: +/- 0.005" (0.127mm)
   Corrugated: +/- 0.0625" (1.5875mm)
   Film: +/- 0.010" (0.254mm)

3. **Dirt & Gels**
   Importance
   This defect appears as apparent dirt on the substrate,

influencing its aesthetic appearance. Dirt and gels can create print defects and voids. Size, frequency, color and location are typical criteria used for measuring dirt and gels visually.

Procedure
Paper: Tappi T437 om-96 & Visual Evaluation.
Paperboard: Tappi T437 om-96 & Visual Evaluation.
Corrugated: Visual Evaluation.
Film: Visual Evaluation.

Specification
Established by the customer.

Tolerance
Film: < $2.0mm^2/m^2$

4. **Formation**
   Importance
   Formation is a structural property of paper products. It is a measure of the uniformity of the fiber distribution in the paper. The higher the number, the more uniform the sheet. The "formation value" is reported as "flocs" (clumps of fiber) and "voids" (areas of less fiber). The flocs will absorb less ink than the adjacent voids, producing a non-uniform, mottled, or blotchy print. This is because flocs usually have a tighter pack than the voids. Increasing impression on press will not resolve the blotchy appearance. In process printing, the image will appear grainy especially when printing with higher line screens.

   Procedure
   Paper: Visual inspection (ie. mottled appearance).
   Paperboard: Visual inspection (ie. mottled appearance).
   Corrugated: Visual inspection (ie. mottled appearance).
   Film: Not applicable.

   Specification
   Established by the customer.

5. **Porosity**
   Importance
   Porosity is a measure of the resistance to airflow through a sheet of paper under pressure. It is an indicator of absorbency (penetration of oil and water) and hence the amount of ink that penetrates into the surface of the sheet. It can influence ink absorption, drying and adhesion.

20.1.2.1b Sheet Formation: *Substrates with poor formation result in non-uniform, mottled or blotchy print.*

20.1.2.2a Brightness: *The brightness of a substrate influences the intensity of the printed color and the perception of print contrast.*

Procedure
Paper: Tappi T460 om-96 (Gurley Method – air resistance of paper); T547 (Sheffield Method – air permeance of paper & paperboard).
Paperboard: Tappi T547 (Sheffield Method – air permeance of paper & paperboard).
Corrugated: Monitor ink color match.
Film: Not applicable.

Specification
Established by the customer and measured using a spectrophotometer.

Tolerance
Paper: +/- 10%
Paperboard: +/- 10%
Corrugated: +/- 10%

6. **Thickness: Caliper & Gauge**

Importance
Caliper is the thickness of a single sheet of paper, paperboard, or combined board. Gauge is the term used to reflect the thickness of a single layer of film. Paper is reported in mils or thousandths of an inch, paperboard in points and film in gauge (1 mil = 0.001", 100 gauge = 0.001", 20 point = 0.020"). Thickness is important because wide variations can cause the final print impression to be uneven. For paper substrates, caliper and smoothness are inversely related. Higher caliper papers tend to be rougher while lower caliper papers tend to be smoother.

Procedure
Paper: Tappi T411 om-89; Tappi T551; ASTM D645; ISO 534.
Paperboard: Tappi T411 om-89; Tappi T551; ASTM D645; ISO 534.
Corrugated: Tappi T411 om-89; ISO 3034.
Film: Tappi T411 om-89.

Specification
Established by the customer and measured using a micrometer.

Tolerance
Paper: +/- 5% of target caliper.
Paperboard: +/- 0.001" (0.0025cm) of target caliper.
Corrugated: +/- 0.005" (0.013cm) of target caliper.
Film: +/- 10% of target gauge.

20.1.2.2 Surface Properties – Influencing Print Quality

1. **Brightness**

   Importance

   Brightness is the measurement of blue light (457nm) reflectance. Higher numbers on a 0-100 scale indicate a brighter surface. Most white papers have a brightness of 60%-90%. The brightness of a substrate will influence the intensity of printed color, and the perception of print contrast. High brightness substrates can improve bar code contrast and scannability. Brightness is influenced by the fillers and pigments added during substrate manufacturing as well as by the addition of optical brighteners.

   Procedure

   Paper: Tappi T452 om-98; ISO 2470: 1999.
   Paperboard: Tappi T452 om-98; ISO 2470: 1999.
   Corrugated: Tappi T452 om-98.
   Film: Tappi T452 om-98.

   Specification

   Established by the customer.

   Tolerance

   Paper: +/- 2%
   Paperboard: +/- 3%
   Corrugated: +/- 3%
   Film: +/- 3%

2. **Color**

   Importance

   Whiteness can be defined as a substrate's ability to reflect all colors of light equally. The most brilliant color reproductions are on substrates with high reflectance values balanced across the visible light spectrum (400–700nm).

   Procedure

   Paper: Tappi T524 om-94; Tappi T562; ISO 11475: 2004.
   Paperboard: Tappi T524 om-94; Tappi T562; ISO 11475:2004.
   Corrugated: Tappi T524 om-94; Tappi T562; ISO 11475:2004.
   Film: ISO 5631; ISO 11745:2004.

   Specification

   Established by the customer and measured using a spectrophotometer.

   Tolerance

   Paper: ΔE < 4.0
   Paperboard: ΔE < 4.0

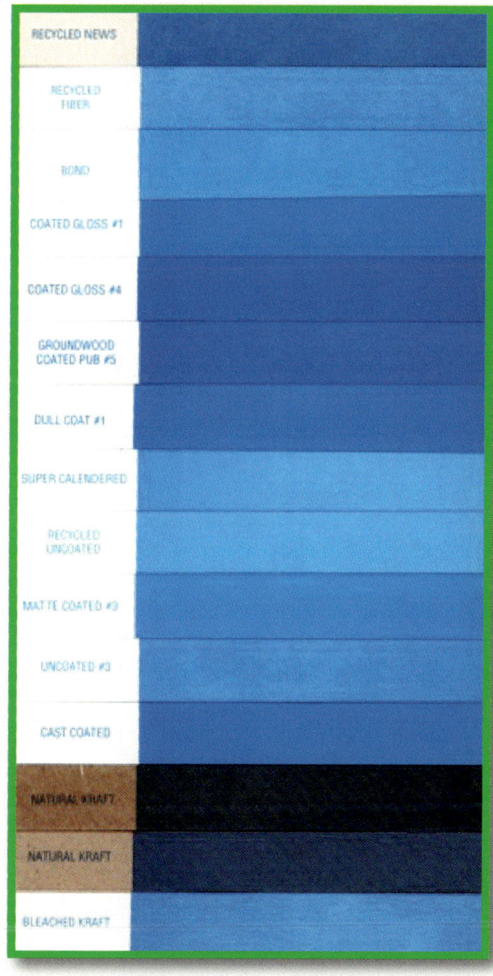

20.1.2.2b Color: *As this ink drawdown over several different substrates illustrates, substrate color significantly influences the color of the printed ink.*

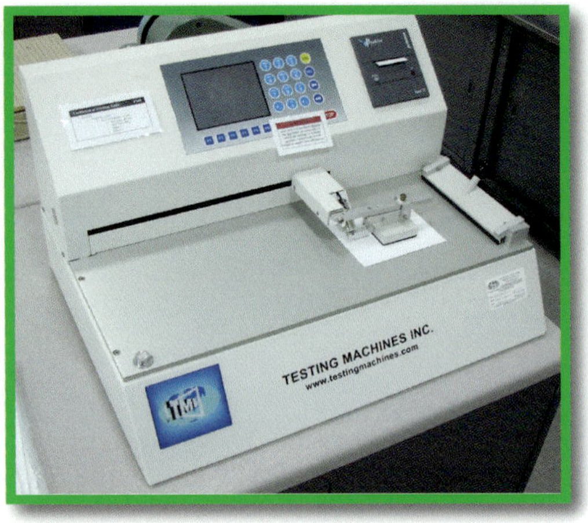

20.1.2.2c CoF: *CoF has limited influence on substrate printability, but it is critical in converting and filling operations and many end-use applications.*

Corrugated: ΔE < 4.0
Film: ΔE < 4.0

3. **Coefficient of Friction (CoF)**
   Importance
   Static CoF measures the force required to initiate movement between two surfaces. Kinetic CoF measures the force required to sustain uniform movement. Modifiers, like waxes, are added to reduce CoF and therefore, increase the ease with which a substrate will move across itself. Colloidal silica is added as an anti-skid treatment to increase CoF. CoF has limited influence on substrate printability, but it is critical in converting and filling operations and many end-use applications. The desirable CoF values vary with product application.

   Procedure
   Paper: Tappi T815 – Inclined Plane Method; T549 – Horizontal Plane Method.
   Paperboard: Tappi T815 – Inclined Plane Method; T549 – Horizontal Plane Method
   Corrugated: Tappi T815 – Inclined Plane Method; T549 – Horizontal Plane Method.
   Film: ASTM D1894-95.

   Specification
   Established by the customer and appropriate print/package application.

   Tolerance
   Paper: None.
   Paperboard: > 18° (inclined plane method).
   Corrugated: > 18° (inclined plane method).
   Film: +/- 30% of target CoF.

4. **Gloss**
   Importance
   Gloss reflects light, like a mirror, and gives the substrate a shiny appearance. It increases with surface smoothness. Printed ink gloss is influenced by substrate gloss. Gloss is reported as a percentage; higher values indicate higher gloss. For most paper substrates, gloss is measured using a 75° angle. A 20° angle is recommended for very high gloss papers. Film substrates typically use a 45° angle for gloss measurement, except for high gloss films which use a 20° angle.

Procedure
Paper: Tappi T480 om-92 (75°); T653 om-98 (20°); ISO 8254-1:1999 (75°); ISO 8254-3:2004 (20°).
Paperboard: Tappi T480 om-92 (75°); T653 om-98 (20°); ISO 8254-1:1999 (75°); ISO 8254-3:2004 (20°).
Corrugated: Tappi T480 om-92 (75°); T653 om-98 (20°); ISO 8254-1:1999 (75°); ISO 8254-3:2004 (20°).
Film: ASTM D2457-08.

Specification
Established by the customer.

Tolerance
Paper: +/- 3% @ 75°
Paperboard: +/- 5% @ 75°
Corrugated: +/- 5% @ 75°
Film: +/- 5% @ 45°

5. **Opacity**
Importance
Opacity obstructs light transmission. Opacity is reported as a percentage; higher values indicate higher opacity. Opacity of paper products is influenced by the degree of fiber refining. Increased refining results in increased opacity. Fillers in paper also increase opacity, or hiding power, of paper products. High opacity minimizes show-through of an image printed on the opposite side of the substrate or the product within the package.

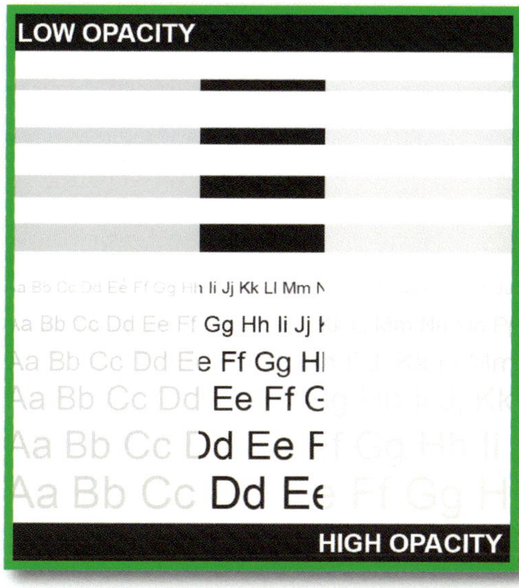

20.1.2.2d Opacity: *High opacity minimizes show- through of an image printed on the opposite side of the substrate or the product within the package.*

Procedure
Paper: Tappi T425 om-96; ISO 2471:1998.
Paperboard: Tappi T425 om-96; ISO 2471:1998.
Corrugated: Not applicable.
Film: None.

Specification
Established by the customer.

Tolerance
Paper: +/- 2% of target opacity.
Paperboard: +/- 2% of target opacity.
Film: +/- 5% of target opacity.

6. **Smoothness**
Importance
Smoothness is a measure of the finish, or texture, of the substrate's surface. It is arguably the most important substrate

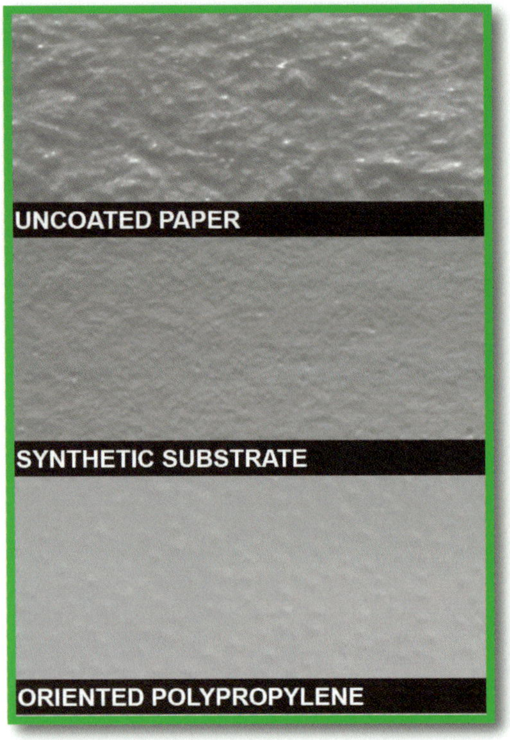

20.1.2.2e Smoothness: *Smoothness is a measure of the finish, or texture, of the substrate's surface. It influences ink lay and ink transfer.*

surface property for print quality because it influences ink lay down and ink transfer. For paper products, the fillers, coating, super calendaring and sizing influence it. With most test methods (air leak methods), the lower the number, the smoother the substrate.

Procedure
Paper: Tappi T538 om-96 (Sheffield Method); ISO 8791-3; T555 om-99 (Print-Surf Method); ISO 8791-4; T575 om-07 (Emveco Method).
Paperboard: Tappi T538 om-96 (Sheffield Method); ISO 8791-3;
T555 om-99 (Print-Surf Method); ISO 8791-4; T575 om-07 (Emveco Method).
Corrugated: Tappi T538 om-96 (Sheffield Method).
Film: None.

Specification
Established by the customer.

7. **Surface Strength / Pick Resistance**
   Importance
   Pick resistance is a measure of the substrate surface cohesive strength versus the force required to split a wet ink film. If the pick resistance is too low, the ink will pull the coating off of the substrate onto the printing plate instead of transferring and adhering to the substrate.

   Procedure
   Paper: Tappi T499 um-591; T459 om-93 (Wax Pick Test); ISO 3783:2006.
   Paperboard: Tappi T459 om-93 (Wax Pick Test); T514 cm-92; ISO 3783:2006.
   Corrugated: Tappi T459 om-93 (Wax Pick Test).
   Film: None.
   Specification
   Established by the customer.

8. **Surface Tension/Treat Level**
   Importance
   Surface tension is a measure of the substrate surface energy that influences ink transfer and adhesion to a substrate. Substrates typically should be 8 to 10 dynes/cm higher than the ink.

   Procedure
   Paper: Polylam = Tappi T552 pm-92 (Mayer Rod Technique).

Paperboard: Polylam = Tappi T552 pm-92 (Mayer Rod Technique).
Corrugated: Tappi T552 pm-92 (Mayer Rod Technique).
Film: Tappi T552 pm-92 (Mayer Rod Technique).

Specification
Established by the customer and ink adhesion requirements.

Tolerance
Paper: Dyne Indicator Standard +/- 2.
Paperboard: Dyne Indicator Standard +/- 2.
Corrugated: Dyne Indicator Standard +/- 2.
Film: Dyne Indicator Standard +/- 2.

9. **Washboarding/Fluting**
   Importance
   Washboarding and fluting are terms used interchangeably. On corrugated substrate, the liners follow the contours of the fluted medium producing alternate ridges and valleys instead of forming a flat, smooth outer surface. Washboarding appears as dark and light parallel lines and often looks worse in images that contain dots. Larger flute sizes, such as C-flute or B-flute, are more prone to washboarding than smaller flutes sizes, such as E-flute. The lighter the weight of the liner used, the more likely that washboarding will occur. Washboarding compromises print quality by creating an uneven surface. Increasing impression on press to overcome washboarding leads to poor print quality, loss of caliper and flute crush.

   Procedure
   Corrugated: Visual evaluation.

   Specification
   Established by the customer.

20.1.2.2f Washboarding: *Larger flute sizes are more prone to washboarding than smaller flute sizes – as are lighter weight liners.*

20.1.2.3 Chemical Properties – Influencing Print Quality

1. **Aging/Fade Resistance**
   Importance
   Aging/Fade Resistance reflects the ability of a substrate to resist changes in its optical, chemical, or structural properties over time. Aging is characterized by a change in the appearance of the substrate, such as yellowing, loss of brightness and/or fading.

Substrate Properties Influencing Print Quality				
Substrate Properties	Substrate Specifications *(Coated and uncoated paper products use similar test methods but typically have different specifications.)*			
	Film	Corrugated	Paper	Paperboard
Aging/Fade	Visual Customer	Visual Customer	Visual Customer T453 sp-97; T544 sp-97	Visual Customer T453 sp-97; T544 sp-97
Brightness	+/- 3%; T452 om-98	+/- 3%; T452 om-98	+/- 2%; T452 om-98 ISO 2470:1999	+/- 3%; T452 om-98 ISO 2470:1999
Caliper/Gauge/ Thickness	Micrometer +/- 10% T411 om-98	Micrometer +/- 0.005" *(0.013cm)* T411 om-98 ISO 3034	Micrometer +/- 10% T411 om-98; ASTM D645 T551; ISO 534	Micrometer +/- 0.001" *(0.0025cm)* T411 om-98; ASTM D645 T551; ISO 534
Clarity/Haze	Customer +/- 10%; ASTM D1003	Visual Customer	Visual Customer	Visual Customer
Color	Customer; Delta E < 4.00 ISO 11745:2004 ISO 5631	Customer; Delta E < 4.00 T524 om-94; T562 ISO 11475:2004	Customer; Delta E < 4.00 T524 om-94; T562 ISO 11475:2004	Customer; Delta E < 4.00 T524 om-94; T562 ISO 11475:2004
Coefficient of Friction	Customer +/- 30%; ASTM D1894-95	Customer; > 18 degrees T815; T549	Customer T815; T549	Customer; > 18 degrees T815; T549
Dimensional Stability	Registration +/- 0.010" (0.025cm)	Registration +/- 0.0625" (0.16cm)	Registration +/- 0.005" (0.013cm)	Registration +/- 0.005" (0.013cm)
Dirt/Gels	Visual < $2.0mm^2/m^2$	Visual Customer	Visual Customer	Visual Customer
Flatness	Go / No Go	Straight Edge < 0.25" / linear foot	Go / No Go ISO 11556:2005	Go / No Go ISO 11556:2005
Formation	N/A	Visual Customer	Visual Customer	Visual Customer
Gloss	Customer +/- 5% @ 45 degrees ASTM D2457-08	Visual Customer +/- 5% @ 75 degrees T480 om-92 ISO 8254-1:1999	Visual Customer +/- 3% @ 75 degrees T480 om-92 ISO 8254-1:1999	Visual Customer +/- 5% @ 75 degrees T480 om-92 ISO 8254-1:1999
Ink Absorbency	N/A	Spectrophotometer Customer; Delta E 4.0	Spectrophotometer Customer; Delta E 4.0	Spectrophotometer Customer; Delta E 4.0
Moisture Content	N/A	5.5%; +/- 2%; T412 om-02	5.5%; +/- 1.5%; T412 om-02	5.5%; +/- 1.5%; T412 om-02
Opacity	Customer +/- 5%	N/A	Customer +/- 2%; T425 om-96 ISO 2471:1998	Customer +/- 2%; T425 om-96 ISO 2471:1998
Porosity	N/A	Customer Ink Color Match +/- 10%	Spectrophotometer Customer; +/- 10% T460 om-96 T547; ISO 8791-4	Spectrophotometer Customer +/- 10% T547
Smoothness	Embossed Smooth	Customer T538 om-96	Customer T538 om-96; T555 pm-94 T575 om-07; ISO 8791-3	Customer T538 om-96; T555 pm-94 T575 om-07; ISO 8791-3 ISO 8791-4
Surface Strength/ Pick Resistance	N/A	Visual Customer	Visual Customer	Visual Customer
Surface Tension/ Treat Level	Dyne Indicator Std. +/- 2; T552 pm-92	Customer +/- 2 T552 pm-92	Polylam = Customer +/- 2 T552 pm-92	Polylam = Customer +/- 2 T552 pm-92
Washboarding/ Fluting	N/A	Visual Customer	N/A	N/A

Table 20.1

Procedure
Paper: Tappi T453 sp-97 (Dry Heat); T544 sp-97 (Moist Heat).
Paperboard: Tappi T453 sp-97 (Dry Heat); T544 sp-97 (Moist Heat).
Corrugated: Visual evaluation.
Film: Visual evaluation.

Specification
Established by the customer.

2. **Moisture Content**
   Importance
   The moisture content of paper directly influences how much ink the paper will absorb during printing. Ambient air moisture will affect the moisture content of the paper. A very moist sheet will require more ink and multiple press adjustments to hold register. Excessive moisture can result in wrinkles and wavy edges making downstream converting difficult. Papers with low moisture content are more susceptible to web breaks on press.

   Procedure
   Paper: Tappi T412 om-02 (Moisture in Pulp, Paper & Paperboard).
   Paperboard: Tappi T412 om-02 (Moisture in Pulp, Paper & Paperboard).
   Corrugated: Tappi T412 om-02 (Moisture in Pulp, Paper & Paperboard).
   Film: Not applicable.

   Specification
   Target = 5.5% Moisture Content for all Paper Products.

   Tolerance
   Paper: +/- 1.5% of target % moisture.
   Paperboard: +/- 1.5% of target % moisture.
   Corrugated: +/- 2% of target % moisture.

3. **Sizing**
   Importance
   Sizing refers to chemicals added to improve the paper's end-use performance. Surface sizing involves applying a light film of starch (or other material) to one or both sides of the web with a size press. Sizing reduces surface linting and increases the resistance of paper to absorbing liquids such as water or ink. It improves ink holdout by slowing the rate of ink

20.1.4 Combination Flute: *When a package requires both increased structural strength and high quality graphics, a combination flute is recommended.*

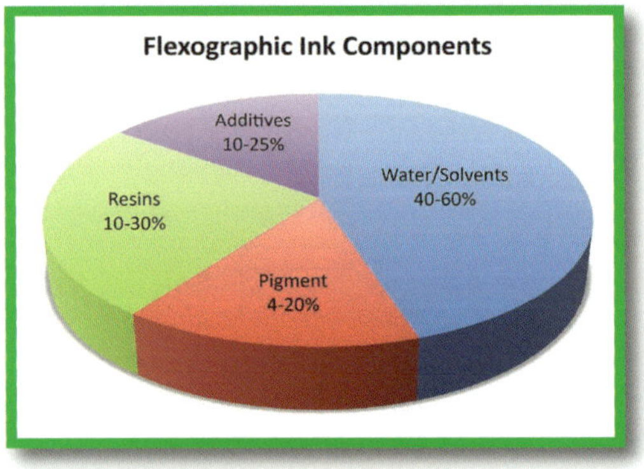

20.2.1a Flexographic Ink Components

absorption into the fiber structure of the sheet. This reduces the risk of wicking, feathering, chalking and print density loss.

Procedure
Paper: Tappi T441 om-98 – Cobb Test; T433 – Dry Indicator Method.
Paperboard: Tappi T441 om-98 – Cobb Test.
Corrugated: Tappi T441 om-98 – Cobb Test.
Film: Not applicable.

Tolerance
Corrugated: <155 grams of pick-up per square meter (30 minute test).

20.1.3 Lamination & Color Matching

Color matching on press, when printing a design that will be laminated in a downstream operation, is particularly challenging. A reverse-printed web, that is part of a lamination, will have a very different color off-press (pre-lamination) then it will after final lamination. There are a couple of ways to simulate the final lamination in order to achieve the desired color match on press:

Option 1: Laminate the reverse-printed web to a similar secondary substrate press-side using a tabletop laminator (designed to laminate paper documents and posters).

Procedure

1. Cut the image to be laminated from the press tear sheet.
2. Place the image on a non-sealable carrier sheet ink side up (paper will work as a carrier in most cases).
3. Cut a piece of secondary substrate (simulate lamination) and place it on top of the printed sample.
4. Cover with a non-sealable carrier sheet and pass through the heated lamination device.

With proper selection of the secondary substrate, press-side lamination should be a close match to the final product and can be used for color matching procedures.

Option 2: Laminate the printed press characterization target on production equipment. By measuring the printed characterization target both laminated and unlaminated, the resulting color shift can be quantified and predicted. The data from the laminated press characterization can then be used to show the expected color shift on a digital proof.

Caution: Dark, wet products, such as soups or meats, can dramatically alter the appearance of the graphics compared to the graphics mounted as a sample on white board stock.

20.1.4 Corrugated Flute Profile Selection

When selecting the optimum flute profile for a package, the strength requirements of the package and the desired graphic quality must be considered. There is an inverse relationship between board strength and print quality. As flute height increases (N to F to E to B to C) board strength also increases. However, print quality decreases. Thinner flutes (ie. E, N and F), tend to minimize the appearance of washboarding or flute lines, resulting in improved appearance of the graphics. When a package requires both increased structural strength and high-quality graphics, a combination flute is recommended. For example, an E-B double wall, in which a printed E-flute board is glued to the outside of a B-flute board, accomplishes both objectives.

20.2 Ink

20.2.1 Components of Ink

Maintaining ink balance on press throughout a production run is critical for a successful and reproducible outcome. Ink chemistry balance determines color, ink transfer and ink lay characteristics, as well as drying, trapping and end-use performance. This Section explains the components of ink and each components function. The components of ink can be generically described as:

- Ink = Pigment + Vehicle
- Vehicle = Resin + Solvent + Additives

Generally, the pigments and additives are not system dependent. With some exceptions, the same pigments and additives can be used across ink systems (ie. water, solvent and UV/EB). The resin and solvent systems are ink system dependent.

20.2.1b Pigments: *Pigments provide the color, or visual identity, of the ink.*

Pigments

Pigments are insoluble colorants that provide the color, or visual identity, of the ink. Pigments also contribute to functional properties such as fade resistance, opacity/transparency and product resistance. The pigment is usually the most expensive component of the ink. A printer's pigment selection determines the achievable color gamut. In Section 20.2.2 *FIRST* recommends pigments by color index name and number for both process and line inks, to maximize the printable color gamut while delivering standard functional properties.

Vehicle

The vehicle components are responsible for carrying the pigment to the substrate and the non-volatile vehicle components provide most of the properties of the finished ink.

20.2.1c Ink Components: *The resins are responsible for binding the pigment to the substrate while the solvents primary function is to convert the other ink components into a fluid form capable of being printed.*

Resins

The resin system is part of the nonvolatile vehicle that is responsible for binding the pigment to the substrate and contributing to the functional properties of the ink (ie. wetting, adhesion, resistance properties, gloss, ink transfer, solvent release properties, flexibility and toughness). Most formulations employ a combination of resins to achieve the desired properties.

Typical Resin/Solvent System Examples

There are many different chemistries that can be used when formulating ink. This table includes some of the more common materials currently in use.

Vehicle	Water Ink	Solvent Ink	UV/EB Ink
Resin	Acrylic	Nitrocellulose & Polyamide	Oligomers (Acrylic Based)
Solvent	Water & Amine	Alcohol & Acetate	Monomers

Table 20.2.1a

Solvents

Solvents comprise the "liquid" part of the ink. The primary function of the solvent is to convert the pigments, resins and additives (all solids) into a fluid form capable of being printed. Solvent selection is critical in determining the performance of the ink. The proper solvent balance will solubilize the resin(s) used, easily evaporate or absorb, impart minimal odor, aid substrate wetting and adhesion, not influence the printing plates or rollers, and interact minimally with other ink ingredients. Most ink formulations utilize a combination of solvents to achieve the desired balance of solubility, rheology and drying speed for a given resin system. An incompatible resin-solvent combination can result in a loss of gloss, reduced adhesion, poor printability, increased ink viscosity and reduced product resistance. When the solvents and resins are incompatible, it is referred to as "ink souring" or "kick out".

Different resin formulations require different solvent combinations and ratios. Different solvents evaporate at different rates. Therefore, it is critical to maintain solvent balance on press. If inks are not managed properly on press, they will gum up in the decks, the print will become streaky (because the pigment is kicking out) and drying problems will occur.

Solvent Ink Systems: Balance is maintained by measuring viscosity and adding the appropriate blend of solvents to

Typical Solvents		
Ink Chemistry	Solvent	Comments
Solvent	N-Propyl Acetate	Solvent and water ink chemistry is designed to work as wet ink only. Failure to remove nearly all in drying will typically result in reduced end use and processing properties.
	Ethyl Alcohol	
	N-Propyl Alcohol	
	Glycol Ether	
Water	Water	
	Ammonia	
	Amine	
UV/EB	Monomers	Mostly non-volatile, reactive component, interacts with the resins/oligomers and influences final properties.

Table 20.2.1b

keep on-press viscosity within the specified range. The solvent blend may need to be adjusted as temperature and humidity increase in the summer to compensate for accelerated evaporation and water absorption.

Water Ink Systems: Ammonia and/or organic amines, combined with water, function as the "solvents" of water based inks. The alkalinity of the water based ink system determines its performance. Generally, maintaining a level of alkalinity between pH 8.5-9.5 is necessary to control the performance of ink on press. Therefore, pH is the primary ink variable monitored and controlled on press for water based ink systems. The ink supplier should provide specific recommendations for optimum pH control. Viscosity should not be adjusted until the proper pH is achieved because these variables are connected and can have an inverse relationship. Higher viscosity = lower pH; therefore, as the ink is adjusted to increase pH the ink viscosity will drop. For uncoated papers and corrugated, inks systems are available that operate in a lower pH range or do not typically require pH adjustment on press.

UV/EB Ink Systems: The monomer (reactive diluent) is similar to a solvent in its ability to thin down the ink. Monomers are typically acrylic based and help to determine the characteristics of the ink such as gloss, hardness and flexibility.

Typical Additives	
Additive	Description
Plasticizers	Make the dried ink more flexible and elastic. They can also improve gloss and adhesion.
Waxes	Improve scratch and rub resistance, reduce blocking or set-off, improve slip and water repellency. Excess wax may result in reduced gloss, poor ink rheology, and reduced transfer characteristics.
Silicones	Used as substrate or pigment wetting agents. Improves scratch, rub and slip resistance, antifoams, and release agents. Excessive use can cause pinholing.
Surfactants	Used to improve wetting and spreading. Too much surfactant can cause problems with foaming, adhesion, and reduced water resistance.
Defoamers	Foaming is a problem that typically occurs in water based inks. Defoamers dramatically reduce surface tension in ink, causing existing bubbles to burst and prevent stable foams from forming.

Table 20.2.1c

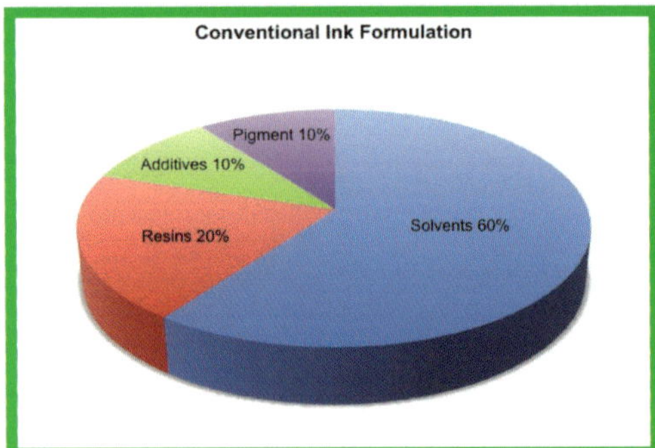

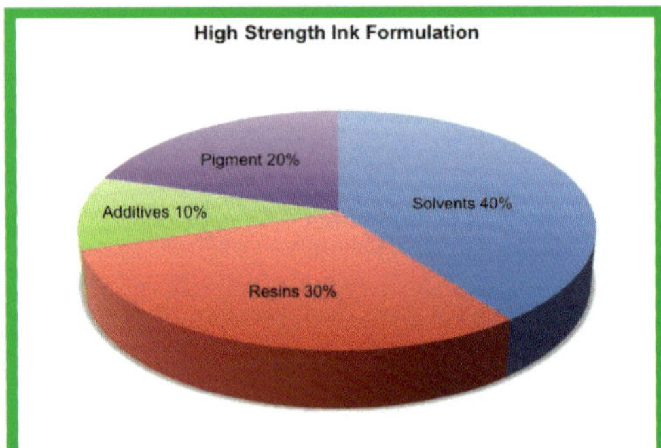

20.2.2a Conventional vs. High-Strength Ink Formulations: *High-strength inks, formulated with a higher pigment load, will increase the printable color gamut – within limits. It is important to conduct print optimization trials to determine the optimum press set-up and ink formula combination*

Additives

Additives enhance, or modify, the ink characteristics to achieve necessary performance attributes. If a vehicle is optimally formulated, the use of additives will be minimal. Over-use of additives can cause secondary problems.

20.2.2 *FIRST* Recommended Pigments

FIRST endorses the use of the color index number system (CI #) to identify and specify pigments used in ink dispersions and formulations. Specifying the CI # of the pigments used to create a finished ink:

- Improves the ability to match color across platforms (proof-to-press, press-to-press, plant-to-plant)
- Reduces the occurrence of metameric color matches
- Results in improved consistency of the printed product

The color gamuts produced by the *FIRST* recommended process ink and line dispenser pigments can be fully quantified for use with a color management workflow. Refer to Section 14.4 for additional information. While the pigment CI # represents a specific and unique chemical structure, some variation in color will occur from vendor to vendor and from grade to grade. Differences in manufacturing equipment and formula content from ink supplier to ink supplier may produce some differences in color as well. These differences are usually negated by using the recommended pigments with slight adjustments in the ratios of pigments in the ink formulation.

While the *FIRST* recommended pigments represent a small number of pigments currently available, they are a good cross-section of the pigments used by the majority of flexographic printers as determined by reviewing extensive data from leading

FIRST Recommended Process Ink Pigments						
Color	Water Based Ink		Solvent Based Ink		UV Based Ink	
	C.I. Name	C.I. #	C.I. Name	C.I. #	C.I. Name	C.I. #
Yellow	Y14	21095	Y14	21095	Y13	21100
Magenta	R57:1	15850:1	R57:1	15850:1	R57:1	15850:1
	R52:1	15860:1	R52:1	15860:1	R52:1	15860:1
Cyan	B15:3	74160	B15:3	74160	B15:3	74160
	B15:4	74160	B15:4	74160	B15:4	74160
Black	K7	77266	K7	77266	K7	77266

Table 20.2.2a

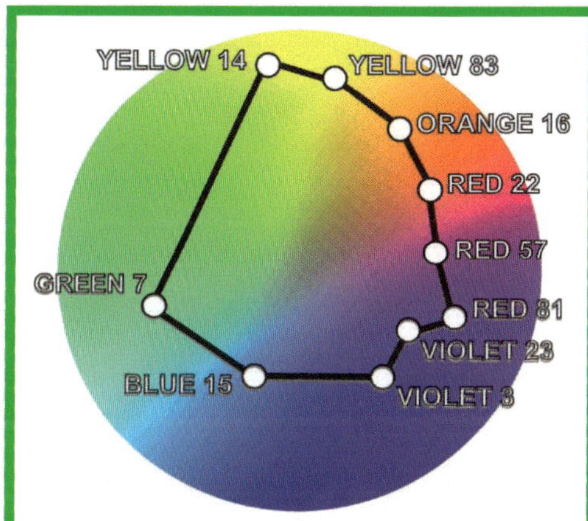

20.2.2b Line Ink Color Gamut: *This graph illustrates the achievable color gamut using FIRST recommended line ink pigments.*

FIRST Recommended Process Ink Pigment Color Values						
Color	Ink Chemistry	L*	a*	b*	C*	h°
Yellow	Solvent	90	- 4	111	111	92°
	Water	92	- 6	101	101	93°
	UV	92	- 5	83	83	93°
Magenta	Solvent	46	72	9	72	7°
	Water	53	70	- 4	70	357°
	UV	50	76	- 1	76	359°
Cyan	Solvent	56	- 41	- 47	62	229°
	Water	59	- 37	- 45	59	230°
	UV	56	- 38	- 45	59	230°
Black	Solvent	21	1	3	3	N/A
	Water	16	0.8	1	1	N/A
	UV	16	0.2	5	5	N/A

Table 20.2.2b

ink suppliers, pigment manufacturers and printers. The *FIRST* recommended line dispenser and process pigments were selected to achieve these objectives:

1. Provide the largest printable color gamut, on various substrates, with a minimum number of pigments.
2. Identify pigments that are compatible with the primary flexographic ink systems (solvent, water, UV/EB).
3. Provide pigment selections with moderate-to-enhanced fade resistance.
4. Identify reasonably priced pigments.

***FIRST* Recommended Link Ink Dispenser Pigments**					
Pigment Name	C.I. Name	C.I. #	Shade	*FIRST* Process Color	Description
Lithol Rubine	Red 57:1	15850:1		Yes	Blue Shade Red
YS Napthol	Red 22	12315		N/A	Yellow Shade Red
YS Rhodamine	Red 81	45160		N/A	Bright, Very Clean, Blue Shade Red
Dianisidene Orange	Orange 16	21160		N/A	Clean Orange
AAOT Yellow	Yellow 14	21095		Yes	Green Shade Yellow
HR Yellow	Yellow 83	21108		N/A	Red Shade Yellow
Phthalo Green	Green 7	74260		N/A	Blue Shade Green
Phthalo Blue	Blue 15:3	74160		Yes	Green Shade Blue
	Blue 15:4	74160		Yes	Green Shade Blue
Methyl Violet	Violet 3	42535:2		N/A	Blue Shade Violet
Carbazole Violet	Violet 23	51319		N/A	Red Shade Violet
Titanium White	White 6	77891		N/A	Opaque White
Carbon Black	Black 7	77266		Yes	Untoned Black

Table 20.2.2c

***FIRST* Recommended Pigment Properties**							
Pigment Name	C.I. Name	Ink Type			Lightfastness	Heat Resistance	Bleed Resistance
		Solvent	UV	Water			
Lithol Rubine	Red 57:1	Yes	Yes	Yes	2.5	4	4
BON Red	Red 52:1	Yes	Yes	No	2.5	4	4
YS Napthol	Red 22	Yes	Yes	Yes	3	4	5
Dianisidene Orange	Orange 16	Yes	Yes	Yes	2.5	2	5
MX Yellow	Yellow 13	Yes	Yes	Yes	2	2	4
AAOT Yellow	Yellow 14	Yes	Yes	Yes	2.5	4	4
HR Yellow	Yellow 83	Yes	Yes	Yes	2.5	5	4
Phthalo Green	Green 7	Yes	Yes	Yes	5	5	5
Phthalo Blue	Blue 15:3	Yes	Yes	Yes	5	5	5
	Blue 15:4	Yes	Yes	Yes	5	5	5
Methyl Violet	Violet 3	Yes	No	Yes	2	3	2
Carbazole Violet	Violet 23	Yes	Yes	Yes	5	4	5
Titanium White	White 6	Yes	Yes	Yes	5	5	5
Carbon Black	Black 7	Yes	Yes	Yes	5	5	5
Ratings Scale: 1 to 5 (5=Best)							

Table 20.2.2d

FIRST recognizes the need for high performance pigments in specialty applications. Section 20.2.3 identifies specific pigments for enhanced light fastness and other properties.

The pigment properties chart (Table 20.2.2d) is a guide for linking pigment performance to end-use requirements. Include the ink supplier in the initial product design discussions to ensure pigment properties correlate to the product requirements. In addition to pigment selection, substrate and ink strength also play a critical role in determining the achievable color gamut. High-strength inks, formulated with a higher pigment load in order to print thinner ink films while maintaining the desired color saturation, will increase the printable color gamut (within limits). A thinner ink film enables the printer to incur less dot gain/TVI, which expands the tonal range and thereby achievable color gamut. However, there are limiting factors; high-strength inks require slower drying solvent blends to compensate for the thinner ink film thickness; otherwise, the ink will dry too fast. Therefore, conduct print optimization trials to determine the optimum anilox engraving and ink strength/dry rate combination. Refer to Section 1.3.1 for detailed information on print optimization.

The consumer product company (CPC) and the design firm should consult with the printer and ink supplier to determine if a particular color is achievable with a specified substrate or lamination structure. Image 20.2.4 illustrates the color gamut achieved with *FIRST* recommended pigments. Color gamuts are created by plotting the hue and chroma values of each pigment on a two- dimensional graphic in CIELAB color space. The addition of the "L" values would produce a three-dimensional model. It should be noted that the two-dimensional color gamuts are an approximation of color.

20.2.3 High-Performance Pigments: *These pigments tend to be considerably more expensive, weaker and dirtier in color. They may result in conditional, metameric color matches.*

20.2.3 *FIRST* Recommended High-Performance Pigments

Alternate pigments are sometimes required to achieve high performance properties, such as: increased fade resistance, enhanced chemical, water and bleed resistance, reduced migration, insolubility in solvents and increased product resistance (grease, fats, etc). There are many different levels of performance available above the standard pigments. These high performance pigments range from slightly better to significantly better. It is important for the printer to discuss the specific application and requirements with the ink supplier to identify the optimum pigments for the application. High performance pigments tend to be considerably more expensive, weaker, dirtier in color and may result in conditional, metameric, color matches.

Color	Water Based Ink		Solvent Based Ink		UV Based Ink	
FIRST Recommended High-Performance Pigments						
	C.I. Name	C.I. #	C.I. Name	C.I. #	C.I. Name	C.I. #
Yellow	Y74	11741	Y74	11741	Y74	11741
Magenta	R184	12487	R184	12487	R184	12487
	R269	12466				
Red	R254	56110	R254	56110	R254	56110
Purple	V23	51319	V23	51319	V23	51319
Orange	O36	11780	O36	11780	O36	11780

Table 20.2.3

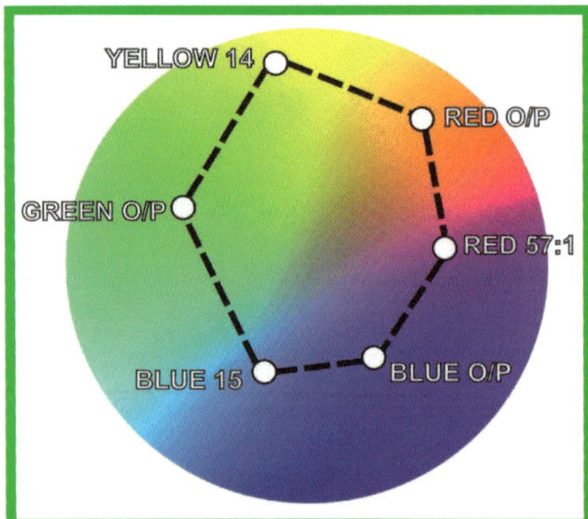

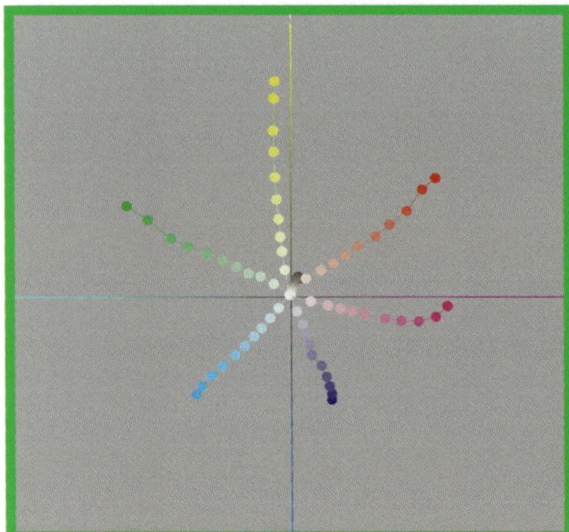

20.2.4 Process Color Gamut: *This graph illustrates the printable color gamut achieved using FIRST recommended process ink pigments.*

Typical colors requiring high performance pigments in specialty applications include: yellow, magenta, red, pink, purple, orange and reflex blue.

Applications that usually require high performance pigments include:

- Wall paper and floor covering
- Shower curtains
- Bumper stickers
- Industrial labels
- Packaging exposed to direct sunlight and/or weather
- Inks in contact with grease, fats, etc.
- Packaging that under goes sterilization, autoclaving, or pasteurization processes

20.2.4 Optimizing the Process Color Gamut

The 4 color process color gamut of a printing press is usually smaller than that of other output devices (proofing systems, displays, etc.). Therefore, maximizing the color gamut of the printing system is critical. Maximize the printable gamut by optimizing the ink formulation (strength), anilox engraving, plate and mounting tape materials to achieve the highest possible densities while minimizing dot gain/TVI and maintaining gray balance. Refer to Section 19.4.1 for detailed information. Additional information regarding process control for expanded gamut can be found in Appendix F Expanded Gamut: Reasonable Measurement for Process Control.

When optimizing the color gamut, the goal is to increase chroma (C*) while maintaining hue angle (h°) and lightness (L*). In practice, all three components change at different rates as the ink strength is increased. The key is to not let hue angle and lightness change too much. Often, as ink strength increases, at some point there will be a "hooking" of the hue angle, resulting in a significant shift in the hue of the ink. Printing at this level presents two problems: undesirable color change (resulting in a reduction in the size of the color gamut) and a condition where as ink strength varies there are large swings in hue and overall color. It is better to operate at an ink strength where normal variations result in minor hue angle changes. Additionally, it is important to evaluate the overprint colors (red, blue and green) to minimize changes that may occur in the overprints while the individual process colors perform acceptably.

Expanded Gamut Ink Color Specifications
For expanded gamut ink specifications the yellow, magenta and cyan inks follow the ISO 12647-6 specification. In that specification only the hue angle is specified. This is an understanding that given the wide variety of flexo printing it is difficult to identify a chroma. The chroma target should be those chroma values specified by ISO 12647-2, but on some substrates it will be difficult to achieve those chroma values. Regardless of substrate the printer should focus on getting as much chroma as possible, while still meeting the hue angle specifications.

ISO 12647-6 hue angle specifications:

Color	Hue Angle
Cyan	233°
Magenta	357°
Yellow	93°

The same approach will be used for specifying the orange, green and violet inks used in expanded gamut printing.

Color	Hue Angle	Ink Type	Recommended Pigment
Orange	54°	Solvent	C.I. Pigment Orange 16 or C.I. Pigment Orange 34
		Water	C.I. Pigment Orange 16
		UV	C.I. Pigment Orange 64
Green	181°	All	C.I. Pigment Green 7
Violet	307°	All	C.I. Pigment violet 23

**Tolerance around all hue angles is +/- 2°*

20.2.5 Printing with an Expanded Gamut

Expanded gamut refers to any process that expands the range of reproducible colors beyond that typically available with standard four color process printing (CMYK). Expanded gamut can be achieved by increasing solid ink density, increasing the whiteness and reflectance of the substrate, or by adding extra ink sets. Expanding the color gamut by increasing ink density is limited, as the hue angle of the ink tends to curve away from

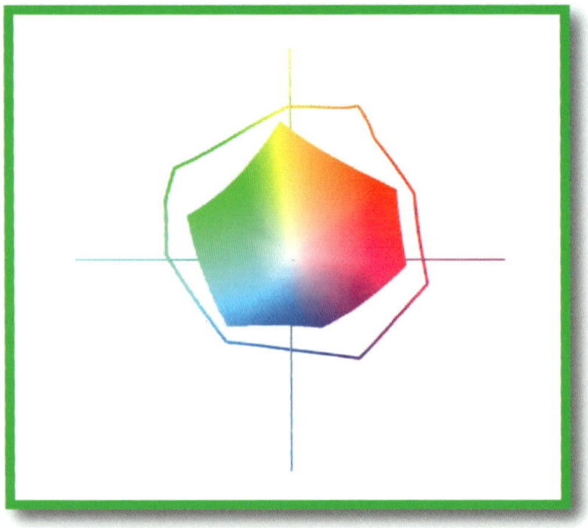

20.2.5a Expanded Gamut: *Adding additional inks to traditional CMYK, 4/color process printing, expands the printable color gamut.*

pure with increased ink film thickness. Adding extra ink sets is what is generally referred to as "expanded gamut" printing. This technology can expand not only to the outer edges of the color gamut with the most saturated colors, but also expand to include the purest pastels and semi-saturated colors that are often missing using traditional CMYK technology.

Following *FIRST* Methodology, a characterization target must be printed. There are no industry standard targets for profiling the expanded gamut color space. Targets are available from specific vendors; they contain solid and tint patches for CMYK as well as for the additional inks. Once the data is analyzed with a spectrophotometer, a color gamut can be mapped. The same target should be used to map the proofing device, and a correlation of the color space between the press and proofer. This correlation provides the information necessary to match the proof to the predicted print outcome. Refer to Section 1.3.4 for more information on Press Characterization and Section 16.6 on Proofing for Expanded Gamut Printing.

When printing with more than four colors, carefully choose the screen angles to avoid moiré problems. One solution is to place each extra ink color on the same angle as the opposing complimentary color. For example, when adding OGV to the traditional CMYK, place the green on the magenta angle, the orange on the cyan angle, and the violet on the black angle. Stochastic screening technology offers another solution for avoiding moiré problems by eliminating screen angles altogether.

Although expanded gamut printing can reproduce a more desirable product, the ability to control color consistency poses many challenges. Printers utilizing this technology must have in place superb process documentation and control systems to be successful. Expanded gamut printing is used to print both expanded process work as well as to create spot colors by trapping multiple inks. Each type of graphic poses unique challenges for measurement and control.

When using expanded gamut technology for process work, measurement equipment limitations pose the primary challenge. Using a spectrophotometer to measure the color (L*C*h° values) of each solid and dot patch is one method of measurement. While a spectrophotometer provides an accurate measure of the color printed, the corrective actions required on press in response to the readings may not be as intuitive to press operators. Press operators understand what press variables influence density and dot gain/TVI and how to adjust accordingly. The correlation between the L*C*h° values of a dot area and the corrective action required on press is less clear. Using a densitometer to measure solid ink density, dot area, etc. is another option. However, this method has some shortcomings as well.

Densitometers use complimentary color theory to measure light reflectance. When measuring magenta ink the green filter is used, the blue filter is used when measuring yellow ink and the red filter is used when measuring cyan ink. Using the complimentary filters to measure light reflectance results in accurate and repeatable measurements of densitometric variables that influence process printing such as: ink density, dot area, dot gain/TVI, ink trap, print contrast, etc. The problem is that the densitometer does not have complimentary filters for the additional ink sets. When measuring density of the additional ink sets, *FIRST* recommends using the "all filters" option on the densitometer. Equipment manufacturers report this setting to provide repeatable density measurements for non-process inks. Control of tone reproduction for non-process inks is more problematic. *FIRST* suggests using the filter for the process ink closest to the additional ink when measuring dot area and dot gain/TVI (ie. use the magenta filter when measuring a red ink). The results may not be accurate but the results should be repeatable; therefore, providing a relative measure that the printer can use to control the job on press.

Both spectrophotometer and densitometer functions should be used to measure spot colors created with "multi- ink traps". First, the printer should check the hue angle of each contributing ink prior to measuring density and dot area. The printer should still use the densitometer functions of density and dot area to control each individual color. These variables must be controlled on press if the spot colors created from "multi-ink traps" are going to match their respective color targets. A spectrophotometer should be used to measure the custom colors created by "multi-ink traps". Control of spot colors created using "multi-ink traps" is critical because the printer must match custom colors, which were previously single inks, with tight DE specifications.

20.2.6 On-Press Ink Control

Ink chemistry is fluid and dynamic. To consistently achieve desirable print results, the ink must be carefully measured and kept in balance throughout the press run. The primary ink variables that must be controlled on press include: viscosity, pH, temperature and foaming.

Additional information regarding Process Control of Spot Colors can be found in Appendix H.

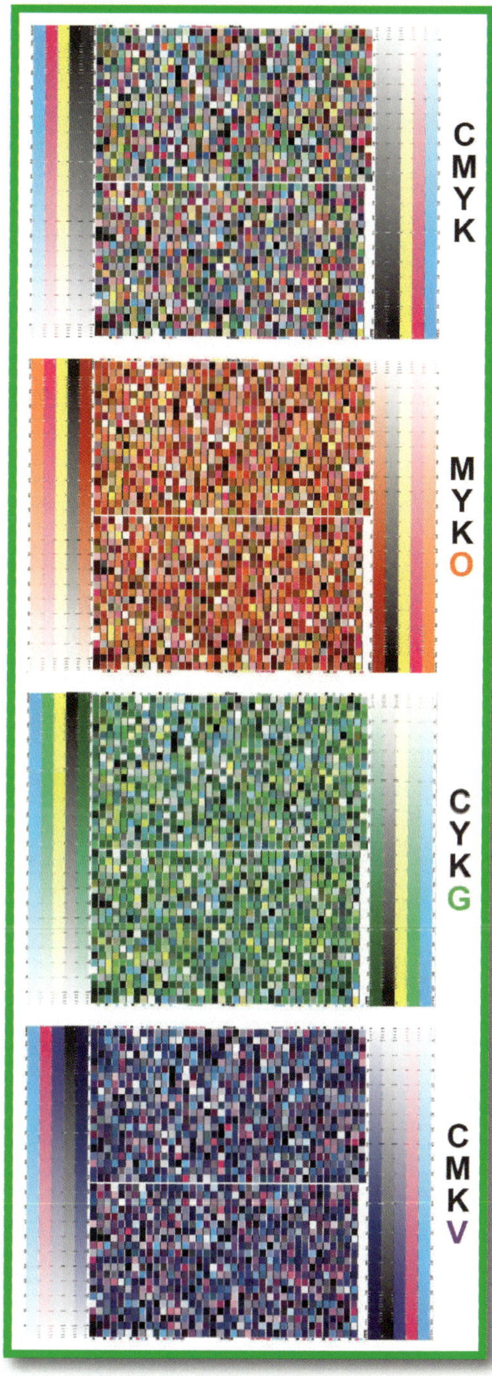

20.2.5b Expanded Gamut Characterization Targets: *This is one example of expanded gamut characterization targets. Due to the printing constraints of this publication, this graphic is a CMYK simulation.*

20.2.6a Viscosity: *In this graphic, viscosity is measured with a Shell cup.*

20.2.6b Press-Side Viscosity Measurement: *A press operator measures the viscosity of a solvent ink using a #2 Zahn cup.*

Viscosity

Viscosity is a measure of resistance to flow at a specified temperature. Ink viscosity is a critical variable influencing the finished print quality. It is dependent upon temperature and must be monitored and adjusted throughout the press run. Controlling viscosity on press is necessary to maintain:

- Color (hue and strength)
- Print quality (ink lay down, dot gain/TVI and trapping)
- Performance properties (coating weight, drying speed, and solvent retention)

Viscosity can be measured both manually, using an efflux cup, or automatically with on-press viscometers. The same method should be used continuously and recorded along with viscosity measurements to avoid confusion between the ink supplier and printer, or between operators from shift to shift. For accurate measurements, the ink should be fully circulating before measuring viscosity. Once the correct viscosity is established, the ink supplier should be able to supply the same ink properties at the ideal viscosity for all subsequent press runs.

The flexographic industry commonly utilizes gravimetric, or dip type, cups; also referred to as efflux cups. The efflux cup, used with the manual measurement method, is a vessel with a hole in the bottom, which is submerged in ink and then removed. The time required for the ink to drain from the cup is measured in seconds using a stopwatch. A larger aperture cup is preferred for water based inks (ie. #3 Zahn); while a smaller aperture cup is preferred for solvent inks (ie. #2 Zahn). The accuracy of efflux cups varies by design, manufacturer, condition and operator. Efflux cup brands include Zahn, Shell, Din and Ford cups. Zahn cups are the most widely used in the flexographic industry due to ease of use and cleaning. While Shell cups provide greater precision, this design tends to be more difficult to keep clean. To ensure accuracy, efflux cups must be kept clean and periodically verified using a "standard" liquid. On press ink viscosity should be maintained within +/- 3 seconds using a Zahn cup or +/- 2 seconds using a Shell cup. Refer to ASTM D4212 (Test Method for Viscosity by Dip Type Viscosity Cup) for measurement procedures.

An automatic viscometer is an electronic tool installed on press at each ink sump to automatically measure and adjust ink viscosity at a programmable frequency. Automatic viscometer units should maintain on-press viscosity within +/- 1 second of the programmed viscosity target. Some units also monitor ink temperature. Refer to the equipment manufacturer's manual for proper set-up and standard measurement procedures.

Temperature will also influence viscosity. Always allow fresh ink to reach temperature equilibrium in the press before measuring and adjusting viscosity. Cold ink will appear higher in viscosity, but will drop as it warms to press temperature.

pH

pH applies only to water based ink systems. It is a measure of the acidity or alkalinity of a water based ink or coating. It is measured on a scale of 0-14.0 (0-7.0 is acid and 7.0-14.0 is alkaline). The neutral point is 7.0. Water based inks rely on precise pH control to maintain resin solubility and stability. The resins used are generally modified acrylic polymers. When the polymers are adjusted with an amine to an alkaline pH range, specified by the ink supplier, the resins perform optimally:

- Dispensing and wetting out pigments
- Transferring and laying smoothly on the substrate
- Imparting the product resistance properties intended

Optimal pH ranges for water based inks vary depending on the chemistry of the ink and the end-use application. The typical pH range for the majority of water based inks (anionic) is 8.5-9.5. There are a limited number of chemistries that are pH neutral (7.0) or acidic (cationic). The ink supplier should recommend the optimal pH range for peak performance of a particular system and the preferred method to maintain the pH level.

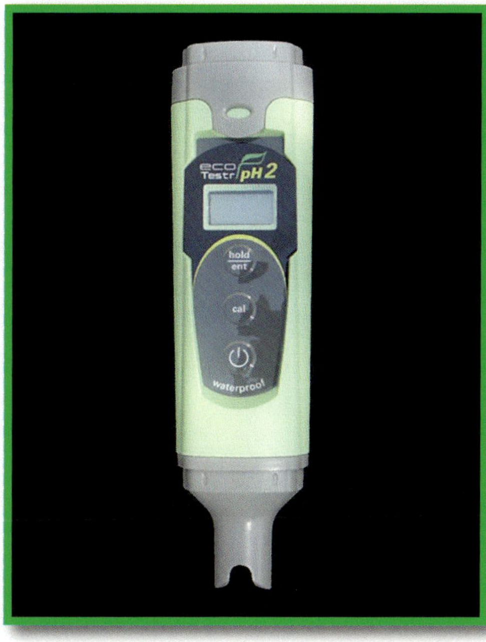

20.2.6c pH Meter: *Water based inks rely on precise pH control to maintain resin solubility and stability.*

20.2.6d Press-Side pH Measurement: *During the press run, water based inks should be measured frequently to maintain a pH of +/- 0.3 units.*

pH meters are used to measure the acidity or alkalinity level of water based inks. A pH meter, which reads to two decimal places and has a variance range of +/- 0.01, is recommended. During a press run, water based inks should be measured frequently to maintain a tolerance of +/- 0.3 units. Refer to ASTM E70-97 (Standard Test Method for pH of Aqueous Solutions with the Glass Electrode). In summer months, it is advisable to monitor and adjust pH hourly. In cooler months, every two to three hours is usually adequate. Increased frequency in summer months is required because as the ink temperature increases, the amine evaporates. As the amine evaporates, the pH of the ink falls and the resin begins to revert back to a heavy-body, higher viscosity state. Adjusting the viscosity with water will not effectively lower the viscosity because it is a formula imbalance and not a physical problem. Therefore, it is very important to adjust the pH of the ink to within the specified range prior to adjusting viscosity.

A blend of water and amine, typically a 9:1 ratio, should be used to adjust the pH of most water based inks. Undiluted amine should never be added to the ink because it can shock the ink and cause it to "kick out". The alkaline blend should be added slowly

pH of Water Based Inks	
pH Range	Print Result
6.0 - 7.0	Precipitate: Ink and resin separate. Result is high viscosity and poor print.
7.0 - 8.5	Ink Unstable: Dirty, fuzzy print. High Viscosity, heavy body and build up on anilox and plate.
8.5 - 9.5	Optimum Flow Characteristics: Good print, good adhesion and excellent wet-out properties. High viscosity does not necessarily mean a low pH; but a low pH usually results in high viscosity.
9.5 - 11.0	Potential Pigment Burn-Out: Excessive foam, corrosive to steel and iron. Lack of water resistance is possible. Strong amine odor on printed material.

Table 20.2.6

while agitating the ink. The viscosity of the ink will decrease as more of the alkaline blend is added. Add only a small amount at a time and measure the pH after each add. It is important not to raise the pH above the specified range (typically maximum 9.5). Once ink exceeds the specified pH range, it cannot be easily adjusted to bring it back into range. It is best to drain it from the press and replace it with virgin ink. After the pH is in the desired range, the viscosity should be checked. If the viscosity is too high, a little water can be added to reduce the viscosity.

Ink Temperature & Ambient Humidity

Temperature and humidity influence an inks dry rate, resin solubility and viscosity. Optimum conditions are high temperature and low humidity for both water and solvent ink systems.

Both ink systems dry by evaporation of the respective solvents from the printed image. Dry air can accept more moisture than humid air. Because of the increased moisture in the air, high humidity results in slower dry rates. High humidity results in the alcohol (in solvent based ink systems) absorbing moisture from the air. The moisture absorption rate increases with faster drying alcohols. This moisture in the ink can cause certain resins, such as nitrocellulose, to become insoluble. The result is blushing of the ink, giving it a flat appearance or causing it to kick out, resulting in dirty print. Because of the increased evaporation rate, water based inks may experience reduced resin solubility in high heat conditions due to a loss of amine in the ink. As the amine evaporates, the pH will decrease and the viscosity will increase resulting in dirty print.

With both water and solvent ink systems, there is an optimum temperature at which the ink will print best. Keeping ink fountains covered and reducing the amount of ink in the sump can minimize moisture absorption by the ink and slow the rate of solvent evaporation; helping to stabilize the ink chemistry.

Operators should have a thermometer to measure the temperature of the ink in the sump. The ink supplier should provide the printer with the temperature range for optimum ink performance.

Foaming

The occurrence of excessive foaming in ink, which is caused by air entrapment, results in print problems such as pinholing or ink starvation. Although usually associated with water based inks, foaming problems arise infrequently with both solvent and energy-cured (UV/EB) systems.

Excessive agitation should be avoided to reduce the occurrence of foaming. To reduce foaming in the sump, ink return hoses should be submerged into the ink sump as low as possible or positioned above the surface so that the return flow runs down the side of the sump container. The return should never be allowed to fall straight down while above the surface of the sump level. If foaming occurs, a "defoaming agent" can be added to the ink. Defoaming agents reduce the surface tension of the ink, causing existing bubbles to burst and preventing stable foams from forming. Defoaming agents should be sprayed into the ink sump. Avoid spraying it on the plates. Too much defoaming agent can result in "fisheyes".

20.3 Specialty Inks & Coatings

Specialty ink and coating applications can be grouped into four basic categories:

1. **Promotional Branding:** Viewed under normal light without artificial aids, intended to attract and hold the consumer's interest.
2. **Interactives:** Require interaction by consumer, no special instruments required.
3. **Brand Security:** Covert systems, special instruments required.
4. **Track and Trace:** Systems used to locate and help manage supply and logistics. Track and trace technologies can also be used for brand security by authenticating products.

Many specialty inks and coatings can be flexographically printed with success. However, ink chemistry variables such as particle shape, particle size and viscosity often differ from the typical flexographic ink. The printer should inquire about these variables and their impact on the printing process prior to printing with a specialty ink. A few ink chemistry variables to consider include:

20.3 Specialty Inks: *When specialty inks or coatings are utilized, the ink supplier should be consulted early in the process to tailor the specialty ink to the printer's operation.*

Particle Size: Many specialty inks have a large particle size which influences the anilox engraving (line count, volume, cell shape) that can be successfully used with the ink. The required anilox engraving may limit the image detail the process is capable of printing with the specialty ink.

Particle Shape: The actual shape of the pigment particles (flat vs. round) influences the rheology of the ink and therefore, the optimum anilox engraving, doctor blade setting, doctor blade material and metering system used.

Viscosity: Many specialty inks print at higher viscosities; impacting the anilox engraving, blade material, blade settings, metering system used, ink pumps and hoses.

Specific Gravity: Many specialty inks have a high specific gravity and may settle. Therefore, it is important to thoroughly mix the ink prior to use and monitor to ensure the particles do not settle during printing. Sometimes hard settling occurs during prolonged storage, making the ink difficult to use.

Specialty inks often require heavy coating weights. Achieving the necessary coating weight can pose significant challenges to the printing process. The printer may need to use a coarse, high-volume anilox engraving, print a double bump of the specialty ink, reduce press speed, or make other adjustments to the equipment or process in order to print and dry these inks successfully. These requirements can limit the type of graphics that can be successfully printed.

When specialty inks or coatings are utilized, the ink supplier should be consulted early in the process to tailor the specialty ink/coating to the printers operation. The ink supplier should also provide guidance to the printer on how the specialty ink may impact the printing process and make equipment recommendations. The printer may choose to also involve the anilox roll and drying system suppliers in discussions on how to optimize the press.

20.3.1 Promotional Branding

Thermochromics

Description: Thermochromics are inks that change with temperature. They are manufactured as either reversible or irreversible formulations. Irreversible formulations change in color only one time; they are not encapsulated. Thermochromics are commonly referenced by their transition temperature (Type 10, Type 27, etc.); this often does not match application/end-use properties. The transition temperature (T°) is different if the process temperature increases or decreases (boiling vs. freezing). The temperature

Thermochromic Inks		
Ink Line	Shift T°	Application Examples
Arctic	- 18°C (0°F)	Deep-Frozen Items
Freeze	- 10°C (14°F)	Frozen Items
Ice	0°C (32°F)	Freezing Indicator
Cold	10°C (50°F)	Cold Beverages
Cellar	15°C (59°F)	White Wine
Room	20°C (68°F)	Red Wine
Touch	28°C (82°F)	Skin Contact
Warm	40°C (104°F)	Hot Beverages
Hot	60°C (140°F)	Depilatory Waxes
Blaze	70°C (158°F)	Burning Prevention

Table 20.3.1

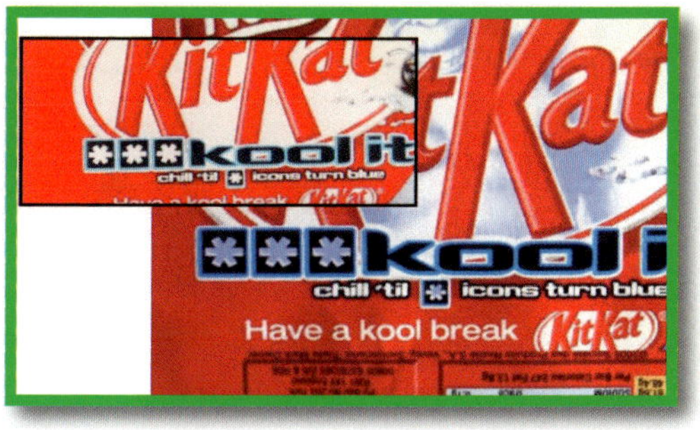

20.3.1a Thermochromics: *Thermochromic inks change with temperature. They can change from colorless to colored, or from one color to another. They can be blended with traditional pigments to create a multi-colored design.*

range for irreversible thermochromics is above 113°F (45°C). Reversible thermochromics range between 59°F (15°C) and 149°F (65°C). Thermochromics can change from colorless to colored, or from one color to another. The activity depends on the medium and somewhat on the surfaces in contact (substrate, additional ink layers). Thermochromics can be blended with traditional pigments, and multi-color designs can be created. Color-shifting technology is available that does not use heavy metals. Functional applications for thermochromics include battery checks, cosmetic depilatory wax and consumption temperature indicators such as: cold beverages, beer, wine, ice cream, hot food, hot beverages and baby food.

Print Considerations: The color strength of reversible thermochromics is much weaker than conventional inks. Heavier ink film thicknesses are required to obtain strong color or full hiding ability. Solvent resistance must be evaluated on a case-by-case basis; however, alcohol should not be used. Small amounts of short-chain alcohol (such as isopropanol) rapidly permeate through the capsule wall and triggers permanent color change. Capsule dispersions are typically stable for 6 months when stored at ambient temperature in a dark environment. Ideal temperature for maximum shelf life is 50°F (10°C). Do not expose to temperatures above 122°F (50°C). Most colors have very limited fade resistance (blue wool scale = 1). Therefore, outdoor applications are not recommended without protection.

20.3.1b Metallic Inks: *Metallic pigments produce a shine and luster that standard inks cannot provide. This property is referred to as brilliance.*

20.3.1c Fluorescents: *Fluorescent inks create eye-catching effects for products on the store shelf by providing bright and vivid colors.*

Metallics

Description: Metallic inks are comprised of pigments that are typically made of aluminum particles for silver and bronze, or aluminum tinted with organic pigments, for gold. Silver tinted with a R/S yellow is also used to create various shades of gold. Metallic pigments produce a shine and luster that standard inks cannot provide. This property is referred to as brilliance.

Print Considerations: Challenges include maintaining water based ink in the proper pH and viscosity ranges for optimum results. Because the inks are typically lower in pH, dirty print can become a problem in screen and highlight areas. pH ranges are lower than standard water based inks due to problems with the pigments gassing. The level of brilliance depends upon applying sufficient coating weight or particle alignment. Inks may tarnish or scuff if the proper pigments/ coatings are not used. An over lacquer is often required to obtain higher rub resistance. It is important to minimize the shear on metallic pigments when mixing or blending. High shear processing will damage the particles and reduce the brilliance.

Fluorescents

Description: Fluorescent inks convert ultraviolet light into color by absorbing both visible and non-visible electromagnetic radiations and then releasing them. Fluorescent inks provide bright and vivid colors, which can create eye-catching effects for products on the shelf.

Print Considerations: Strength is the major challenge on press. Coarse, high-volume, anilox rolls are recommended to achieve bright colors; double bumps are sometimes necessary. Agitation on press, to avoid settling and separation of the colorant, is also recommended. Fade and bleed resistance is generally less than conventional pigments.

Pearlescents

Description: Pearlescent inks use pigments based on titanium dioxide (TiO_2) or natural mica. These pigments scatter light rays to provide depth, creating a metallic sheen when viewed at an angle. The degree of sheen is dependent on the ink film thickness. Pearlescents are sometimes considered an alternative to metallic inks and are primarily used for novelty applications, personal care products and cosmetic packaging.

Print Considerations: Inks and coatings containing pearlescent pigments should be mixed thoroughly before and during print production to avoid settling and separation. Additionally, heavy coating weights are recommended to achieve optimal results. The anilox roll should have a cell opening at least 4 times the pigment particle size. Generally, these inks have larger pigment particle sizes than standard inks, requiring a higher volume anilox engraving. Pearlescents can be formulated as either water or solvent based inks.

Invisible Fluorescents

Description: Invisible fluorescents are invisible under normal light but glow, a variety of colors, under UV light. The typical example is a special dye that glows under black light. A black light can be added to displays, or the item can be promoted with products that produce black light. Invisible fluorescents can be used for both promotional branding purposes such as promotional games and product verification. They are also used for low-level security functions such as anti-counterfeiting and anti-tampering.

Print Considerations: Most invisible fluorescents are dye based; therefore, they are prone to bleed and fade. Solvent-based inks tend to work best on dull substrates. Since these inks are not visible under normal lighting conditions, a press-side UV or black light is required to ensure proper coverage and register on press.

20.3.2 Interactives

Phosphorescent

Description: Phosphorescent pigments (glow in the dark) absorb light energy and then slowly release it in the form of visible light. It is a clear, or semi-transparent, ink that is charged by exposure to UV light and emits an afterglow for a limited time. The strength and duration of the afterglow is influenced by the intensity of the light source, period of exposure and amount of ink applied. Phosphorescent inks are now available in long-life versions that can be printed flexographically. Phosphorescent inks are often used for promotional branding and are particularly popular in candy applications. They are also used as low-level security features such as anti-counterfeiting and anti-tampering.

Print Considerations: Because of the large pigment particle size, it is difficult to achieve a suitable afterglow unless a heavy ink film thickness is applied. Therefore, a high volume anilox roll, metered with a doctor blade, is recommended. Inks should be mixed thoroughly before and during use due to the heavy density of the pigment. Phosphorescent inks can be formulated either as water or solvent based. Since phosphorescents are invisible on press, a UV, or black light, and a darken viewing area is required to ensure proper coverage and register on press.

20.3.1d Pearlescents: *Pearlescent inks create a metallic sheen when viewed at an angle. They are primarily used for novelty applications, personal care, and cosmetic packaging.*

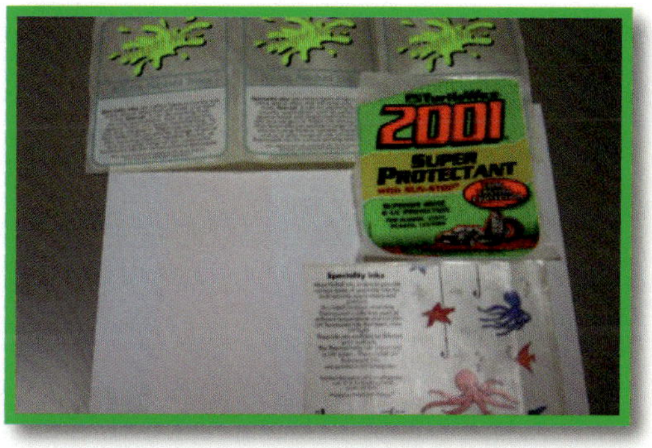

20.3.1e Invisible Fluorescents: *Invisible fluorescent inks glow under UV light but are invisible under normal light. They are used for novelty applications and low-level security features.*

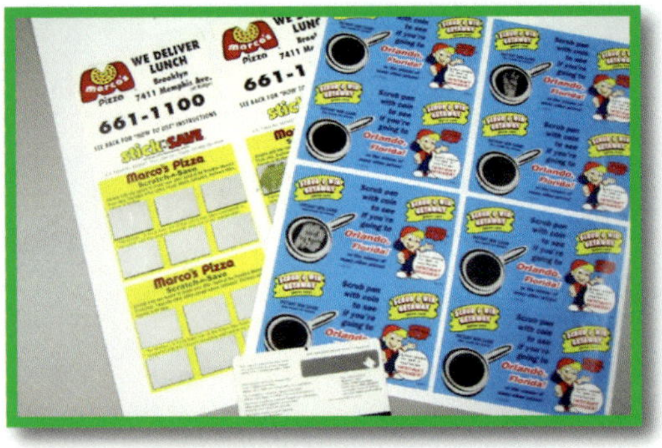

20.3.2 Coin Reactive: *Coin reactive inks are invisible until rubbed with a metal object to reveal a hidden message. They are used as a low-level security feature as well as promotional branding applications.*

Encapsulated Fragrance

Description: Encapsulated fragrance inks can be made by encapsulating an aroma in a resin carrier. As the microcapsules are scratched, they are broken; thereby, releasing the fragrance.

Print Considerations: The printer should use a very light nip pressure (anilox-to-plate and plate-to-substrate) to ensure the capsules are not broken. Avoid recirculating the ink through the ink fountain when the press is not printing to reduce the risk of breaking capsules. The printer should consult the anilox roll supplier to determine the optimum anilox engraving for the capsule size/shape being utilized.

Metameric Pairs

Description: Metameric pairs (hidden message method) use two inks that are closely color matched over the visible spectrum but diverge at a critical point. The metameric ink pair appears as a single ink on a product in normal lighting conditions but, with the use of a suitable color filter, reveal two inks. It is used for promotional branding (such as gaming), product verification and low-level security features (such as anti-counterfeiting and anti-tampering).

Print Considerations: Proper viscosity control is critical to provide consistent color strength and coating weight. Tight color tolerances are required to ensure proper performance. Inks should not be toned on press with components outside of the formulation.

Coin Reactive

Description: Coin reactive inks are invisible until rubbed with a metal object, such as a coin, revealing a "hidden message". The pigment reacts with nickel, producing a gray color when scratched. The ink must be highly pigmented and white. A matching overprint is required to achieve the desired effect; the gloss must match perfectly. Because the inks cannot be scanned or copied, they are used as low-level security features such as anti-counterfeiting and anti-tampering. Coin reactive inks are also used for promotional branding applications such as gaming, promotional and product verification.

Print Considerations: pH, viscosity and coating weight should be carefully monitored throughout the press run. Typically, anilox engravings between 200-400cpi are used. Coin reactive inks work best on low gloss, white papers.

20.3.3 Brand Security

Photochromics

Description: Photochromics change color when exposed to UV light, such as sunlight, and quickly revert back to the original color when the UV light is removed. Photochromic ink can be colored or colorless. Print applications include promotional branding (ie. product verification and novelty applications) and security features (ie. anti-counterfeiting and anti-tampering).

Print Considerations: The color strength of photochromic ink is much weaker than conventional inks. Therefore, apply a heavier ink film thickness for more intense color. An anilox roll engraved for coating applications, such as 150cpi or less with a high cell volume, and metered with a doctor blade is recommended to achieve a suitable ink coating weight. It is important to monitor the anilox to prevent plugging. Because of the coating weight required to achieve reasonable color strength, detailed designs are not recommended. Use a harder mounting tape and very light nip pressures (anilox-to-plate and plate-to-substrate). Inks should be stirred thoroughly before and during use. Water based inks require increased drying capability because of the higher coating weight required. Photochromics should be printed only on paper substrates; avoid prolonged exposure to excessive light and heat. Usually, higher viscosities will yield better results, more saturated color. The recommended viscosity range is 35-55 seconds in a #2 Zahn cup for narrow web applications. pH should be kept around neutral (7.0). Use only distilled water to reduce viscosity of water based inks. All solvents should be avoided due to the sensitivity of the pigment. Since these inks are invisible under normal lighting conditions, a UV light, or black light, is required to ensure proper coverage and register on press.

20.3.4 Track and Trace

RFID: Radio Frequency Identification Device – Printed Method

Description: RFID is similar to bar code technology but uses radio waves to electrically capture data from tags rather than optically scanning a bar code. Conductive inks are inks that conduct electricity, allowing a circuit to be printed onto a variety of materials, including paper. Conductive inks, typically silver and carbon-based, are used with conventional printing platforms to fabricate RFID antennas directly onto labels and packaging. RFID is used for both brand security and track-n-trace functions for high-end products such as: cosmetics, fragrances, pharmaceutical and food supply chain. It allows for improved inventory management, enhanced anti-theft control and anti-counterfeiting. Current research within the flexographic community addresses printing the functional microchip, and assembly of the RFID devices.

20.3.4 RFID: *The printed method of RFID utilizes conductive inks which allow a circuit to be printed onto a variety of materials, including paper. This enables RFID antennas to be printed directly onto labels and packaging.*

Print Considerations: RFID inks are applied at a specified ink film thickness. The ideal ink film thickness often requires multiple passes, usually with special high-temperature drying considerations. They can be formulated using water, solvent, or UV ink chemistries. These conductive inks can be printed on a variety of substrates, including paper. Particle sizes range from 0.5 microns to 8.0 microns in diameter. Ink film thickness and particle alignment are key variables to ensure conductivity. Additionally, the anilox engraving, ink metering, ink temperature and the ability to dry/cure the ink are also critical considerations.

20.4 Ink Metering System

The ink metering system is comprised of the anilox roll, the fountain roll, the doctor blade and the doctor blade chamber. The function of the ink metering system is to control the amount of ink being transferred onto the printing plate.

20.4.1 Doctor Blades

The primary function of the doctor blade is to uniformly remove ink from the surface of the anilox roll without damaging the anilox roll. When the doctor blade is properly installed in the chamber, and the chamber is correctly aligned with the anilox roll, there should be a dull sheen uniformly across the anilox roll. The consistent dull sheen across the anilox roll confirms the doctor blade is properly metering the ink onto the anilox engraving.

Doctor Blade Variables

Material: Doctor blades are manufactured using steel, plastic, or composite materials. Regardless of the material used, the doctor blade must be flat and straight with a controlled thickness. The doctor blade surface should be finished to provide clean, uniform metering of the anilox roll.

1. **Steel blades** are made of high purity ores to provide consistent metering and minimize damage to the anilox roll. Steel blades provide the cleanest wipe of the anilox surface. Carbon steel, either bright or blue, is the most common blade material. Stainless steel is often used in water based applications where corrosion is a concern. Tool steel is used when long life is needed or abrasive inks are required.
2. **Plastic blades** provide long life and corrosion resistance. They are safer to use because they do not sharpen like steel or composite blades. Plastic blades are thicker to provide necessary support. Because of the thickness, plastic blades do not wipe the anilox surface as well and therefore transfer excess ink to the plate. Containment

blades in a chambered doctor blade assembly are often polyester to reduce anilox wear and scoring. The corrugated print segment often uses plastic blades.

3. **Composite blades** provide long life and corrosion resistance. While sharper than plastic blades, they are not as sharp as steel blades. Because of the thinner edge, they provide a cleaner wipe than plastic with only a small amount of surface ink on the anilox. Composite blades are typically run in corrugated and flexographic newsprint applications.

Edge Profile: There are three types of edge profiles available for reverse angle doctoring.

1. **Beveled Edge:** Blades with a beveled edge are available from 2° to 45°. The bevel should be installed away from the anilox roll so that the small tip is in contact with the anilox.
2. **Lamella or Stepped Edge:** The stepped edge maintains a constant tip thickness as it wears.
3. **Round or Radius Edge:** Blades with a round edge were designed for flexographic printing; they provide a faster run-in on press.

Typical Doctor Blade Thickness	
Material	Measurement
Steel	0.004" (0.010cm); 0.006" (0.015cm); 0.008" (0.020cm); 0.010" (0.025cm); 0.012" (0.030cm)
Plastic	0.020" - 0.125" (0.051cm - 0.318cm)
Composite	0.015" - 0.035" (0.038cm - 0.089cm)

Table 20.4.1

Contact Angle: The contact angle of the doctor blade-to-anilox, and its consistency across the anilox, determines the ink transfer from anilox-to-plate. The dynamic contact angle, also referred to as the "wear angle", should be between 25°-40°. A contact angle greater than 40° can result in doctor blade chatter. Blade chatter prints as equally spaced lines across the width of the web in varying ink densities. A contact angle less than 26° leads to insufficient metering of the anilox, resulting in an increased ink film thickness transferring from the anilox to the plate. A contact angle less than 26° also increases the potential for scorelines on the anilox roll. The inconsistent, increased ink film thickness will appear as increased dot gain/ TVI and dirty print. Loading pressure of the doctor blade chamber to the anilox influences the contact angle of the

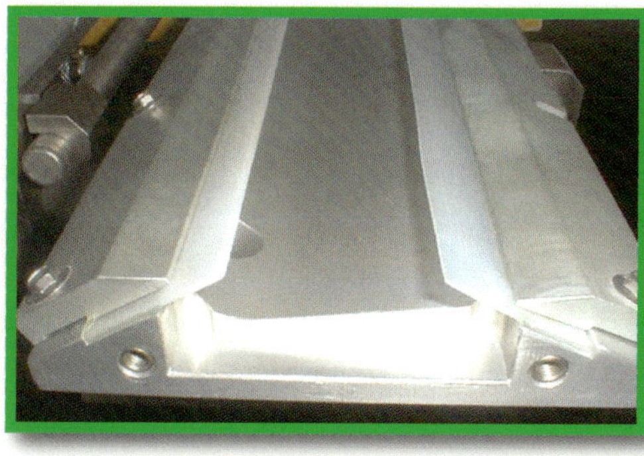

20.4.1a Plastic Blades: *The corrugated print segment often uses plastic blades because of their long life and safer handling.*

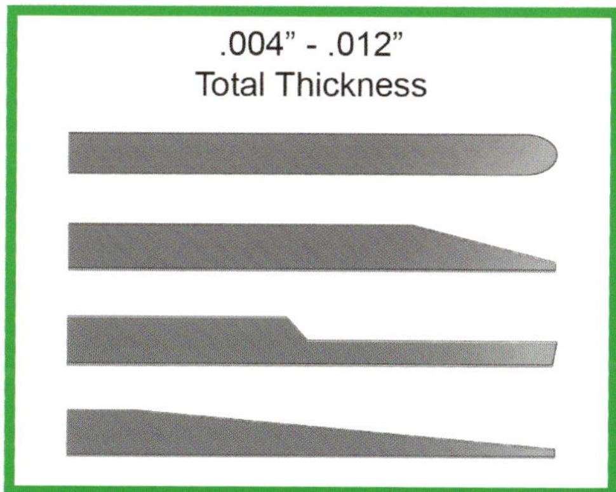

20.4.1b Edge Profile: *Each edge profile is designed to provide specific advantages to the printing process.*

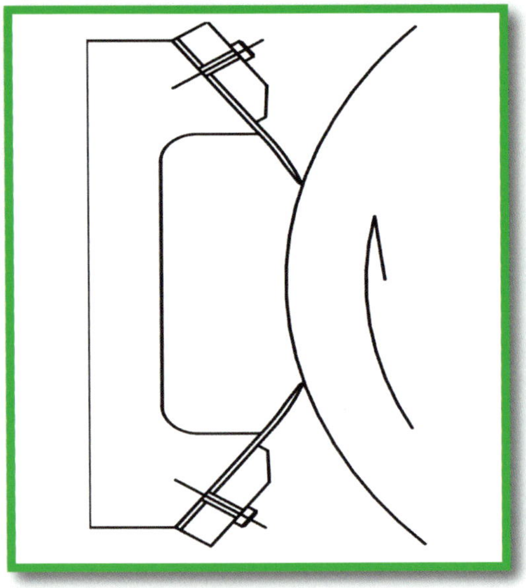

20.4.1c Contact Angle: *When the doctor blade is properly installed in the chamber, and the chamber is correctly aligned with the anilox roll, there should be a dull sheen uniformly across the anilox roll.*

blade to the anilox. Increasing the loading pressure results in a decreased contact angle. Excessive loading pressure can curl the tip of the blade printing inconsistent, increased ink film thickness, which results in increased dot gain/TVI and dirty print. Additionally, metal slivers of the doctor blade can appear in the ink because of excessive blade pressure.

Chamber Loading: When observing the ink sheen on the metered anilox roll, if the dull sheen is not consistent across the anilox, the doctor blade may not be loaded parallel to the anilox, causing a shift in density across the web. When the blade is engaged with the anilox, if it only meters on one side, with minimal loading pressure, this suggests the doctor blade chamber is out-of-alignment (not parallel to the anilox). When the entire chamber is overloaded, a wet shiny anilox can result because there is surface ink on the anilox roll. Surface ink on the anilox results in back doctoring or ink build-up on the containment blade along with increased dot gain/TVI and inconsistent print.

Deflection: When running at high speeds, a tremendous amount of turbulence is created inside the doctor blade chamber, placing pressure on the blade, which can cause it to hydroplane on the anilox. This allows excess ink to remain on the surface of the anilox, resulting in an inconsistent transfer of ink to the plate. Increasing the thickness of the doctor blade helps to minimize deflection. However, a thicker blade tip creates a larger contact area on the anilox, which results in less effective ink metering. Therefore, use a thicker blade, 0.010"-0.012", to reduce deflection. However, it is also important to reduce the metering tip with the thicker blade by specifying a lamella or 10°-15° bevel edge profile. Blade deflection can also cause ink to leak from the chamber. Use a 0.0075"-0.030" thick polyester containment blade, that is 1/16"-1/8" wider than the metering blade, to minimize the risk of leaking and damage to the anilox roll.

20.4.2 Anilox Rolls

20.4.2.1 Anilox Roll Selection

An anilox roll engraving consists of three primary variables:

1. **Cell Count:** The cell count represents the number of cells (or lines) per linear inch (or centimeter); typically denoted as cpi, lpi, or l/cm, cpcm. The cell count is sometimes referred to as anilox line screen, which should not to be confused with the line screen of the separation.
2. **Cell Volume:** Cell volume is determined by the size (depth) of each cell. Measured in Billion Cubic Microns per square inch (BCM/in^2) or cubic centimeters per square meter (cm^3/m^2).

3. **Engraving Angle:** The engraving angle describes how the cells are arranged on the roll in relation to the axis of the roll. It influences the number of cells per square inch. The engraving angle is measured in degrees.

The optimum combination of cell count/volume/angle varies based on substrate and ink properties as well as the graphics to be printed. Follow *FIRST* Methodology to obtain the optimum anilox engraving and ink chemistry combination for a particular job.

Printers may opt to run a banded anilox roll to determine the optimum combination of ink chemistry, print components, press components and anilox roll engraving. A banded anilox trial determines the impact of ink film thickness on the print quality characteristic(s) being evaluated and identifies the optimal anilox engraving. A banded anilox roll is a single anilox that is engraved with several bands of different line screens and volumes across the width of the roll. The number of bands engraved is determined by the press width and test target width. Generally, use the largest cylinder repeat available to maximize test design. An identical test target is run on each band. On each end of the roll, include impression/control bands (bands of the same engraving) allowing the press operator to accurately set the same anilox-to-plate and plate-to-substrate impressions. Monitor the impression settings on each end of the roll by including control targets on the control bands. The lowest volume band achieving slightly higher than target density while incurring minimum dot gain/TVI is the optimum anilox specification. A banded anilox can be run with several ink strengths to identify the optimum combination of anilox engraving and ink strength.

The printer should work closely with the anilox manufacturer, ink supplier, prepress provider, press team and other test component suppliers when designing the banded anilox optimization trial to ensure critical variables can be isolated and evaluated. After the optimization trial, the printer should thoroughly clean the banded anilox and store it for future optimization trials. The same banded anilox can be used to determine the optimum anilox engraving and ink chemistry under different conditions (ie. new substrates, new ink chemistries, new mounting tape, new plate material, etc.). A banded anilox optimization trial could also be designed to determine the optimum engraving for line decks as well. This would require different engravings on the anilox and a different test design. Refer to Section 1.3.1 for additional information on press optimization trials.

20.4.2.1a Anilox Roll Selection: *The optimum combination of cell count, volume and angle varies based on substrate and ink properties as well as the graphics being printed.*

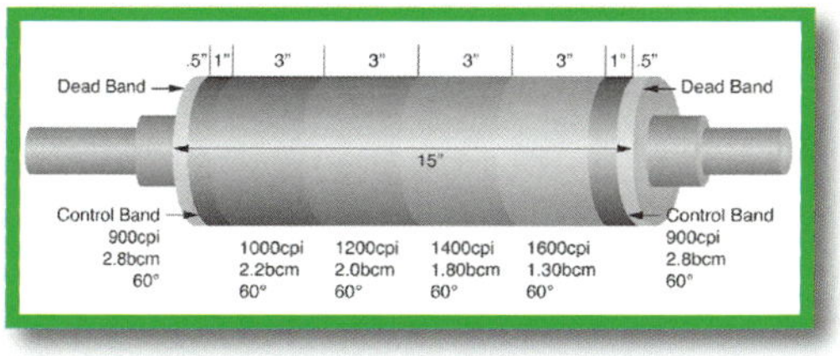

20.4.2.1b Banded Anilox Optimization Trial: *A well-designed anilox trial determines the impact of ink film thickness on the print quality characteristics being evaluated and identifies the optimal anilox engraving.*

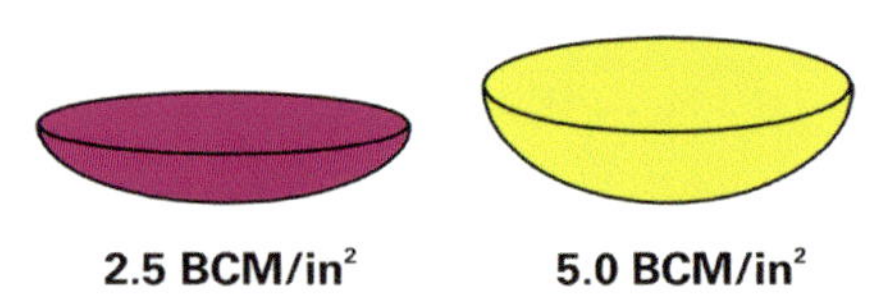

20.4.2.2a Cell Volume: *The cell volume of an anilox roll determines the amount of ink supplied to the printing plate.*

20.4.2.2 Cell Volume

The cell volume of an anilox roll determines the amount of ink supplied to the printing plate and, therefore, the resulting solid ink density. The anilox roll volume is measured in billion cubic microns per square inch (BCM/in^2) or cubic centimeters per square meter (cm^3/m^2).

The conversion of anilox volume between BCM/in^2 to metric cm^3/m^2:

(BCM/in^2) x 1.55 = cm^3/m^2 OR (cm^3/m^2) x 0.645 = BCM/in^2

Cell Volume Measurement

Printers measure the cell volume of an anilox roll for two primary reasons: first, to confirm a new roll conforms to the specifications ordered and second, to monitor anilox wear and ink plugging in the cells throughout the life of the roll. As an anilox roll is used in the press, miniscule particles of ink dry in the bottom of the cells influencing the print volume of the roll. Also, the surface of the roll can become worn over time. Both of these variables will negatively impact the volume of ink delivered to the printing plate. By periodically measuring the volume of an anilox roll, the printer is able to determine when an anilox roll should be replaced.

It is important to note that there is limited correlation between the volume measurements being generated by various anilox roll manufacturers due to the absence of an industry standard for measuring anilox cell volume. Anilox volume measurements are a relative measure and are only useful when done using the same procedures under controlled conditions. Therefore, *FIRST* recommends the printer and anilox manufacturer agree upon the methods and procedures used to measure cell volume and correlate the methods or use the same method exclusively. By using only one method, confusion between parties is minimized. The printer can accurately track changes to roll volume both over time and before/after cleaning. The printer can also compare volumes between multiple rolls. Once a measurement method is agreed upon, all newly manufactured rolls should be supplied with a Certificate of Analysis (CoA) confirming the roll conforms to the specifications. Manufacturer tolerance on cell volume, regardless of measurement method or volume, is ±5%. The two methods commonly used to determine anilox cell volume are:

1. Scanning interferometry
2. Liquid volume measurement

Liquid Volume Measurement

The most important variable to control when using liquid volume measurement is consistency, as this method is very operator dependent. The liquid volume measurement method can introduce as much as ±10% measurement variability if run by an inexperienced technician. However, the measurement variability of this method can be reduced to ±3% to 5% when conducted by a properly trained operator using properly calibrated equipment. The liquid volume measurement method typically measures cells within a 2 to 3 square inch area and calculates cell volume based on this sample. A trained technician should take multiple measurements across the roll. Below is a summary of the steps required to use the liquid volume measurement method:

1. Draw an exact amount of a slow drying ink, typically 10 to 25 microliters (one microliter equals 1 billion cubic microns or BCM), into a micropipette.
2. Spread the ink across the anilox roll using a metering device. The ink is spread with a "tractor" or some other device where a metering blade can be held stationary and drawn down the surface of the roll. This causes the ink to fill the cells with a known volume of control liquid over the surface of the roll.
3. The coverage area is then transferred to a paper. The area of the blot is measured using a planimeter.
4. The amount of ink (BCM or cm^3) is divided by the area to determine the volume of the roll.

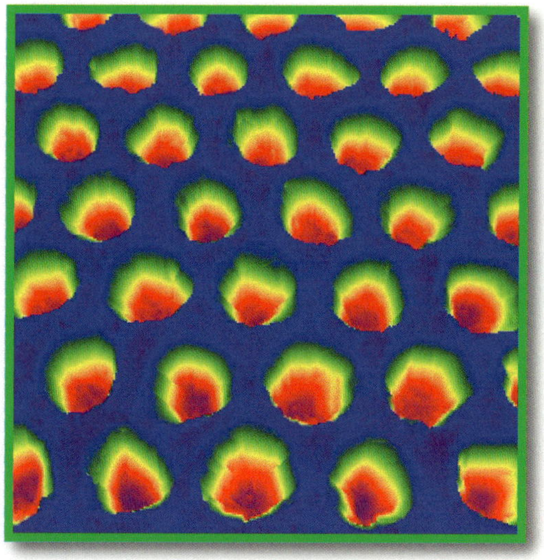

20.4.2.2b Scanning Interferometry: *Appropriate sample size is critical for the accuracy of this measurement method.*

Scanning Interferometry Measurement

The scanning interferometry measurement method utilizes a microscope and computer to produce a three-dimensional view of an anilox cell. This method determines:

- Volume of the measured cells
- Cell depth
- Depth-to-opening ratio of the cells
- Percent of cell opening
- Cell count
- Engraving angle

There are two primary disadvantages to this measurement approach. First, each reading of the scanning interferometer microscope consists of roughly 30 to 300 cells, dependent upon cell count. This represents an extremely small sample of the anilox roll and can result in misleading conclusions if the number of readings from a particular roll is too small. Appropriate sample size is critical for the accuracy of this measurement method. The second disadvantage of this approach is the upfront cost of the equipment. The upfront cost is a barrier for many

20.4.2.4 Engraving Angle: *The higher the number of cells per area, the more uniform the dispersion of ink across the anilox and the better the print quality. Both the 30° and 60° engravings have 15% more cells per unit area than the 45° engraving angle.*

printers. An alternative to purchasing the scanning interferometry equipment is to work with the anilox roll manufacturer. The printer can make an impression of an area of cells on the anilox roll using a very thin metal foil strip (the foil adds variability to the measurement). *FIRST* recommends sampling several areas across and around the roll. Label each sample with a code to identify the roll and the location on the anilox roll. Carefully place the metal foil strips in containers that will prevent them from being damaged during shipment. The accuracy of this approach is dependent upon the size of the area sampled, the accuracy of the impression and preventing damage to the foil after the impression has been pulled.

20.4.2.3 Cell Count (CPI/LPI)

The terms cells per inch (CPI) and lines per inch (LPI), are used interchangeably when referring to the anilox roll cell count. *FIRST* uses the term CPI for consistency. Generally, the cell count (CPI) chosen should be the optimum engraving specification that will accommodate the cell volume necessary to achieve the desired print characteristics and can be properly maintained on press under the given operating conditions throughout the press run. When screen or process work is being printed, the industry standard is to select an anilox cell count that is, at a minimum, capable of supporting a 2% dot, which requires at least 4 times the value of the plate screen (LPI). For example, printing a process job with a plate screen of 120lpi would require an anilox roll with a minimum cell count of 480cpi.

Higher cell counts typically have lower volumes; therefore, in order to achieve desired densities (or color matches) the ink formulation must be adjusted. There is a limit to how much pigment an ink chemistry can hold in suspension and still achieve the necessary performance properties (such as rub resistance). An additional challenge of very low volume rolls is getting the "pigment particle size" of the ink small enough to fit within the cells. Ink dry-rate is another challenge to consider with very low volume rolls. The printer, anilox manufacturer and ink supplier must work together to optimize the ink and anilox combination for the job.

To convert between cells per inch (CPI) and the metric cells per centimeter (CPCM), use the following equations:

$$CPCM = CPI \div 2.54 \quad \text{or} \quad CPI = CPCM \times 2.54$$

20.4.2.4 Engraving Angle

Theoretically, any angle between 0° and 90° can be used to engrave an anilox roll. However, the most common engraving angles are 30°, 45° and 60°. The most prevalent engraving angle used today is 60° whether for solids and type or screen and process work. Both the 30° and 60° engravings have 15% more cells per unit of area than the 45° engraving angle. The higher the number of cells per area, the more uniform the dispersion of ink across the anilox and the better the print quality.

For some high viscosity radiation cured inks, 30° channeled or open-cell engravings are preferred because they appear to reduce pinholing and striations when printing solids. When working with radiation cured inks, it is always best to discuss the anilox requirements with the anilox roll manufacturer and the ink supplier, as well as the press manufacturer.

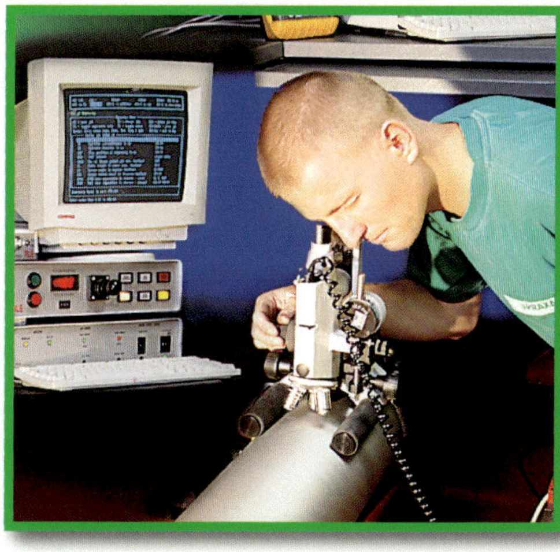

20.4.2.5a Anilox Roll Inspection: *A microscope is used to inspect the engraving quality of the anilox roll. Variables such as land area, cell shape, and channeling are evaluated.*

# Cells Per Square Inch		
CPI	45° Engraving Angle	30° or 60° Engraving Angle
200	40,000	46,000
400	160,000	184,000
600	360,000	414,000
800	640,000	736,000

Table 20.4.2.4

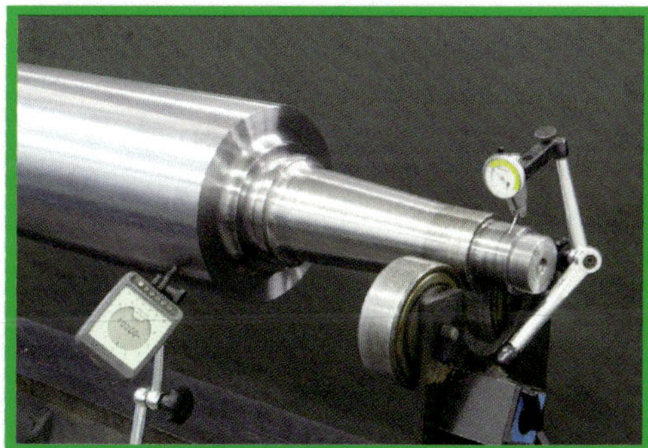

20.4.2.5b TIR Measurement: *The total indicated reading of the roll face and journals should be compared to the anilox roll specifications.*

20.4.2.5 Inspection of New Anilox Rolls

The anilox roll supplier should issue a certificate of analysis (CoA) with each anilox roll. The CoA should include both the engraving characteristics and the roll dimension characteristics of the anilox roll being shipped. The actual roll data should be compared to the specifications ordered. A printer may elect to have the CoA replace an in-house engraving and dimension inspection for incoming anilox rolls. Some printers conduct "random" audits of incoming rolls to confirm the CoA accurately describes the roll being received.

If the printer performs an incoming dimensional and engraving inspection, it should be completed as soon as possible after the roll is received and before it is used. An inspection table for general, visual inspection and to check total indicated runout (TIR) is required. A microscope with a minimum 10X eyepiece and objectives of at least 20X, 40X and 80X are also needed to inspect the overall appearance of the engraving. TIR measures the roundness of the anilox roll. Rolls that are out-of-round will negatively impact the consistency of the ink lay characteristics on press.

20.4.2.6a Ink Wear & Damage: *To minimize wear, run proper doctor blade pressures and install & maintain magnets and filters in the ink-delivery system to reduce the potential for score lines.*

20.4.2.6b Anilox Cleaning: *Cleaning rolls to maintain consistent cell volume is critical to achieving specifications. When dried ink plugs anilox cells, the ink carrying volume of the roll decreases erratically.*

Total Indicated Reading (TIR)	
Ceramic Rollers Under 65" Wide	+/-0.0005" (0.0013cm)
Ceramic Rollers Over 65" Wide	+/-0.001" (0.0025cm)

Table 20.4.2.5

Anilox Roll Inspection Procedures

1. First, check the roll for any damage that may have occurred in shipping, or in the press, by looking at the face of the roll and the shafts.
2. Next, check the TIR of the roll face and journals. Compare to the roll specifications.
3. Using a microscope, inspect the engraving for general quality.
   Evaluate:
   - Excessive land area between the cells
   - Cell shape
   - Unspecified channeling, excessive amounts of channeling or excessive channel depth indicate an inferior engraving
4. Measure the cell volume using the agreed upon volume measurement method (Section 20.4.2.2). Usually three readings across the roll face are sufficient to determine the average cell volume of the roll.
5. Photograph the engraving or request that the supplier provide a photograph of the anilox roll engraving.
6. All anilox roll information, both CoA's and in-house inspections should be documented and maintained for reference.

20.4.2.6 Anilox Roll Maintenance

To maximize the life of an anilox roll, regular inspection and maintenance procedures should be implemented. Anilox rolls should be routinely inspected to monitor the cleanliness and wear of the engraving. Maintain a data sheet for each roll recording: manufacturer, cell count, cell volume, engraving angle, age, location, cleaning and inspection history.

Wear and ink dried in the cells reduces the actual print volume of the anilox roll. Press wash-ups are critical to keeping the anilox roll clean, which will help maintain the print density and extend the life of the anilox roll. To minimize wear, run proper doctor blade pressures and install/maintain magnets and filters in the ink delivery system to reduce the potential for score lines. If a decrease in density occurs when printing with a clean roll, the roll should be inspected to verify the current cell volume agrees with the actual cell volume when the roll was new. Anilox roll suppliers should assist in designing and implementing an anilox roll monitoring and maintenance program.

Anilox Roll Cleaning Methods
Even with on-press cleaning during press wash-ups, invariably anilox cells plug with dried ink. When plugged, the ink carrying volume of the roll decreases erratically. This plugging causes a change in the rolls printing characteristics. Many systems and techniques exist for cleaning anilox rolls. Regardless of the system used, cleaning rolls to maintain consistent cell volume is critical to achieving specifications. Cleaning methods include:

- Cleaning compounds and brushes
- Ultrasonic
- Bicarbonate of soda or poly bead bombardment
- Caustic soak with high-pressure water rinse

20.5 Plate Package

The combination of plate, sleeve and mounting material, referred to as the "plate package", interact to influence print quality. The plate package must be standardized to achieve optimum print results. Changing any of these components will influence the print outcome and therefore require re-characterization. *FIRST* recommends optimizing the combination of plate, sleeve and mounting material prior to fingerprint trial. The optimum combination of components (plate, sleeve, mounting material) should be recorded on the job set-up form. The same combination should be used in the characterization trial and each time the job is printed.

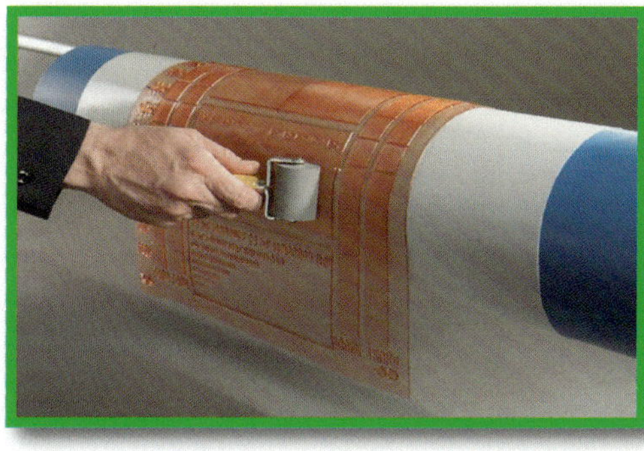

20.5 Plate Package: *The combination of plate, sleeve, and mounting material, referred to as the "plate package", interact to influence print quality.*

20.5.1 Plate Type

Flexographic printing plates are produced using many different materials and parameters. Different plate materials produce different ink release characteristics. Refer to Prepress Section 17.0 for more information on platemaking specifications and tolerances. The primary plate variables influencing print quality include:

Plate Material: Ink release and imaging traits vary widely among the many materials available. Flexographic plate materials can be grouped into two broad categories: rubber and photopolymer. Even within a category of material, such as photopolymer, the print characteristics vary dramatically by type of material:

1. Polymer or combination of polymers.
2. Material format: sheet, sleeve, liquid.
3. Material imaging method: conventional film, laser ablation, or laser engraving.

20.5.1 Plate Quality: *The plate imaging method, plate exposure, plate wash, plate drying, and post exposure all influence finished plate quality characteristics such as: dot structure, shoulder angle, and plate relief.*

20.5.2 Mounting Tape Density: *Print quality characteristics, such as dot gain and pinholing in solids, are directly influenced by the density of the mounting tape.*

Plate Durometer: Plate types vary in durometer (hardness). Durometer influences impression pressure, dot gain/TVI, plate durability and the effectiveness of mounting materials. Hardness is measured using a Shore A gauge on a scale of 0 (soft) and 100 (hard). ASTM D2240 durometers allows for a measurement of the initial hardness, or the indentation hardness after a given period of time. The basic test requires applying the force in a consistent manner, without shock, and measuring the cured photopolymer plates are generally between 25 to 70 Shore A. The optimum durometer is determined by substrate and graphics. Most film and paper printing use plate materials with a cured plate durometer of 45 to 60 Shore A. Rough and uneven substrates, such as corrugated board, require lower-durometer materials of 25 to 40 Shore A.

Capped Plates are "dual-durometer"; a thin (0.004"-0.010") harder image surface lies atop a lower-durometer polymer base. Some contain a dual durometer and others modify the surface texture to increase the quantity of ink collected on the plates surface. The advantages of capped plates include wider exposure latitude, less distortion on the printing surface, a deeper relief and less dot gain/TVI from plate-to-substrate impression.

Plate Caliper: Thickness (caliper) influences imaging traits and tonal range. Usually thinner plates cause less image distortion. The industry trend is toward thinner plate materials for this reason. Generally, a thin plate is defined as having a caliper of 0.112" or less.

Plate Uniformity: Plates must have uniform caliper to avoid over impressing. Monitor photopolymer exposure times to ensure consistent imaging caliper. The more accurate the plate caliper, the longer the plate life on press. A caliper tolerance of +/- 0.0005" (12 microns) or better should be maintained.

Plated Imaged Relief: Reference Table 17.1a in Prepress Section 17.1 for general guidelines on plate and sleeve relief specifcations.

Plate Processing Method: Controlling the plate imaging process is critical to producing a quality flexographic plate. All of the variables in the imaging process directly impact print quality. Photopolymer plate quality is influenced by the method used for imaging the polymer, plate exposure, plate wash, plate drying and the degree of post exposure.

20.5.2 Mounting Tape

All mounting materials that adhere plates to the print cylinder, sleeve, or carrier sheet have inherent gauge variation that can significantly influence print quality. Some types and brands have more variation than others. Similarly, the individual characteristics of the mounting material can change the overall hardness of the plate/mount combination, significantly influencing plate performance. Print quality characteristics influenced by mounting tape include:

- Solid ink density and uniformity (mottle)
- Dot gain/TVI and dot structure
- Slur and banding

There are two general categories of mounting materials: hard tapes and compressible backings (foam tapes). Hard tapes are very thin double-sided tapes that do not cushion the impression of the plate to the substrate. Compressible backings cushion the impression of the plate to the anilox and substrate, reducing dot gain/TVI by compensating for imperfections in the plate and cylinder TIR. Compressible backings also cushion (like a shock absorber) any mechanical bounce created by certain image designs or press components. Mounting tape variables include:

Thickness: The thicker the cushion backing, the more "give" it has on press. Thicker tapes have greater variances in caliper because they are made of a foam material. Thinner tapes provide less plate cushioning and less caliper variance.

Density: Foam tapes are generally listed as high, medium and low density. Medium density tapes cushion the plate less than low density tapes. Although low density tapes normally produce less dot gain/TVI, they may also cause pinholing in the solid areas or on the line portion of a combination plate. Medium density tapes print solids smoother but create more dot gain/TVI. High density tapes are most appropriate for large solids.

Resiliency: Foam tapes are constructed as a closed cell or open cell design. Closed cell tapes generally have a better memory or ability to rebound after the impression pressure is relieved. This helps the tape last longer on long production runs, therefore it has greater resiliency.

20.5.3 Sleeves

Sleeve Materials

Nickel Sleeve: A nickel sleeve is a thin seamless metal tube made to varying diameters, lengths and thicknesses. It is the thinnest sleeve available, permitting the printer to become a sleeve user with minimum modifications to existing equipment. The only modifications required are:

1. Air holes drilled in plate cylinder.
2. Thinner mounting tape or plate used to compensate for the sleeve thickness.

20.5.3a Sleeve Mounting: *Typically, sleeves are mounted onto mandrels or print cylinders using compressed air.*

Nickel Sleeve Specifications	
Surface Hardness	40 Rockwell C
Tensile Strength	200,000psi (13790 Bars)
Diameter	3" - 24" (76.2mm - 610mm)
Length	Up to 145" (3683mm)
Thickness Range	0.003" - 0.015" (0.076mm - 0.381mm)
Standard Thickness	0.005" (0.127mm)
Surface Finish	>2Ra

Table 20.5.3a

Composite Sleeve: A composite sleeve is made of a polymer (plastic), which is reinforced with a fiber, such as fiberglass or carbon fiber. Various polymers and fibers are combined to obtain specific characteristics and properties and to achieve the desired wall thickness.

1. **Fiber:** Fiberglass is the most widely used fiber, primarily because of its low cost. Additionally, tensile strength and strain to failure, heat and fire resistance, chemical resistance, moisture resistance, thermal properties and electrical properties are also advantages of fiberglass.
2. **Polymer:** Epoxies, Polyesters, Vinyl Ester and Polyurethanes are the most widely used polymers. Polymers are used as the matrices to bind together the reinforcement material (fiber) in some composite sleeves. However, most sleeves now utilize polyurethane materials to build wall thickness and achieve a smooth mounting surface.

Sleeve Design

Parallel/Cylindrical Sleeves: Cylindrical sleeves are the most common sleeves used in flexography today. They have constant and parallel inner and outer diameters and are designed to mount on existing plate cylinders or cantilevered sleeve presses riding a cushion of compressed air.

Tapered/Conical Sleeves: Conical sleeves have a tapered inner diameter and constant outer diameter and are designed to mount on matched, tapered mandrels or print cylinders. Tapered sleeves are most often used in applications where nip pressure is much higher than in standard flexographic printing environments, such as solventless laminating.

Laser-Engraved Rubber-Covered Sleeves

Material: Rubber can be vulcanized onto a nickel or composite fiberglass sleeve and then laser engraved (Section 17.4)

Composite/Mylar Sleeve Specifications	
Surface Hardness	Varies by Manufacturer. Approximately 75 Shore D.
Diameter Range	2" - 24" (51mm - 610mm) Varies Depending on Print Repeat Specifications.
Length	Up to 145" (3700mm)
Wall Thickness Range	0.015" - 4"+ (0.51mm - 102mm+)
Standard Thickness	Does Not Exist. Sleeves are Made to a Variety of Wall Thicknesses.
Outside Diameter Tolerance	Varies by Manufacturer; Generally +/- 0.0005" - 0.0008" (0.0127mm - 0.02032mm)
Total Indicated Runout (TIR)	Varies by Manufacturer; Generally < 0.001" (0.0025mm)

Table 20.5.3b

to create a continuous printing design roll or used as a full coverage coating/tint roller. Fiberglass sleeves can be provided with either a smooth or rough finish for the vulcanization process depending on what the rubber requires. Urethane covered/polyester sleeves are not suitable for the vulcanization process because most urethanes/polyesters cannot be exposed to the high temperatures required to vulcanize rubber.

Cleaning Laser-Engraved/Coating Sleeves: Clean the surface of the sleeve using a horsehair brush and a solvent recommended by the ink supplier for the proper breakdown of the ink. Remove all ink from the sleeve and let it dry completely before storing. Do not clean sleeves with rags, nylon, cheesecloth, etc. as these products will wear the edges of the plates and destroy process or screened plate dots.

Sleeve Mounting & Demounting

Typically, sleeves are mounted onto mandrels or print cylinders using compressed air. Compressed air is used to expand the sleeve in order to slide it onto the mandrel or print cylinder. Once in position, the air is disconnected and the sleeve clamps tightly into place.

It is necessary to conform to the air pressure and volume requirements as described by the manufacturer. The recommended range in pressure and volume required for the safe mounting and demounting of the different sleeve materials varies by manufacturer and sleeve thickness.

Sleeve Storage

1. Vertical storage of sleeves is highly recommended. Store sleeves vertically with adequate space between each sleeve so the sleeves are not in contact with one another. Several commercial sleeve storage systems are available to accommodate most printers' needs for safely storing sleeves.

20.5.3b Sleeve Storage: *Store sleeves with adequate space between each sleeve so the sleeves are not in contact with one another.*

20.5.3c Sleeve Care: *Sleeves are classified as a semi-consumable; they can last for many years if properly maintained and stored.*

2. Sleeves can be stored horizontally, but should be supported internally using as much surface area for support as possible. Printers should exercise caution when horizontally storing large diameter, thick wall sleeves, due to the heavier weight.
3. If sleeves are stored with plates mounted in lighted areas, they should be wrapped with a barrier material, such as dark plastic, so that oxidation is minimized.
4. Sleeves should ideally be stored under climate-controlled conditions to eliminate exposure to extreme temperatures (heat and cold). Sleeves that have been exposed to excessive cold or heat should be allowed to return to room temperature prior to mounting.

Sleeve Care & Maintenance
Flexographic printing sleeves are classified as a semi-consumable; they are not designed to last forever, nor should they fall apart after a few uses. The primary determinant of how long a sleeve lasts is how well it is cared for. Sleeves can last for many years if properly maintained and stored. In general, printers should adhere to the following guidelines. For more detailed care and handling instructions, contact the sleeve manufacturer.

1. Although nickel sleeves are cut and scratch resistant, heavy cutting on the surface will shorten the life of the sleeve.
2. Ink should be removed from the surface of the sleeve as soon as possible. Ink that is allowed to dry on the sleeve can create mounting and/or diameter issues.
3. When cleaning the surface of the sleeve, use as small amount of cleaning solution as possible. Solvents such as acetate, acetone, etc., should be in concentrations less than 25%. If acetate or acetones are used to clean solvent ink, it is critical the sleeve surface is cleaned with alcohol to remove any residual chemicals that would prevent bonding between the mounting tape and the sleeve surface.
4. After cleaning the surface, allow the sleeve to completely dry before re-mounting (usually 5-10 minutes under normal pressroom conditions).
5. Periodically clean the inside diameter of the sleeve to ensure a good, smooth mounting surface. Dirt, ink and other materials can build up on the inside diameter of the sleeve creating interference when mounting under compressed air; resulting in a tight fit.

Register Pin Standards: Sleeve Pre-Positioning	
Mandrel Location	Gear/Drive End
Pin Diameter	6mm (0.236")
Pin Location: Varies by Press Type / OEM	7 - 18mm (0.276" - 0.709") from Gear/Drive End
Pin Height	2mm (0.787") +/- 0.1mm

Table 20.5.3c

20.5.4a Plate Mounting: *The maximum allowable mounting registration tolerance is less than half the dimension built into the art for trap.*

20.5.4 Mounting Methods

Mounting Objectives

The mounting and proofing process verifies a job is ready to be printed and is capable of both accurate impression and accurate register.

Mounting For Accurate Impression: A quality print result is dependent upon even impression; therefore, each component that contributes to total caliper variation must be evaluated prior to plate mounting.

1. **Printing Plate:** Before a printing plate is mounted, it should be inspected for defects and measured with a micrometer to ensure the caliper tolerance is within specifications. Refer to Section 17.9 for Printing Plate Measurement and Control. It is helpful to mark the caliper reading on the plate for review by the mounter while laying out the work.
2. **Print Cylinder:** The bare cylinder face and journals should be measured for total indicated runout (TIR) as well as consistent diameter before the mounting procedure begins.
3. **Sleeve:** The TIR and diameter of the sleeve must be measured after the sleeve is applied to the base mandrel. It is wise to mark the position of the sleeves on the plate mandrel if a demounted sleeve is to be re-used for multiple runs.
4. **Mounting Material:** Although it is not currently possible for a plate mounter/press operator to accurately measure mounting tape for gauge, foam density and consistency, inspection for visible defects is recommended.

Mounting for Accurate Register: Each job must have the correct copy mounted on the correct cylinder in register to the other cylinders. The mounting process should strive to minimize its contribution to registration inaccuracies. The registration tolerance built into the artwork is intended to allow for on-press, mechanical variation. The maximum

Air Actuated Sleeve Mounting Mandrel & Cylinder Specifications	
Mandrel/Print Cylinder	Seamless Tubing Preferred
Header Journal	Either Machined from Solid Round Stock or Welded Plate Construction
BCD (Bare Cylinder Diameter)	Zero to + 0.0006" (0.015mm)
TIR (Total Indicated Runout)	0.0006" (0.015mm)
Cylinder Dimensional Stability	Thermally Stress Relieved
Air Hole Placement: Cylinder Diameter Less than 4.676" (121.074mm)	4 - 3/32" (2.38mm) Diameter Holes Equally Spaced Circumferentially
Air Hole Placement: Cylinder Diameter Between 4.767" - 9.779" (121.074mm)	6 - 3/32" (2.38mm) Diameter Holes Equally Spaced Circumferentially
Air Holes: Cylinder Diameter Greater than 9.779" (248.398mm)	8 - 3/32" (2.38mm) Diameter Holes Equally Spaced Circumferentially
Air Hole Distance: Measured from the Top of the Bevel/Leading Edge to the Center of the Air Hole (Referred to as the "Seal Face")	0.39" - 0.47" (10 - 12mm)
Bevel / Leading Edge	0.16" – 0.25" with 15° Bevel (4mm - 6.35mm with 15° Bevel)
Downstream Air Holes (As Needed)	2 - 0.78" (2mm) Diameter Holes Located Opposite Each Other at Zero & 180°; Usually Placed at Half the Cylinder Face Length
Air Inlet	1/4" (0.64cm) NPT, 1 Hole Either Through Journal or Header
Air Pressure: Composite Sleeves (Nickel, Fiberglass, Kevlar, etc.)	87 - 116 psi (6 - 8 bars)
Air Pressure: Mylar Sleeves	60 – 100 psi (4 - 7 bars)
Chrome Plated Cylinders	Minimum 0.0005" (0.0127mm) Chrome
Chamfer	5°
Chamfer Length	4mm (0.157") from Mandrel End
Internally Piped Cylinders	Optional

Table 20.5.3d

allowable mounting registration tolerance is less than half the dimension built into the art for trap. Refer to Section 19.4.7 for general image trap recommendations by print segment. On critically registered jobs, the pitch diameter of each mounted print cylinder should be verified within the allowable tolerance. Differences in print height around the cylinder, from color to color, will cause registration problems.

Mounting Equipment

Optical Mounting: Optical mounters, the oldest of the three types of mounting equipment, operate on the principle of mirrors that reflect a proofed image from the impression drum onto the plate cylinder. It is critical when using this type of mounter to verify that the mirror alignment is accurate and the cylinders are parallel before beginning to mount a job. This method of mounting is the only one that allows visual inspection of the total image; this is critical when mounting rubber plates without a stable backing. It is possible to achieve specifications for mounting using an optical mounter; however, it is dependent on operator skill. A mounter proof should be pulled to verify the concentricity and fit of the mounted plates before going to press.

Pin Register Mounting: This method mechanically attaches the plate to the cylinder. Pinholes are either drilled or punched in the negatives and in the raw plate material to ensure accurate register. Finished plates are then mechanically attached to a pin bar or table which applies the plate squarely to the plate cylinder. This equipment can only be used with stable backed plates, such as photopolymer or rubber with a stable backing. With this type of mounting, each color is mounted as a single plate. Therefore, stepped negatives of multiple images can be mounted. The accuracy of this mounting technique is dependent upon the ability to accurately place the pinholes in the plate material. Verify the parallelism of the mounting pin bar to the plate cylinder at the point of contact to ensure accurate register color-to-color. The speed and repeatability of mounting is usually an improvement over optical mounting; however, the degree of accuracy is not always consistent.

Video Mounting: Video mounting requires stable-backed plates and is the newest approach to mounting. Registration is achieved through the use of microdots (typically 0.01" diameter) that are accurately imaged in the film or placed in the file and remain permanently on the printing plate. Mounting is done by aligning video cameras into datum positions based on the location of the dots and their position on the web. Magnification of the dots in the video monitor varies by manufacturer. A magnification of at least 16X should be used. Magnifications of 16X and higher ensure superior accuracy in mounting registration and repeatability color-to-color. Dot-to-dot registration can be achieved within ±0.001" (0.025 mm). In narrow web applications, registration cross hairs are used instead of microdots. Typically, the crosshairs are located in the matrix and are 0.0625" (1.6 mm) tall.

20.5.4b Video Mounting: *Registration is achieved through the use of microdots that are accurately imaged in the film or placed in the file and remain permanently on the printing plate.*

Mounter's Proof

Ultimately, it is the responsibility of the printer to print within agreed-upon specifications. A mounter's proof is an actual proof of the printing plate created by rolling ink onto the plate, which is mounted on a plate cylinder, and rolling it against a drum with

20.6 Contract Proof: *The contract proof represents the customer's expected content and appearance of the printed product. It is the basis for negotiations on project performance.*

a substrate attached to it. A mounter's proof is not profiled and not used for matching color. It is used to confirm plate cylinder (TIR, concentricity, fit) and image accuracy (register, impression, placement, contents, etc.) prior to going to press. In narrow web, typically the press serves as the proofing press. A mounter's proof is often used in wide web applications.

20.6 Contract Proof

The contract proof represents an agreement between the printer, prepress provider, designer and the customer regarding the expected content and appearance of the printed product. It is an important quality control tool and communication device and is the basis for negotiations on project performance. The contract proof does not have to be a dot-for-dot reproduction, but it must be an overall visual simulation of the expected print result. Therefore, it must simulate the dot gain/TVI, color attributes, detail and contrast of the printed image. Either analog or digital proofing technologies can be successfully used to create a contract proof to *FIRST* specifications.

Before a contract proof can be accurately used, the entire reproduction system must be characterized to calibrate the proof to match the printed result. The proof is profiled using a color management system (CMS) and is prepared based upon profiles provided by the specific printer or prepress provider. It must be made according to *FIRST* specifications. Once the proofing system and press have been profiled, both must be maintained for consistency and predictability. Refer to Section 16.1 for additional information on contract proofs.

Proofing Compliance Cover Sheet & Certificate of Result

A "Proof Compliance Cover Sheet," or label, must accompany the contract proof submitted for color match at press and approved by the customer. It should identify the proofing product or system used, the company supplying the proof (contact name, telephone/fax numbers), as well as operator, date, job number and customer. The cover sheet must also contain information required to verify the proof's compliance to the technical attributes required for that proofing type (Section 16.5). It is a best practice approach for all proofs to include a "Certificate of Result". It should include all pertinent measurements: density, dot area, Delta E (@ 100% & 50%), trap, print contrast, bar code scan analysis, etc. Proof densities should be within the printer's on-press density specifications (Section 19.4.4). The Proof Compliance Cover Sheet and Certificate of Result can be combined into one document.

Proofing Pigments/Dyes
FIRST has identified process and line ink pigments by color index number (CI #). The *FIRST* recommended process ink pigments are further defined by CIELAB color values. The pigments used in proofing inks or dyes should match, as closely as possible, the spectral response of the inks being used to print the commercial job. It is the printer's responsibility to communicate the ink pigments used on press to the proofer.

Using the same pigment (referenced by the color index number) will minimize the risk of metamerism. If the proofing method employed does not utilize pigments, the CI # of the pigments being used by the printer will facilitate the identification of the best match available for the given proofing colorant. Refer to Sections 20.2.2 and 20.2.3 for *FIRST* recommended pigments.

21.0 Press Component Print Variables: *Press components such as dryers, registration controls, tension controls, and press mechanics can have a direct impact on print quality.*

21.0 PRESS COMPONENT PRINT VARIABLES

21.1 Press Dryers

Drying Solvent & Water Based Inks & Coatings
Flexographic press dryers utilize hot air to remove the volatile portion of the ink from the printed substrate. As the hot air contacts the printed substrate, the temperature of the substrate, ink solids, and ink solvents increases, resulting in evaporation of the ink solvents. There are two types of flexographic press dryers:

1. **Between Color/Interstation Dryers:** Interstation dryers remove a sufficient amount of the volatile solvents from the printed substrate to allow for a dry trap of the next down color.
2. **Main Tunnel/Overhead Dryer:** The overhead dryer removes the volatile solvent from the last print station and completes the heat setting of all the inks applied to the substrate.

The basic dryer components include a supply fan, burner and exhaust fan. The airflow schemes vary with press design. These components are critical in controlling the three primary variables that influence dryer efficiency:

Air Temperature: Higher temperatures result in faster drying. However, an excessive dryer temperature can damage the substrate (especially films but paper substrates as well). Too high temperatures can also dry the top surface of the ink, trapping solvent in the ink. Trapped solvents result in odor problems as well as creating pockmarks or "fisheyes" on the printed surface. Optimum dryer temperature is influenced by ink chemistry (solvents used and solvent load), substrate and

21.1 UV Dryers: *For better, deeper, or faster curing of some colors, or of some specialty formulations, other materials may be added to the mercury in the lamp to alter the wavelength of the emitted light.*

press speed. Measure the temperature of the printed web as it exits the dryer section of the press using a pyrometer. Analog and digital pyrometers are available which measure the surface temperature by either a probe or infrared beam.

Air Velocity & Volume: The greater the velocity and volume of heated air directed on the substrate, the quicker the volatile ink components evaporate. The ink must reach the ink-vapor temperature for evaporation to occur. The moving air helps carry the volatile components away from the substrate, speeding additional evaporation. The horsepower of both the supply and exhaust fan determine dryer velocity and volume. The dryers must be clean, the nozzles clear and the airflow balanced for maximum velocity. Unbalanced dryers may cause air to escape and blow onto the plates. This causes dirty print and reduced color density. Adjusting the airflow (velocity) down too low causes poor ink trap and drying. General definitions of air velocity:

1. Low Velocity = < 5,000fpm
2. Medium Velocity = 5,000-10,000fpm
3. High Velocity = > 10,000fpm

Exhaust volume should be greater than the supply volume. This imbalance keeps the dryer under a slight negative pressure, preventing volatile air from escaping into the pressroom. Design and maintain the exhaust airflow volume from the dryer to prevent the vapor concentration in the dryer from exceeding 25% of the lower explosive limit (LEL).

Dwell Time: Sufficient dwell time along with appropriate air velocity and temperature, allows the printed substrate to get hot enough to vaporize the volatile component of the ink without overheating the substrate. The length of the dryer and the press speed determines the dryer dwell time. Interstation dryers are typically </= 1 foot long. The printed web passes through the interstation dryer in a fraction of a second. Main tunnel dryers are typically ~15 feet or longer. The actual length varies by press design. The printed web spends one second or more in the main tunnel, significantly longer than in the between deck dryers.

Radiation Curing of UV & EB Inks & Coatings

UV and EB curing utilizes ultraviolet light or electron beam, respectively, to polymerize a combination of monomers and oligomers in the ink or coating.

For UV curing, ultraviolet radiation (light) is generated using either mercury lamps, pulsed xenon lamps, or lasers. With mercury lamps, either an electrical current or microwave radiation vaporizes the mercury into a gaseous state, naturally emitting

Print Defects Related to Dryer Temperature & Air Volume	
Insufficient Drying	Ink Pick-Off
	Poor Ink Trap
	Poor Adhesion: Fails Tape, Crinkle, and/or Rub Test
	Bleed
	Blocking: Due to Insufficient Drying
Excessive Drying	Blocking: Due to Web Re-Wound Too Warm
	Brittleness: Ink Film or Substrate Breaks or Cracks When Flexed

Table 21.1

radiation in the ultraviolet frequency. Polished reflectors direct the light onto the web. For better, deeper, or faster curing of some colors, or of some specialty formulations, other materials may be added to the mercury in the lamp to alter the wavelength profile of the emitted light.

Electron beam accelerators generate the electron stream for curing EB inks and coatings. Unlike photons of light, which tend to be absorbed mainly at the surface of the material, electrons have the ability to penetrate through matter. Primary variables for radiation curing:

1. **Dwell Time:** Controlled by press speed and lamp length.
2. **Radiation Intensity:** Influenced by the power setting of the lamp, lamp bulb life, cleanliness of the lamp reflectors, accuracy of the lamp focus, lamp shutter opening, and ink film thickness.

Reducing Web Temperature

Controlling the temperature of the printed web is a fundamental challenge for radiation curing systems. Both dwell time and radiation intensity influence web temperature. There are three basic methods for reducing web temperature:

1. **Remove Heat From The Web:** Chill rollers are cooled with air or water to maintain their capacity to transfer heat. The location of the chill roller in relation to the dryer determines its effectiveness.
2. **Avoid Heating The Substrate:** Run the lamp at a low power setting to significantly reduce web temperature. Increasing press speed will also reduce the temperature of the web by decreasing exposure time. Design parameters, such as lamp diameter and reflector shape, impact web

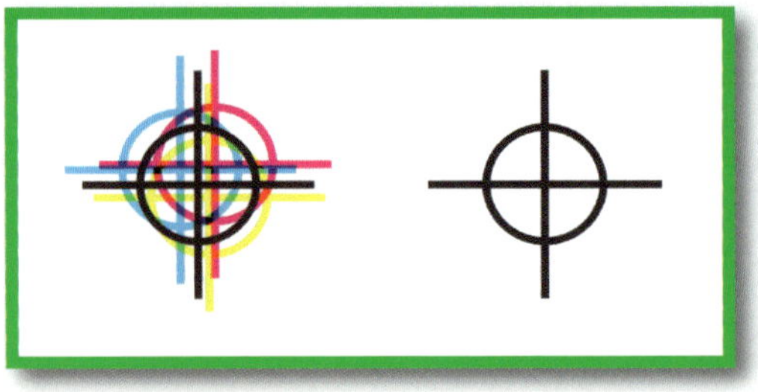

21.2 Traditional Registration Target

temperature as well by controlling the amount of infrared (IR) radiation emitted and the concentration of the IR radiation on the web.

3. **Use IR Filters:** IR filters restrict the amount of IR energy reaching the web by altering some wavelengths of energy and not others; thereby, reducing the peak temperature of the web.

21.2 Registration Control

In order for a job to print successfully, the printer must achieve and maintain both color-to-color registration and print-to-substrate registration throughout the pressrun. Many variables influence print register such as:

- Location and movement of the web
- Web tension (stretching)
- Print cylinder TIR
- Gear condition
- Plate/sleeve mounting method and accuracy
- Press design (CI, Inline, Stack)
- Press maintenance (play in the decks)

Print registration is achieved, monitored, and maintained throughout the pressrun using three primary technologies: side and circumferential registration methods, web guides and web viewers.

Side & Circumferential Registration

Side and circumferential registration can be achieved using either mechanical or hydraulic/electrical devices.

Mechanical Methods: Registering a job quickly and accurately using mechanical methods requires more operator experience and skill. Side registration is adjusted using a hand wheel (or other mechanical device), which when connected to the plate cylinder, laterally moves the plate cylinder. Circumferential register is adjusted using a hand wheel (or other mechanical device), connecting the print cylinder gear to a helical gear, to slide the cylinder forward or backward. This adjusts the circumferential register without adjusting the side register of the plate cylinder.

Hydraulic/Electrical Methods: Automatic deck positioning systems allow both side and circumferential registration automatically, resulting in much closer registration upon rack-in and gear mesh than mechanical systems. These devices have the capability to be pulsed, allowing for incremental movement, which provides the operator with fine register capability. Pulse registration affords the operator the ability to adjust and fine-tune registration from a remote area, such as a web viewer or video monitor.

Web Guiding
Web guiding systems control the position of the web as it moves through the press. The typical web guiding system consists of several components integrated into a closed control loop. Components include a sensor, controller, hydraulic actuating cylinder and the web. There are three locations on a flexographic press where guiding is typically applied: the unwind, prior to printing and the rewind.

Unwind Guiding: The unwinding roll shifts laterally. The web guide sensor is fixed so that the edge of the substrate is aligned with the press. It corrects for misalignment of mill roll position, roll defects such as telescoping and poorly wound rolls.

Intermediate Web Guides: If there is a long span between the unwind and the first print station, an intermediate web guide is often installed immediately ahead of the first print station, either in lieu of or in addition to the unwind guide. The intermediate guide corrects for any web position errors that may occur in the span between the unwind and the first print section.

Rewind Guiding: The rewind guide is not truly a lateral control of the web; it is actually a chasing control system. At the rewind, guiding is accomplished by attaching the sensor to the rewind stand and shifting the stand to chase the web as it comes from the press. If there is not sufficient frictional engagement between the web and the last fixed idler roll, a finished roll with ragged edges will be produced.

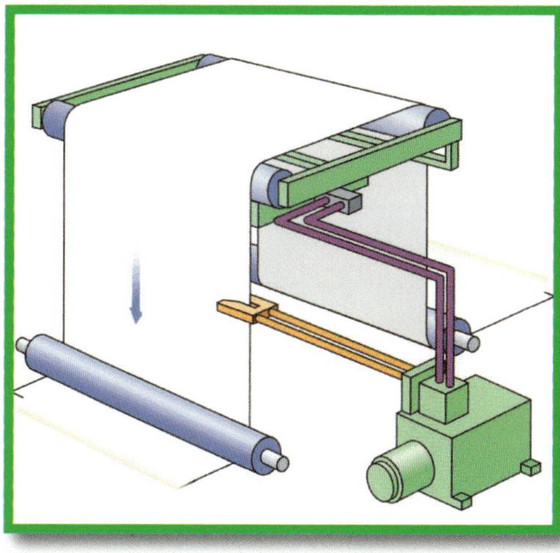

21.2a Web Guiding: *Web guiding systems control the position of the web as it moves through the press.*

The range of substrates printed, the press geometry, and auxiliary equipment layout combined with various guide arrangements and costs determine the choice of specific guide types and locations. There are four types of web guide systems:

1. **Edge Guiding:** The sensor detects the web edge and the guide system maintains this edge at the desired lateral position throughout the pressrun.
2. **Fixed Sensor Center Guiding:** Two sensors are held in a fixed position; they detect both edges of the web. The guide system maintains the centerline in an exact position and accommodates small web width variations throughout the pressrun.
3. **Moving Sensor Center Guiding:** The sensors continuously reposition themselves automatically to detect both web edges and to maintain the centerline of the web in an exact position. This method is used in

21.2b Rotating Drum Mirrors: *When printing transparent or translucent substrates, it is advantageous to utilize a rear lighting fixture for improved web illumination.*

applications where there are large web width variations during a production run.

4. **Line or Pattern Guiding:** The sensor detects a printed line, pattern, or some distinguishable feature on the web. The system then maintains the printed line, pattern, or feature in an exact lateral position regardless of the web edge position.

Web Viewers

Web viewers enable the operator to evaluate the printed web while the press is running. How efficiently an operator is able to evaluate the printed web, depends upon the capability of the web viewer to provide a stable image of high enough quality that precise definition of detail, registration, color, etc. is readily visible at all printing speeds. There are four basic technologies used for web viewing:

1. **Stroboscope:** This is the simplest web-viewing device; it consists of a lamp and voltage pulse source. It provides a brief instant flash in a frequency synchronous with the print repeat length. While it is the least expensive method, it has severe limitations:
   - Small viewing area
   - Poor image definition
   - Operator eye fatigue and discomfort
   - Viewing frequency is dependent on repeat length and press speed
2. **Oscillating Mirrors:** This method consists of a rectangular mirror oscillating in synchronization with the print repeat length. The viewing area is generally 18" to 20". It can be mounted on side-motion tracks allowing it to traverse the web width. Only a portion of the print repeat length is visible through the viewing area at any given time. The advantages over the stroboscope technology include:
   - A constantly illuminated web, reducing operator fatigue
   - Increased visual dwell time, improving the operator's ability to evaluate print detail

   The primary disadvantages of oscillating mirrors technology include:
   - Speed limitations
   - Lack of image stability, must continually refocus on the image each time a print length is viewed
   - Only able to view web repeat lengths of 6" to 12"
3. **Rotating Drum Mirrors:** This method uses a rotating drum of mirrors to offset the speed limitations inherent in the previously discussed web viewers. The multi-sided

drum of mirrors is rotated in synchronization with the viewed web segment of the given repeat length. This method generally has a fixed viewing width of 18" to 20". It is mounted on an overhead track assembly to traverse the total web width. It can provide suitable web scanning at speeds up to 2,000fpm. When printing on transparent or translucent substrates, it is advantageous to utilize a rear lighting fixture for improved web illumination. Because of the image stability of a rotating device, scopes are available for print magnification in the range of 5X to 10X. Critical inspection of small details is realistic.

4. **Video Scanning:** These systems are the most popular form of on-line print inspection methods. Video inspection systems are capable of viewing the printed image at high press speeds without jitter and distortion. They are also capable of high magnification and can detect print defects normally seen only with the aid of a magnifying glass on a stationary web. They are particularly helpful in detecting registration errors, doctor blade streaks, dot gain/TVI and bar code verification. They are capable of monitoring preset tolerances and warning the operator when the press exceeds the tolerances for registration, color variance, repeat variation, bar code scannability, etc. By changing the illumination, they also have the capability of viewing clear varnishes without using additives. The basic components of a video web inspection system include:
   - **Camera Assembly:** The high-resolution color camera with motorized zoom lens and close-up lens determine the overall magnification of the system. The zoom lens also controls the focus and iris. The iris controls the amount of light reflected into the camera and hence the brightness of the image.
   - **Camera-Positioner Assembly:** Manual positioners are primarily used on narrow web presses. Motorized positioners provide basic left-right jog capability. Programmable positioners offer multiple, user-programmable positions to be stored and then activated in a continuous loop. For example, the operator is able to store the position of the bar code, TM symbol, fine reverse type element and registration marks for a given job and recall it every time the job is printed. The camera-positioner assembly is installed after the last line operation that requires inspection, typically after the dryer but prior to the rewind.
   - **CPU and Software:** The software gives the computer instructions to follow. Examples include: instructions

21.2c Video Scanning: *Video inspection systems are capable of viewing the printed image at high press speeds without jitter or distortion.*

for capturing and displaying images, automatic color monitoring, register monitoring and bar code verification.

- **Signal Input Device/Press Timing Device:** Provides the inspection system with speed and timing information. The timing signal enables the image on the screen to be advanced or retarded, in the machine direction.

21.3 Tension Control

Accurate web tension control is required to achieve production efficiency and product quality. The ability to control tension directly influences print register especially on extensible film. The inability to effectively control tension will result in:

- Slower press speeds to accommodate web handling problems
- Increased waste
- The inability to run a wide range of web thicknesses, widths, or materials

A typical flexographic press has more than one tension zone; each zone may require a different tension level. A tension zone is a length of web extending from one tension-affecting device (TAD) to the next. Typical TADs include: unwind and rewind core shafts (with attached motor, clutch, or brake), driven rolls, braked rolls, nip rolls (where at least one roll is driven or braked) and drag bars. There are three tension zones in a typical flexographic press: the unwind, intermediate and rewind tension zones.

Tension Zones

1. **Unwind Tension Zone:** The objective is to maintain constant tension from full roll to core. The unwind tension level should be equal to or less than the tension used at the rewind. Greater unwind tension than rewind tension can cause the roll to telescope and to tighten on itself (results in blocking). The problem is more serious with smooth, low friction materials than with rough or sticky webs. Extensible webs, such as film, are run with much lower tension than non-extensible webs, such as paper or foil, to prevent wrinkling, stretching and width reduction.
2. **Intermediate Tension Zone:** The objective of this tension zone is to maintain constant tension throughout the roll and press run. The ideal tension level may be higher or lower than the unwind tension. The process, web material, web thickness and width determine optimum intermediate tension. Extensible films must be run with low tension to prevent stretching, which causes

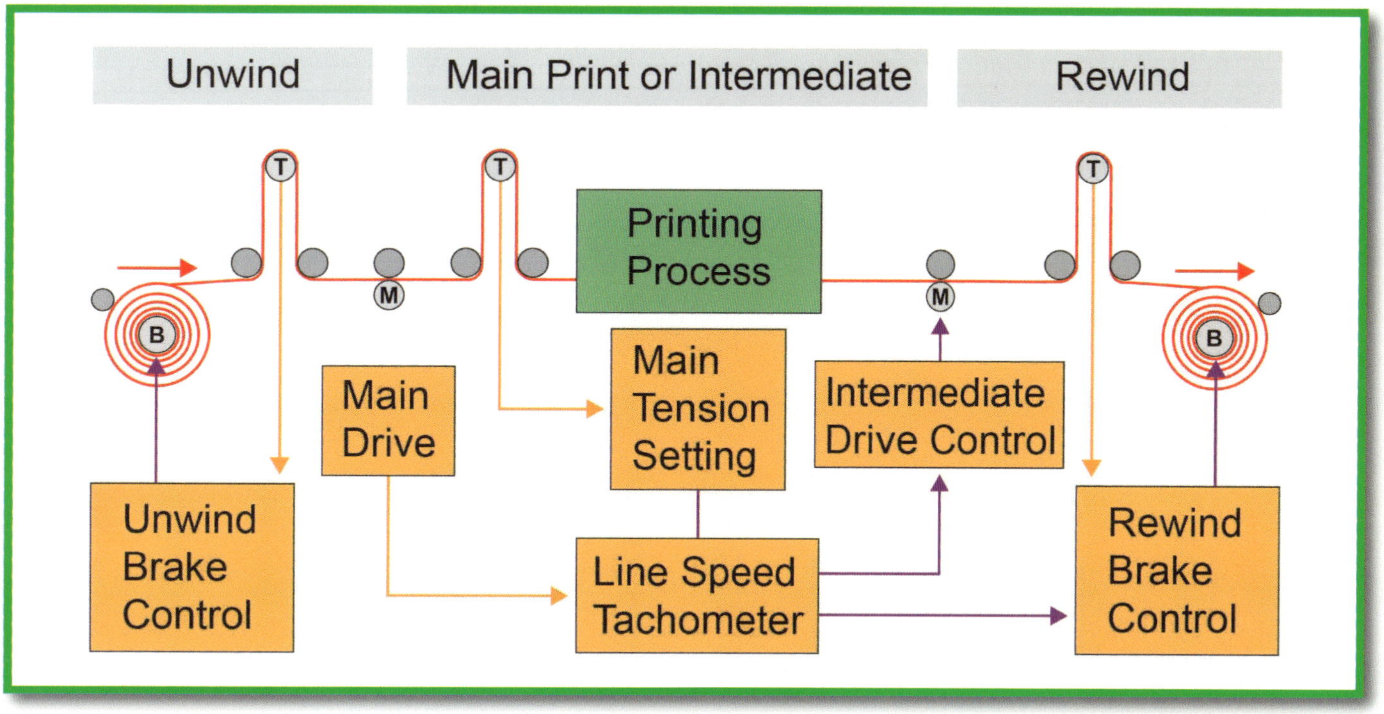

21.3 Tension Zones: *There are three tension zones in a typical flexographic press: the unwind, intermediate, and rewind tension zones.*

short print lengths and curling when tension is released.

3. **Rewind Tension Zone:** Either constant or tapered tension is used in this zone. The web material, the buildup ratio, and the tension capability of the rewind drive determine the optimum type of tension. The buildup ratio is the full roll diameter divided by the core diameter. Usually, buildup ratios (full roll:core size) of more than 5:1 require "tapered tension". A "tapered tension" has less tension at the full roll than at the core (ie. a 40% taper has a full roll tension that is 60% of core tension). The rewind tension determines the quality of the finished roll, which is the priority when setting rewind tension. This priority is only possible if the rewind tension zone is effectively isolated from the tension in the preceding zone by an efficient nip-roll system. If the rewind zone is not isolated, the ideal rewind tension must be compromised to accommodate the intermediate zone requirements. In this situation, finished roll quality usually suffers. Extensible webs, such as film, are wound with low taper or constant tension. Low-friction web materials, such as plastics and high gloss papers, are normally wound with high taper, 50% or more. Webs requiring high tension and a large buildup ratio need high taper to keep from exceeding the capability of the rewind drive.

Typical Tension Settings			
Film	Unwind Tension *(lb. per mil per inch of width)*	Rewind Tension *(lb. per mil per inch of width)*	Preferred Winding
Acetate	0.25#	0.50#	Taper 1.25:1.0
Foil (Aluminum)	0.25#	0.50#	Taper 1.5:1.0
Cellophane	0.375#	0.75#	Taper 1.25:1.0
Nylon	0.125#	0.25#	
Polyethylene	0.06#	0.12#	Constant or Taper 1.25:1.0
Polyester	0.375#	0.75#	Constant or Taper 1.25:1.0
Polypropylene	0.125#	0.25#	Constant or Taper 1.25:1.0
Polystyrene	0.50#	1.00#	Taper
Vinyl	0.025#	0.05#	Constant
Paper *(3,000 sq.ft. ream)*	Unwind Tension *(per inch of width)*	Rewind Tension *(per inch of width)*	Preferred Winding
20#	0.25#	0.50#	Taper 1.5:1.0
40#	0.625#	1.25#	Taper 1.5:1.0
60#	1.00#	2.00#	Taper 1.5:1.0
80#	1.50#	3.00#	Taper 2:1
Paperboard	Unwind Tension *(per inch of width)*	Rewind Tension *(per inch of width)*	Preferred Winding
8 pt.	1.5#	3.0#	Not Applicable: The majority of these substrates are in-line converted.
12 pt.	2.0#	4.0#	
15 pt.	2.25#	4.5#	
20 pt.	2.75#	5.5#	
25 pt.	3.25#	6.5#	
30 pt.	4.0#	8.0#	

Table 21.3

Adjusting Tension To Correct Print/Press Problems

Tension is often used to correct web handling problems, such as "loose edges" or the web not tracking properly. The problem with adjusting tension to compensate for web handling issues is that the "solution" usually creates other problems such as web breaks, stretching, wrinkling and print length variation. *FIRST* recommends addressing the root cause of the web handling problem rather than creating additional problems by increasing tension. However, there are several print problems that can be caused by inadequate tension control. These problems are best corrected by adjusting/controlling tension:

- Web breaks
- Wrap-ups around driven rolls
- Loss of color-to-color registration

- Deformation of web due to stretching or wrinkling
- Print length variations
- Web shifts side-to-side
- Curling or wrinkling of laminated webs
- Variation in coating thickness
- Core crushing: unwind or rewind

21.4 Press Mechanics

Parallelism

The impression cylinder, plate cylinders, anilox rolls, drive rolls and idler rolls must be aligned and parallel. When the anilox roll or impression cylinder is not perfectly parallel with the plate cylinder, one side of the print cylinder will begin to print before the other.

Another consideration is alignment of drive, nip and idler rolls in the web or sheet path. When not aligned, baggy web edges, wrinkles, drifting substrate and inconsistent register occur. The anilox roll and doctor blade chamber (or rubber roll) must also be in alignment with each other, and the print cylinder, for uniform ink metering.

Balancing Rolls & Cylinders

Balancing is a process whereby the distribution of mass in a roll or cylinder is altered to eliminate vibration at the bearings. The vibration is eliminated by "dynamic balancing". Weights are placed in two different planes of the roll or cylinder, perpendicular to the axis of rotation. Imbalance is caused by the lack of homogeneity in the roll or cylinder material, such as uneven wall thickness. The following problems are a direct result of vibrations caused by unbalanced cylinders:

- Excessive plate wear
- Excessive bearing wear
- Excessive roll wear
- Uneven print impression
- Resonant vibration of other parts of the press

There are two types of roll/cylinder imbalance that must be corrected:

1. **Static Imbalance:** Only gravity or weight force is involved; balancing can be accomplished without rotation. This type of balancing is not adequate for flexographic rolls or cylinders due to the rotational speeds at which they operate and the ratio between their diameter and axle length.
2. **Dynamic Imbalance:** Caused by centrifugal force when the roll or cylinder is rotating. The magnitude of the

21.4 Gear Condition: *Gears should be cleaned and regularly examined for wear.*

centrifugal force produced by an unbalanced condition is a function of the speed of rotation. Therefore, the accuracy of balance required increases with press speed.

Roll & Cylinder Deflection

All rolls and cylinders deflect. The amount of deflection experienced is determined by several variables:

1. The weight of the roll or cylinder (heavier = more deflection).
2. Web tension (higher tension = more deflection).
3. Impression loading (increased impression = more deflection).
4. Roll or cylinder imbalance (greater imbalance = more deflection).
5. Length of the roll or cylinder (longer = more deflection).
6. Material used to construct the roll or cylinder.

Gear Condition

Gear trains must have a small amount of space between meshing gear teeth to allow them to mesh correctly, with sufficient lubrication, and to become engaged and disengaged easily. Backlash is a condition in which the space between the gear teeth is increased. It is typically caused by tooth wear or incomplete meshing. Backlash allows the position of the driven cylinder to change in reference to the drive cylinder, causing misregister. The number of gears in the drive train multiplies the register error. Excessive backlash makes it impossible for an operator to maintain accurate multicolor register. Vibrations associated with backlash increase plate bounce. This results in uneven/over impression, which appears as excessive dot gain/TVI, slurred dots and barring (gear marks) in the image, especially in vignettes. The gear teeth should be cleaned and regularly examined for wear. Wear often appears as a sharpening of the top of the gear tooth.

The impression drum/cylinder, plate cylinder and anilox roll must rotate at exactly the same speed the web is traveling. If a roll or cylinder turns at a slightly different rate, excessive plate wear and severe printing problems (including slurred dots) will occur. The diameters of all impression, printing and anilox rolls and cylinders must match the pitch diameter of the gear that propels them. It is important for the gears to always operate on pitch diameter. If they do not, they create a speed mismatch and vibrations resulting in uneven/over impression (causing excessive dot gain/TVI), slurred dots, barring (gear marks in the image) and excessive plate wear. If one gear is set past the pitch diameter, it will cause that roll to rotate at a slower speed than desired. Conversely, if one gear is meshed before the pitch diameter position, the roll that it drives will rotate faster than desired.

Deck Screws & Ways and Cylinder Bearings & Bushings
Worn, dirty deck screws and ways cause impression setting and repeatability problems. These vital parts must be kept clean and oiled for smooth operation. Plate and anilox cylinder bearings or bushings in poor repair also do not allow proper impression setting. Plates must often be over-impressed to avoid the bounce caused by play in worn bearings. The result is slurred dots and banding. It is necessary to replace all worn bearings for optimum press performance.

Central Impression Drum/Impression Cylinders
The two primary variables that must be monitored and controlled on the central impression drum or impression cylinders are temperature and diameter accuracy (TIR).

1. **Temperature:** Central-impression (CI) drums, if not properly cooled, will expand and contract with changes in temperature preventing optimum impression setting. Press dryers and pressroom temperature can influence the CI drum (CID). High CID temperatures result in high drum centers due to expansion. Uneven CID temperature can result in: high/low spots (typically, the image located at the center of the drum will print first) and poor ink transfer from the plate to the substrate. To control the temperature, make sure water is flowing through the cooling coils. Prolonged running of an uncontrolled drum may cause permanent warping. A pyrometer measures the surface temperature of the CID. Analog and digital models are available which measure a surface temperature by either a probe or infrared beam. The press manufacturer establishes the CID target temperature and tolerance range.
2. **Total Indicated Runout (TIR):** The TIR of the central impression drum, or impression cylinders, influence the ability to achieve consistent optimum impression. If a drum or impression cylinder has excessive runout, part of the print will either skip or be over-impressed. For presses with individual impression cylinders, TIR problems come mainly from worn bearings or bushings. A dial indicator measures the TIR of the drum/cylinder. The recommended increment is 0.0001" (0.025cm). The press manufacturer specifies the TIR specification for the CID/impression cylinders.

22.0a Bar Code Symbology: *The type of bar code depends on many factors including where it will be scanned and how it will be printed.*

22.0 BAR CODE PRINT CONSIDERATIONS

Formerly, the Uniform Code Council (UCC) was responsible for managing the bar code system in the USA. The UCC is now the GS1 US organization. GS1 US manages the GS1 system and assigns GS1 company prefixes to companies/organizations in the USA. The most common use of a GS1 assigned company prefix is the creation of UPC's (Universal Product Codes), which contain a 12-digit Global Trade Item Number (GTIN).

The GS1 US publishes the following electronic data interchange guidelines based on the ANSI ASC X12 standard:

- **Industrial/Commercial EDI**
- **Uniform Communication Standard (UCS):** used in the grocery industry
- **VICS EDI:** used in the general merchandise retail industry

The GS1 US is also the code manager for the United Nations Standard Products & Services Code (UNSPSC). The UNSPSC provides an open, global, multi-sector standard for classification of products and services. Identify applicable commodity codes on the UNSPSC website (www.unspsc.org).

For detailed information on instrumentation used to verify bar code quality, refer to Section 19.1.3 Bar Code Verifier. For additional information to consider when determining minimum bar code size and optimum BWR, refer to Section 19.3.4. For more information on design and prepress considerations for bar codes, refer to Sections 4.3 and 12.4. To learn more about correcting symbol dimensions to accommodate the addressable output resolution of the output device refer to GS1 US "Guidelines for Producing Quality Symbols". The GS1 US and UNSPSC contact information as well as information regarding 2D Bar Codes is included in Appendix A.

22.1 Bar Code Specifications

Bar code print specifications are produced by combining three types of related specifications:

1. Application Standards are published by accredited standards organizations. Bar codes are used in many different applications; for example, one bar code application is bar coding products for retail checkout lanes and another application is bar coding shipments for conveyor lane routing in distribution centers. The specifications for bar codes used in these two applications are different because the conditions for scanning the bar codes are very different. Accredited standards

organizations provide specifications in the form of guidelines and standards to assist in:
- Selecting bar code type to be used
- Structuring the data inside the bar code
- Defining the printed human-readable information that is inside the bar code
- Selecting bar code size within the acceptable range
- Understanding where the bar code should be placed on the product or container
- Defining the minimum print quality requirements

2. *FIRST* Print Specifications prescribe a minimal level of capability for all compliant printers. These specifications fall within the acceptable specification limits of the appropriate application standard for the bar code being printed and will assist in:
    - Determining the minimum size for a bar code depending on the printing press and substrate
    - Identifying the preferred bar code orientation given the direction the web or sheet will travel
3. Job Specifications should be published for film or plate output. This type of specification should assist in:
    - Identifying optimum film/plate output resolution
    - Determining bar width reduction (BWR) required by the specified print conditions

22.0b 2D Bar Code: *This is an example of a Quick Response (QR) code.*

Because scanners read bar code graphics, they must be produced according to the appropriate application standard, printing press type, substrate type, and film/plate output conditions. There are many considerations that apply when producing any bar code.

22.2 Printer Responsibilities

The designer, prepress provider and printer all bear responsibility for producing quality bar code symbols. The printer is responsible for inspecting incoming materials to verify conformance to predetermined specifications. Additionally, the printer is responsible for identifying and communicating variables that influence printability, such as: the minimum printable size, the optimum bar width reduction, the correct distortion factor and proper symbol orientation. The printer is also responsible for working with the designer to confirm the substrate and ink selected will provide adequate symbol contrast. Once a job is on press, the printer must monitor the symbol throughout the pressrun to ensure it is scannable. Refer to Section 19.1.3 to review bar code verification recommendations.

Evaluating Printability and Symbol Contrast

Substrate Considerations: The printer is responsible for reviewing the proposed substrate and design to determine

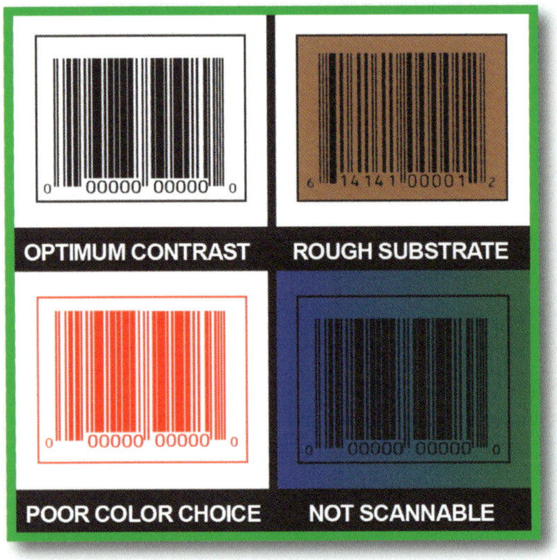

22.2a Symbol Contrast: *If a combination other than black bars and a white background must be used, the symbol contrast should be verified during print optimization trials.*

the impact on bar code scanning. The printer must consider variables such as symbol contrast and symbol size as well as necessary press conditions: anilox volume, mounting tape selection, or the need for a double bump of background color to achieve the opacity required for successful scanning.

Bars and spaces are most accurately reproduced on smooth substrates with high ink holdout characteristics. The rougher, more textured, and more porous a substrate, the greater the potential for printing bars with voids and ink in the spaces. Textured and more porous substrates also tend to increase bar edge roughness, bar growth, and bleeding. Any of these characteristics can reduce scan rates.

Bar codes scan most successfully with an opaque white background that provides white spaces and quiet zones with the maximum reflectance possible. When printing on a transparent or colored substrate, a solid, light-colored background (white is optimum), with maximum opacity, is recommended in the area where the bar code is located. Special considerations for the background ink formulation and press set-up (anilox, double bumps of background color, mounting tape selection, etc.) may be necessary in order to achieve maximum opacity.

Ink Considerations: The colors specified for the symbol background and bars have different reflectance values. It is important to remember that colors with acceptable ANSI/ISO Symbol Contrast (SC) grades on opaque substrates may be completely unacceptable on translucent or transparent substrates. In order to obtain an accurate prediction of a bar code's SC, the reflectance of both the bar color and background should be measured. Translucent or transparent substrates must be measured against a background that is representative of the final, filled package contents.

Symbol contrast is a mathematical expression for the difference in reflectivity between the symbol's dark bars and light background. It is based on scan reflectance profiles specified by ISO/IEC 15416:2000 (Information technology – Automatic identification and data capture techniques – Print Quality Test Specification – Linear Symbols) and ISO/IEC 15415:2004 (Information technology – Automatic identification and data capture techniques – Print Quality Test Specification – Two-Dimensional Symbols). The minimum SC specified varies depending on the bar code symbol selected and where it is used. For example, the EAN/UPC symbol used in retail point-of-sale applications, requires a minimum 40% SC on the final filled product according to GS1 US "UPC Printed-Symbol & Quality Specifications". To meet this 40% specification, it is recommended to target achieving a SC minimum of 55% for the UPC at the print stage (before the packaging process).

Bar code scanners typically use red light; therefore, red, brown and orange inks with a high percentage of red pigment should be avoided when printing the bars in a bar code. However, these colors may be used for the symbol background. The optimum combination is opaque black ink for the bars and opaque white ink, or paper, for the symbol background. Bars printed in black, dark blue, or dark green and backgrounds (spaces and quiet zones) printed in yellow, orange, pink, peach and red generally scan successfully.

Ink color specifications should be evaluated individually for particular substrates. No matter what color is specified, symbol contrast should be monitored closely, and ink densities controlled carefully. A relatively small change in density, on a marginal bar/space combination, can result in unacceptable symbol contrast. Bar codes require bars with sharp edges in order for the scanner to correctly read the code. Therefore, the bars comprising a bar code must be printed on a single print station.

Predicting Adequate Symbol Contrast: Some commercially available ANSI-based verification devices have a mode that allows the user to make static or spot reflectance readings of sample ink and substrate combinations (called a "reflectometer" mode). Two reflectance readings, one of the substrate, or background color (RL), and one of the ink printing the bars on the substrate (RD), may be used to estimate whether the sample color combination will produce an adequate symbol contrast (SC) for the bar code.

Determining Symbol Size

Minimum Magnification/Size: Bar code symbol sizes are specified in an acceptable range for the optics found in the scanning system and vary with symbology. A printer should determine the minimum size symbol that can be printed repeatedly while meeting minimum print quality specifications. Refer to Print Section 19.3.4 Bar Codes Minimum Size and Bar Width Reduction, for information on how to determine the minimum acceptable magnification. Printing a bar code below the minimum size specified by

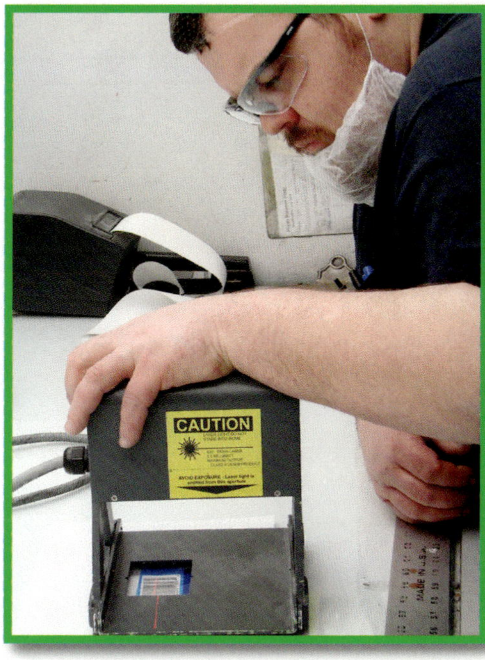

22.2b Press-Side Verification: *The press operator should continually monitor the printed bar code to confirm scannability throughout the production run.*

the applicable symbol specification is not supported by *FIRST*. Compliant printers will be able to meet the minimum bar code size specifications in Table 22.2 while meeting the minimum print quality specifications found within appropriate application section standards. Table 22.2 is only applicable to symbols that are printed with the bars running in the machine direction.

Minimum Bar Code Magnification: General Guidelines *Bar Code magnification is print system dependent; determine optimum magnification with press fingerprint trial (ref. 1.3.2)*		
Print Segment		Machine Direction
Wide Web	Preprint Linerboard	100%
	Combined Corrugated (*flute dependent*)	UPC: 110% - 200% ITF-14: 100%
	Folding Carton	100%
	Multiwall Bag	115%
	Film Products	100%
Narrow Web	Paper Products	80%
	Film Products	100%

Table 22.2

Combined Corrugated – Size Considerations

UPC Symbol: A 200% symbol is the maximum size specified by the GS1 US system. Generally, this represents the minimum size recommended for direct printing on combined corrugated. However, many printers with newer presses, thinner printing plates, smaller flutes and improved white substrates are successfully printing in the 110-160% range. A minimum magnification of 150% is specified by the GS1 system for all containers that will be scanned on automated conveyor lines during the distribution stage.

ITF-14 Symbol: The nominal size specified by the GS1 system for ITF-14 symbols carrying the GTIN-14 number, is based on an X-dimension (narrow bar width) of 0.040" (1.0mm) and a height of 1.250" (31.8mm). Generally, the 100% specification is recommended as a minimum size for direct printing on combined corrugated. If the carton size prohibits the 1.250" (31.8mm) height, (ie. the height restrictions of a tray) it may be preferable to print a slightly truncated symbol while leaving the X-dimension at 0.040" (1.0mm). The current magnification range for ITF-14 symbols has a

GS1 system specification between a minimum of 70% and a maximum of 120%. These specifications are changing in two ways:

1. If the container is large enough to be scanned on an automated conveyor line scanner during distribution, then the minimum height is fixed at 1.25" (31.8mm) regardless of the X-dimension used for the symbol.
2. To improve scanning efficiency, packages marked with ITF-14 symbols that are scanned in a conveyor-based environment should have a maximum X-dimension of 0.040" (1.0mm) not 0.048" (1.2mm). While current packages marked with ITF-14 symbols with X-dimensions between 0.040" (1.0mm) and 0.048" (1.2mm) remain acceptable based on historical GS1 system specifications, new packages should begin using 0.040" (1.0mm) X-dimensions as the maximum.

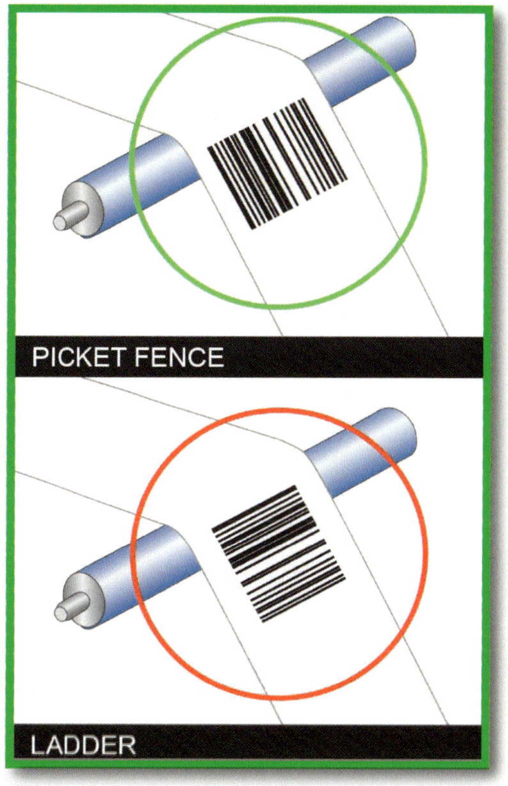

22.2c Bar Code Orientation: *The left figure illustrates the bar code traveling in the preferred machine direction. The right figure illustrates the bars printing perpendicular to the direction the web is moving; this is not supported by FIRST.*

Optimizing Bar Width Reduction

Bar widths increase in flexographic printing in a manner similar to dot gain/TVI. As the bar widths increase, the corresponding space widths between the bars decrease. Just as a dot gain/TVI curve can be applied to a process image to account for expected dot gain/TVI on press, a BWR is typically applied to a bar code prior to film/plate output to account for the bar growth expected on the press. Printers are required to specify the amount of BWR appropriate to the anticipated printing conditions (ie. press type, plate material, mounting material, substrate, anilox roll, etc.). The specified amount of BWR should be corrected at the symbol design stage for digital bar codes files (Section 4.3). Bar code design specifications outside the BWR specified by the printer are not acceptable. Refer to Section 19.3.4 Bar Code Magnification and Bar Width Reduction, to determine optimum BWR for a given magnification and press condition.

Communicating Symbol Orientation and Distortion Factor

Orientation: The bar code should be oriented in the direction of the web going though the press to avoid slurring.. Printers need to advise designers of their requirements regarding the bar code's orientation based on the printed web direction. If the printed symbol must be printed with the bars perpendicular to the direction the web is moving, the printer should provide the designer with the minimum size symbol they can print and continue to meet the minimum print quality specifications within the appropriate application standards. If print slur occurs with the symbol printing in the machine direction, the bars grow in length only and are still

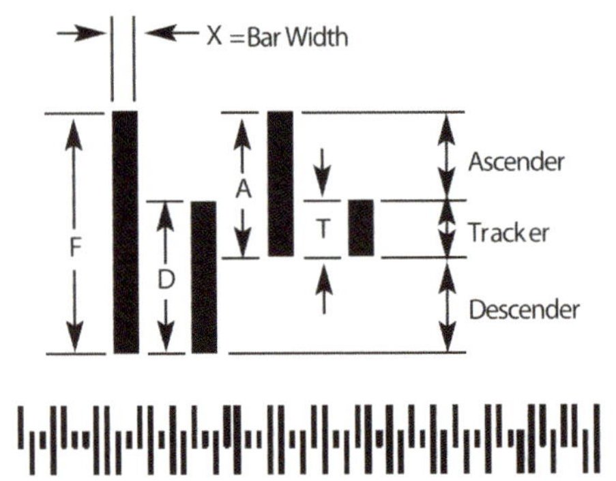

22.3 USPS CB4 Bar Code: *The Intelligent Mail Bar Code (CB4) is a 4-state bar code that consists of 65 bars.*

scannable. However, if the symbol is printed in the transverse direction the bars will grow in width, likely causing the bar code on the printed product to fail to meet specifications. Therefore, *FIRST* does not support printing bar codes perpendicular to the direction the web moves through the press.

Distortion: Printers must specify a distortion factor (or plate-roll circumference) to prepress providers for any bar code whose bars are printed perpendicular to the direction the web will travel. Printing symbols in the cross-direction is not recommended by *FIRST*.

Confirming Correct Encodation

A good manufacturing practice for each phase of the production process is to confirm the information printed below the bar code matches the information encoded within the symbol. It is the responsibility of the supplier providing digital files, or final films for platemaking, to verify that the bar code is properly encoded. However, it is in the printer's best interest to check the encoding for accuracy. A spreadsheet for calculating check digits can be found in Appendix A and on the GS1 US website.

22.3 USPS Intelligent Mail Bar Code

The Intelligent Mail Bar Code (CB4), used by the United States Postal Service (USPS), is a 4-state bar code that consists of 65 bars. The information in this section was obtained from the United States Postal Service Intelligent Mail Bar Code specification USPS-B-3200C. For additional information, reference the USPS-B-3200C specification from the US Postal Service. Contact information is included in Appendix A. Refer to Sections 4.4 and 12.4 for additional information.

Printing Considerations

Background Reflectance: The area of the mail piece where the bar code is located shall be uniform in color. When measured with a Postal Service envelope reflectance meter or equivalent, the minimum reflectance from the background area shall be 50% in the red portions and 45% in the green portions of the optical spectrum.

Print Reflectance Difference (PRD): The PRD is the difference between light reflected from the printed bars in the bar code and the background, when measured with a Postal Service envelope reflectance meter or equivalent. The minimum PRD is 30% in the red and green portions of the optical spectrum.

Quiet Zone Reflectance: Within the quiet zone around the bar code, background patterns, envelope insert "show through", and all other printing shall be limited to a maximum print contrast ratio (PCR) of 15% when measured in the red and green portions of the optical spectrum using a Postal Service envelope reflectance meter or equivalent.

Inking Issues: Over inking can cause the bars to exceed their maximum dimensions and prevent successful scanning. Print variables that can lead to over inking include: insufficient bar width reduction, too high volume anilox, over impression, too hard mounting tape and improper ink metering of the anilox roll. Under inking can cause the bars to fail to meet the minimum dimensions and prevent successful scanning. If bars fail to meet the minimum dimensions, then too much bar width reduction was applied during prepress for the current print system (combination of: substrate, ink, anilox, mounting tape, etc.).

Bar Code Tilt: When printing height-modulated bar codes, two types of tilt can occur:

1. **Bar Code Skew:** The entire bar code is skewed with respect to the bottom edge of the mail piece. Bar code skew for letter mail pieces is limited to +/- 5 degrees.
2. **Bar Code Rotation:** The individual bars are rotated with respect to the centerline of the bar code. Bar rotation for letter mail pieces is limited to +/- 5 degrees.

The combined result of bar code skew and bar rotation is limited to a maximum of +/- 10 degrees from a perpendicular to the top or bottom edges of the mail piece for horizontally oriented bar codes and left or right edges of the mail piece for vertically oriented bar codes.

Physical Dimensions: The printer should verify conformance to the USPS size specifications for all designs containing CB4 bar codes.

USPS 4CB Size Specifications		
Parameter	Minimum	Maximum
Overall Bar Code Width	20 bars per inch	24 bars per inch
Overall Bar Code Hight	0.134" (3.4mm)	0.23" (5.84mm)
Quiet Zone: Above & Below Symbol	0.040" (1.02mm)	N/A
Quiet Zone: Either Side of Symbol	0.125" (3.18mm)	N/A

Table 22.3

23.1 Color Matching: *Use a spectrophotometer to compare the drawdown to the color target. Confirm color accuracy using a numeric color target and tolerancing system such as CIE94, CMC or CIE2000.*

23.0 INK ROOM PROCEDURES & TESTING

23.1 Color Matching

When creating an ink formulation for press, *FIRST* recommends these color matching procedures:

1. Identify any required high-performance properties: fade, heat, chemical, bleed, or grease resistance.
2. Select appropriate pigments per ink supplier recommendation, experience, historical data or formulation software.
    - Utilize *FIRST* recommended pigments
    - Select pigments that achieve necessary performance properties
    - Formulate with the fewest pigments possible for accurate color control
    - Use lowest cost combination of pigments that achieve above criteria
3. Prepare a small lab batch (typically 100 grams) using the selected ink formulation.
4. Adjust viscosity and strength to match the press when proofed on correlated proofing device.
5. Create an ink proof on a proofing device correlated to the press using the same substrate as the print job. Refer to Section 23.2 Ink Proofing for additional proofing information.
6. Compare the proof to the color target using both visual and spectrophotometric methods.
    - Confirm color accuracy using a numeric color target and tolerancing system
    - Typical specifications are: CIE or CMC 2:1 with ΔE 0.5 lower than the finished print tolerance (customer expectation)
    - The color target should accurately represent the printed sample with respect to: substrate, ink and press. It should be comprised of either a single ink or multi-trapped inks to create the custom color.
7. Revise the formula as necessary to achieve an acceptable color match prior to manufacturing the production ink batch.

23.2 Ink Proofing

The purpose of proofing an ink is to generate accurate and repeatable color matches, which minimize or eliminate downtime on press caused by color matching. The ink should be proofed on the same substrate as the print job and compared to the customer-approved standard using a spectrophotometer. Current proofing methods can be divided into three general categories:

hand proofers, mechanical bar proofers and laboratory-scaled flexographic proofing machines.

To control for color accuracy, the proofing procedure should be:

- Correlated to the print station
- Operator independent
- Reproducible over time
- Based on a durable mechanism
- Easy to clean and maintain

Ink Proofing Methods

Hand proofers consist of a rubber roll and anilox roll mounted in a frame. The ink is applied to the nip formed by the two rolls. The proof is created by the operator pulling the hand proofer over the substrate at an even speed and pressure. Changing the anilox volume or rubber roll durometer can alter the amount of ink transferred. This is the simplest of the proofing methods; it is durable and easy to maintain. Unfortunately, this approach creates the most inherent variability. Operator variation in speed and pressure create unpredictable results. *FIRST* does not recommend the hand proofing method because of the limitations of accuracy, predictability and operator influence.

Bar proofers are mechanically driven devices that control speed, pressure and ink deposition film thickness. A wire-wound rod draws ink onto the substrate. The wire thickness determines the ink film thickness proofed onto the substrate. Different coating rods, or proofing bars, correlate with specific press conditions. A thicker wire deposits more ink onto the substrate. Bar proofers offer many advantages: they can be correlated to the print station, they are operator independent, durable and easy to clean. The primary disadvantages of bar proofers are the inability to lay ink down as smoothly as the press on uneven and highly absorbent substrates and the inability to determine transfer differences with the ink.

Laboratory-scaled flexographic proofing machines simulate an actual print deck with an anilox, doctor blade, plate and impression cylinder. Some designs utilize photopolymer plates that can proof both solids and dots. They are mechanical, gear-driven machines with precise impression setting capability, both anilox-to-plate and plate-to-substrate. The proofer can be accurately correlated to specific press conditions. Some designs proof on a narrow web, affording the ink supplier the opportunity to create multiple proofs of the same color at ~100fpm. They can proof water, solvent and/or UV flexographic inks. This approach offers the

23.2a Hand Proofers: *Changing the anilox volume or rubber roll durometer can alter the amount of ink transferred-- as can operator variation in speed and pressure.*

23.2b Proofing Machines: *The proofing machine method offers the highest correlation to the press and is the most reproducible over time -- independent of the operator.*

23.3 Ink Functionality Testing: *The ink properties addressed in this section have a significant impact on the printability and finished properties of the ink.*

highest correlation to press and is the most reproducible over time, independent of the operator. However, these proofers are generally more mechanically involved, more difficult to clean and more expensive. Flexographic proofing machines represent the future direction of proofing; this is the category experiencing the most innovation.

Proof-to-Press Correlation
Regardless of the proofing method selected, the correlation to press must be verified. Correlation is easily confirmed by collecting from the press (while printing within standard operating conditions) a sample of ink from the ink sump, unprinted substrate and a press tearsheet. Proof the ink, from the press ink sump, on the unprinted substrate. Measure and create a standard of the solid color on the press tearsheet using a spectrophotometer. Measure the proof sample and compare it to the standard created by the printed tearsheet. If the difference in lightness (ΔL or DL) is less than 0.5, using CIE color tolerancing, the proof-to-press correlation is acceptable. If the ΔL is greater than 0.5, the proofer settings must be adjusted until the difference in lightness is 0.5 or less. A ΔL of 0.5 is a general recommendation and may vary for different ink colors, processes, or customer requirements.

23.3 Ink Functionality Testing

The functional properties of flexographic printing inks are divided into two distinctively different categories: physical virgin ink properties and printed ink properties. The properties addressed in this section have a significant impact on the printability and finished properties of the ink. Test methods, specifications and tolerances are also described. A variety of procedures may be used for any given test. Where possible, ASTM or industry standard methods have been adopted. Refer to Appendix A for information on how to obtain ASTM standard test methods.

Core physical virgin ink properties include:

1. Fineness of grind/Particle size
2. Viscosity
3. Specific gravity
4. pH (for water inks only)
5. Color (proofing to check reproducibility, not correlation to press performance)
6. Foam
7. Solids
8. Drying rate

Core printed ink properties include:

1. Adhesion to nonporous substrates
2. Color correlation to press results
3. Gloss on surface printed product
4. Block/Set-off resistance
5. Rub/Abrasion
6. Opacity/Transparency
7. Printability and surface defects

Critical performance tests for packaging applications include:

1. Coefficient of Friction (CoF)
2. Heat resistance
3. Crinkle Resistance/Flexibility
4. Odor and taint
5. Lamination bond strength
6. Water resistance
7. Chemical or product resistance
8. Lightfastness

23.3.1 Virgin Ink Properties – Wet Ink

1. **Fineness Of Grind/Particle Size**

Importance

Grind is a critical factor for color strength, opacity/transparency, gloss and abrasion characteristics. In general, finer grind implies higher transparency, stronger color, higher gloss and less abrasion.

Procedure

ASTM D1316 (Fineness of grind of printing inks by the NPIRI grindometer)

Specification

Based on production specification with a NPIRI grind measuring from 0-10 microns.

Tolerance

For Dispersions: 0-1 microns = maximum for all pigments. Pigment "White 6" may be coarser for desired opacity. For Finished Inks: Additives used to provide functional properties (such as wax) may lead to false readings. Even with additives, maximum grind should not exceed 3 microns.

2. **Viscosity**

Importance

Viscosity is a measure of resistance to flow at a specified temperature. It plays a vital role in determining print quality of the ink. Controlling viscosity on press is necessary to maintain color (hue, strength), print quality (ink flow out, dot gain/TVI, trapping) and performance properties (coating weight, drying speed, solvent retention). Refer to Section 20.2.6 for additional information.

One of the largest issues with using efflux cups is the actual reproducibility of the test. It is in general a significant process improvement to cut fresh inks by a standard weight, and then measure the viscosity. Typically, inks from suppliers will be substantially more reproducible than a viscosity measurement.

Another large issue is correlation among cups. Many wasted hours have been spent over the years by failing to calibrate cups on site and among sites.

Procedure
ASTM D1200 (Test Method for Viscosity of Paints, Varnishes, and Lacquers)
ASTM D4212 (Test Method for Viscosity by Dip Type Viscosity Cup)
ASTM 2196 (Test Method for Rheological Properties of Non-Newtonian Materials by Rotational (Brookfield) Viscometer)

Test Method
The flexographic industry commonly utilizes gravimetric or dip type cups. The accuracy of these cups varies with design, manufacturer, condition and operator. Zahn cups are the most widely used in the flexographic industry due to their ease of use. RIT laboratory studies have shown Shell Cups to provide greater overall precision. The cups should be tested with a calibration liquid when new and on a regular basis.

Tolerance
Dip Cups: Zahn = ±3 seconds, Shell = ±2 seconds, Brookfield viscometer = ±10% centipoise

3. **Specific Gravity**

Importance
The specific gravity, or density, of an ink is indicative of ink mileage, as well as the solids/volatile content of an ink or coating.

Procedure
ASTM D 1475 (Standard Test Method for Density of Paint, Varnish, Lacquer, and Related Products)

Test Method
Specific gravity is the weight of a unit volume of the material at 25° C. In the printing industry, it is typically expressed in pounds/

gallon or grams/milliliter. When expressed as specific gravity, it is the ratio of the density of the product tested to the density of water at 25° C; therefore, it is not expressed in any units. Specific gravity is normally measured on the virgin ink prior to extending or reducing.

Tolerance
The allowable tolerance for density is ±5%.

4. pH for Water Inks
Importance
pH is a measure of the acidity or alkalinity of a water based ink. Water based inks rely on good pH control to maintain resin solubility, which in turn directly impacts printability, drying speed and viscosity. This test applies only to water based ink systems. Optimal pH ranges for water based inks vary, depending on the chemistry of the ink and the end-use application. The typical pH range for the majority of water based inks (anionic) is 8.5-9.5 (>7.0 = alkaline). There are a limited number of chemistries that are pH neutral (7.0) or acidic (cationic). The ink supplier should recommend the optimal pH range for peak performance of a particular system and the preferred method to maintain the pH level. Refer to Section 20.2.6 for additional information.

Procedure
ASTM E70 (97 Standard Test Method for pH of Aqueous Solutions with the Glass Electrode)

Tolerance
±0.1 pH units

5. Color – Ink Batch Reproducibility, not Press Correlation
Importance
Color is what the consumer sees when viewing a package. It identifies the product and, if not consistent from package to package, results in the perception of poor quality.

Procedure
ASTM D1729 (Visual) (2003 Standard Practice for Visual Appraisals of Colors & Color Differences of Diffusely Illuminated Opaque Materials)

ASTM D2244 (Instrumental) (Standard Practice for Calculation of Color Tolerances and Color Differences from Instrumentally Measured Color Coordinates)

Test Method
A wet ink sample, at a pre-determined viscosity, is printed on a controlled substrate with a mechanical proofer set up to achieve equal film thickness. This is compared to a print of the accepted standard that was produced identically. The color standard should also be stored as a digital file for long-term reference and to prevent color drift.

In the case of reverse printed laminations, it is sometimes necessary to view the color in the finished lamination structure. It is extremely important to make the lamination exactly as the laminated color target. Refer to Print Section 20.1.3 Lamination and Color Matching for more information.

Color comparisons should first be made visually, and if necessary, adjusted until acceptable by eye. Following this, a spectrophotometer should be used to measure the sample and compare it to the accepted standard. The measurement is reported as a difference between the new batch and the accepted standard. This is recorded as a difference (Δ or D) in lightness (L*), saturation or chroma (C*), hue (h°) and total color (E). Refer to Section 19.1.1 for more information on spectrophotometer settings, including alternative settings.

Tolerance
The allowable tolerance for virgin ink is DE 0.5 less than the finished print tolerance (customer expectation).

Condition of Tolerance

- Instrument: 0°/45° or 45°/0° geometry and 2° standard observer
- Aperture: Either 30mm (fixed) or 10mm (portable)
- Illuminant: D50
- Color Space: LCh°
- Tolerancing: CIE or CMC (2:1)

6. Foaming
Importance
The occurrence of excessive foaming in ink, which is caused by air entrapment, leads to handling and print problems, such as pinholing or ink starvation. Although typically associated with water based inks, problems arise infrequently with both solvent and energy-cured (UV) systems. The test method described is a useful guide to determine the tendency of a non-flammable ink or coating to produce foam under high shear conditions by measuring the volume of an ink or coating before and after subjecting it to high shear forces.

Procedure
ASTM D3519 – (88: Standard Test Method for Foam in Aqueous Media)

Test Method
The foaming observed, after the test, is recorded as an increase in height from the original liquid level in the graduated cylinder. The foam height should be quoted in millimeters (mm).

Tolerance
Allowable tolerance limit ≤5% of the original ink volume.

7. **Ink Solids – Non-Volatile Content**

Importance
Solids content measures the percentage of non-volatile matter in a compound or mixture, based on the entire weight of the mixture.

Procedure
ASTM D4713 – (Standard Method for Non-Volatile Content in Printing Inks, Resin Solutions and Vehicles)

Tolerance
Allowable tolerance is ±2.0%.

8. **Drying and Curing Rates**

Importance
This test determines if drying rates of subsequent batches of inks and coatings are similar. In addition, this test isolates and identifies drying problems while on press.

Procedure
ASTM D1640 – (Standard Test Method for Drying, Curing or Film Formation)

Test Method
For solvent and water inks: On a clean NPIRI grind gauge, place a small amount of the test ink on the top right-hand side and a similar amount of the standard ink on the top left-hand side. Hold the cursor blade vertically, with both hands, and draw the inks down the full length of the gauge scale while applying even pressure. The comparative drying rate can be observed and confirmed by drawing a finger across both samples at various points starting from the bottom of the scale and working up. Once dry, the ink film will not be smudged by the finger. This is a very subjective test.

For UV inks: The drying, or cure rate, of UV inks and coatings can be readily measured. A wet-ink sample and a wet ink standard are printed on a controlled substrate with a mechanical proofer then cured under stated conditions (lamp power, speed).

- MEK rub resistance (used for inks and coatings): A cotton swab soaked in MEK is rubbed (light pressure) in a single motion across the print. This is repeated as per the specification set by the printer/converter. The print is examined for deterioration of the ink/coating. The swab should be examined for color transfer.
- Stain test using potassium permanganate (KMNO4) (only for clear coatings): One drop of a 10% solution of potassium permanganate is placed on the surface of the cured coating. The solution is allowed to sit for 30 seconds. It is wiped off with a clean towel. The discolored area is measured for color density. The level of discoloration is proportional to the degree of cure.

Tolerance

The standard and test samples should dry at equivalent rates. For UV inks/coatings both the standard and test samples should achieve a similar level of cure as judged by MEK resistance. For the stain test, the density of the test areas should not vary by more than ±10%.

23.3.2 Printed Ink Properties – Dry Ink

For the core printed ink properties and specific application properties listed below, the printer/customer determines the specification based upon the end-use application of the printed material.

1. **Adhesion to Nonporous Substrates**

Importance

Ink provides both functional and decorative properties. In order to meet most of the functional needs (flexibility, chemical resistance) the ink must adhere to the primary substrate.

Procedure

ASTM D3359 – (Test Method for Measuring Adhesion by Tape)

Test Method

The adhesion strength of an ink or coating to a substrate is frequently determined by a tape test. The most common version for water and solvent inks uses 3M 610 tape. For UV/EB ink, 3M 810 tape is preferred. Determining the results is subjective and based upon the level of ink removal from the test sample.

Tolerance
For the majority of inks, less than 10% ink removal from the test sample, or equal to an established standard, is acceptable.

2. **Color – Proofing for Ink Approval**

Importance
Ink color consistency on press from run to run is vital in maintaining the consumer's perception of product quality. Color adjustment on press can be a lengthy and costly process. Predictive color proofing methods are critical for achieving color consistency.

Procedure
ASTM D1729 (Visual) (2003 Standard Practice for Visual Appraisals of Colors & Color Differences of Diffusely Illuminated Opaque Materials)

ASTM D2244 (Instrumental) (Standard Practice for Calculation of Color Tolerances and Color Differences from Instrumentally Measured Color Coordinates)

Test Method
An ink batch produced under standard conditions is compared visually and instrumentally against a hard copy and digital file of the customer approved standard. The preferred instrument is a spectrophotometer. The measurement is reported as a difference between the new batch and the accepted standard. This is recorded as a difference (Δ or D) in lightness (L*), saturation/chroma (C*), hue (h°) and total color (E). Refer to Section 19.1.1 for more information on spectrophotometer settings, including alternative settings.

Specification
Customer-based specification.

Tolerance
The allowable tolerance for ink supplied to press is: DE 0.5 less than finished print tolerance (customer expectation).

Condition of Tolerance

- Instrument: 0°/45° or 45°/0° geometry and 2° standard observer
- Aperture: Either 30mm (fixed) or 10mm (portable) aperture
- Illuminant: D50
- Color Space: LCh°
- Tolerancing: CIE or CMC (2:1)

3. **Gloss of Surface Printed Product**

Importance
Gloss level is a critical decorative property of the printed piece. It is often associated with the perception of product quality.

Procedure
ASTM D523 – (Standard Test Method for Specular Gloss)
ASTM E97 – (Standard Test Method for 0 degrees/45 degrees – Direction Reflectance of Opaque Specimens)

Test Method
Gloss is a measure of the ratio of the reflectance of the printed sample to the reflectance of a "standard surface" under the same measurement conditions. The "standard surface" is polished glass.

Reflectance measurements are made with 20°, 60°, or 85° geometry. For most printing applications the 60° geometry is used. 20° is normally used if the printed sample, when measured at 60°, has a gloss reading higher than 70. 85° is used if the printed sample, when measured at 60°, has a gloss reading below 10.

Tolerance
The allowable tolerance for gloss is ±5%.

4. **Blocking or Set-Off**

Importance
Blocking resistance measures the ability of a printed substrate to withstand heat and pressure without sticking or transferring ink. This is extremely critical in roll-to-roll or stacked print operations.

Procedure
ASTM D2793 – (Standard Test Method for Block Testing)

Test Method
This test utilizes a Koehler IC block tester, which uses calibrated springs to exert pressure.

Tolerance
Based on customer specification. Pressure, time and temperature must be specified. Usually no visible ink transfer or significant cling is allowed.

5. **Rub/Abrasion Resistance for All Printed Substrates**

Importance
Printed packaging is subjected to many abrasive forces during printing, converting, shipping and end use. Rub or abrasion

resistance of the ink/coating is important to the aesthetics of the final package. Printers use different terms such as scuffing, marring, blotching, rub off and scratching to describe abrasion.

Procedure
ASTM D5264/92 – (Standard Test Method for Sutherland Rub Test)

Equipment
The Sutherland Rub Tester is the instrument most commonly used. Less commonly used alternatives include the Taber Abraser and the Gavarti CAT tester.

Specification (Industry Expectations)
Corrugated Postprint: 40 rub cycles, 4-lb. block, print to board.
Preprint Linerboard: 40 rub cycles, 4-lb. block, 400° F print to heat platen.
Multiwall Bags: 100 rub cycles, 4-lb. block, print to paper.
Folding Carton: 200 rub cycles, 4-lb. block, print to board.
Surface Print Film: 100 rub cycles, 4-lb. block, print to film.

Tolerance
No observable transfer of ink/coating to the rubbing sheet and no visible damage to the print surface under the specified testing conditions.

6. **Opacity/Transparency**

Importance
The level of transparency or opacity (hiding power) of an ink governs the quality of the printed image. For example, process inks require high transparency to provide good graphic reproduction while high opacity may be preferred for white inks and line colors.

Procedure
TAPPI T425 (15/d geometry, illuminant A/2*, 89% reflectance backing and paper backing)
ASTM D589-97 (Standard Test Method for Opacity of Paper – 15° Diffuse Illuminant A, 89% Reflectance Backing & Paper Backing)

Test Method

Visual Comparison: The relative opacity of the test sample compared to an accepted standard can be made. Proof the inks (standard and test) side by side, at a pre-determined viscosity, with a mechanical proofer to achieve equal film thickness. For visual tests, a controlled substrate featuring a pre-printed black bar, for example a Leneta card, should be used.

Instrumental Comparison:

1. **Black Trap Method:** The opacity of a print, produced either in the laboratory or on press, can be measured and compared against a known print standard (produced under identical conditions) using a spectrophotometer and black trap.
2. **Contrast Ratio Method:** The relative transparency of an ink can be gauged from a measurement of the contrast ratio where an ink is compared spectrophotometrically over white and black.

Tolerance

Visual: The test samples should not exhibit significant visual difference in opacity or transparency when compared to the accepted standard.

Instrumental: Opacity (Black Trap Method) allowable tolerance: ±5%. Contrast ratio allowable tolerance: ±10%.

23.3.3 Performance Properties

1. **CoF/Slide Angle**

Importance

Friction is the force that opposes the motion of two surfaces in contact. In packaging terms, it provides an indication of the level of slip or lubricity of the surfaces. Low CoF denotes high slip and high CoF denotes low slip. Coefficient of friction is a measure of the relative difficulty with which one material will slide over an adjoining surface of itself, or another material. It is the ratio of the frictional force to the gravitational force acting perpendicular to the two surfaces in contact. Flatbed testing equipment provides direct CoF data. Two measurements typically quoted are static CoF and kinetic CoF. Static CoF is a measure of the force required to begin movement of the surfaces relative to each other. Kinetic CoF is a measure of the force required to sustain this movement.

Procedure

ASTM D 1894/95 – (Standard Test method for Static and Kinetic Coefficients of Friction of Plastic Films and Sheeting).
ASTM D4518 –(Standard Test Methods for Measuring Static Friction of Coating Surfaces)

Test Method

The two most common measurement methods used are the inclined plane and flat bed equipment. The inclined plane provides results as a slide angle. Slide angle and static CoF can be related as follows: static CoF = 1/tan (slide angle).

Tolerance
Slide Angle: ±2° on the third reading.
Static and Kinetic CoF: ±10% of specification.

2. **Heat Resistance**
Importance
Ink/coating heat resistance is vitally important to maintain the functional and aesthetic quality of various types of packaging that may be heat-sealed, or subjected to heat during the life of the package.

Procedure
No known ASTM/TAPPI Standard.

Test Method
Generally, heat resistance is determined and tested using a heat seal testing machine. These test machines have either one or two jaws, which can maintain a constant set temperature. The pressure with which the jaws come together and the duration of the test are also controllable. Testing temperature, pressure and dwell time are specified by the printer or converter depending on the processing or packaging equipment used. Printed samples may be tested:

- Face to face
- Face to back
- Face to specified substrate

Tolerance
There should be no visible transfer of ink from one test surface to another. The level of cling between two test samples should be less than 15 grams per linear inch if measured on a tensile tester.

3. **Crinkle or Flexibility**
Importance
The crinkle test indicates the flexibility of an ink and the level of ink adhesion after subjecting the print to mechanical forces.

Procedure
No known ASTM/TAPPI Standard.

Test Method
Crinkle test is performed by firmly grasping a piece of the printed substrate between the two thumbs/forefingers. Place hands approximately 2" apart and rotate rapidly in a circular motion (similar to bicycling motion) for ten cycles. The print is observed for cracking, crazing, flaking, or ink removal. This test cannot be used on rigid or heavy gauge substrates. Caution

should also be exercised in interpreting results when the test is performed on coated substrates or materials subject to high levels of extensibility.

Tolerance
This test is subjective. Ink removal from cracking, crazing, or flaking should not exceed 5% of the crinkle test area.

4. Odor and Taint

Importance
Many materials (solvents, amines, monomers, oligomers) commonly used in solvent, water and UV/EB inks have characteristic odors. Residual levels of such materials, when retained within the printed film, can alter the odor or taste of the packaged product. Low odor/taint levels are necessary for food packing applications.

Procedure
ASTM E-462 (1989 Test Method for Odor & Taste Transfer from Packaging Film)

Test Method
Many of the materials that cause odor are volatile and can be quantitatively determined by gas chromatography. Unfortunately, the accuracy of such tests is entirely dependent on the test protocol followed. Odor levels can also be determined subjectively by an odor panel. Experienced odor panels rate samples 0 (no odor/taint) to 6 (strong odor/taint).

Specification
Customer-based specification.
Odor Panel: Acceptable rating 2 or below.
Gas chromatography: For food packaging, the total level of all retained solvents should not exceed 7,500 mg per ream. Levels for individual solvents should never exceed 25% of this limit.

5. Lamination Bond Strength

Importance
A wide variety of packaging relies on multi-layered structures bonded together to provide the necessary performance properties. Package integrity and performance relies on maintaining this bond.

Procedure
No known ASTM standard.

Test Method
Cut laminated print into 1" strips. Carefully separate lead edge. If difficulty is encountered, the edge can be immersed in solvent (THF, MEK, Toluene, Ethyl Acetate) or add slip-sheets during the lamination. Measure the bond value using a tensile strength tester. Bond values are quoted as grams per linear inch. If the substrates tear before achieving an acceptable bond reading, it is sometimes necessary to back the sample with tape. Note on results if tape backed.

Specification
Customer based specification.
As a general guideline:
  Extrusion: 50 g per linear inch (minimum).
  Adhesive: 100 g per linear inch to film destruction.

Tolerance
Customer defined tolerance is normally quoted as a minimum requirement. The mode of failure should be noted, ink split is generally most desirable.

6. **Water Resistance**

Importance
Water resistance is a critical property required for all packaging subjected to water contact in some manner. Examples of such packaging include frozen food, boil-in-bags and microwaveable food packaging.

Procedure
No known ASTM standard. This section explains three tests for water resistance commonly used:

1. Resistance to bleeding in water at room temperature.
2. Adhesion/flexibility when subjected to ice water or to freezing and thawing.
3. Resistance to boiling water.

Test Method
To test for bleeding in water at room temperature, place the printed film in contact with a damp filter paper or towel for 1-5 hours and note any bleed onto the paper. Also check the print for color fastness. The freeze-thaw test involves immersing the printed film in ice water for a specified time and examining the loss of adhesion and/or flexibility. Alternatively, freeze the film for a specified time and then test adhesion and/or flexibility upon thawing. To examine resistance to boiling, place the printed sample in boiling water for the specified time. Remove and inspect the print for ink adhesion.

Tolerance

- For water bleed: No color bleed or transfer onto the towel or filter paper should be observed.
- For ice water and freeze/thaw resistance: Allowable tolerance is less than 5% loss of tape adhesion or crinkle (See previous tests).
- For boiling water: Tolerance is less than 5% ink removal.

4. **Chemical or Product Resistance – Spot Test**

Importance

The ink on a printed package must be able to resist the product being packaged, especially when chemicals are in the package. While most chemical packaging has barriers to prevent migration of the material through the substrate, the chemical may contact the outside of the package during the filling operation.

Procedure

ASTM D1647 (1996 Standard Test Methods for Resistance of Dried Films of Varnished to Water & Alkali)
ASTM D2248-01 (Standard Practice for Detergent Resistance of Organic Finishes)

Test Method

Several drops of the specified chemical, or product, are placed on a flat, printed portion of the package. The chemical, or product, is left on the print, under ambient or otherwise specified conditions.

Tolerance

Any visible deterioration of the print, or rub off of color onto the swab, could be cause for failure.

5. **Lightfastness**

Importance

Resistance to fading, or discoloration, is important if an ink is to be used on a printed piece that will be exposed to light for extended periods (wall coverings, labels, window displays). Some pigments have much greater resistance to fading, or discoloration, than others. The primary pigments specified by *FIRST* are designed for general use. Higher performance pigments are available. Refer to Section 20.2.3 for additional information.

Procedure

ASTM D3424-01 – (Standard Test Methods for Evaluating the Relative Lightfastness and Weatherability of Printed Matter)
Other References: ASTM D4459, ASTM G23, ASTM G26.

Test Method
The fade test typically utilizes equipment with a Xenon light source. The test sample is partially exposed to the light source in order for a visual or instrumental comparison to be made. Lightfastness can be measured and specified either as a time period (hours) or on a scale (blue/gray wool scale). A spectrophotometer can be used to measure the difference in color between the sample exposed to light and the control sample.

Tolerance
Visual: No visible difference after the stated test time.
Instrumental: Customer based tolerance.

1-Bit TIFF files — A file format that contains all image and screening information at final output resolution with values of either zero or one (on or off) typically for writing engines that utilize a laser to image a medium (plate, film, proofing media, etc.). Also see TIFF (tagged image file format).

Abnormal Color Vision — One of several kinds of defective color vision, inaccurately referred to as color blindness. See Protanopia, Protanomalous, Deuteranopia, Deuteranomaly, Tritanopia and Monochromatism.

Abrasion — Process of wearing away the surface of a material by friction.

Abrasion Resistance — A printing ink's or substrate's ability to resist scuffing and scratching with increased handling, an important consideration in the printing of packaging and other materials destined to be subjected to abrasive forces. Also called scuff or rub resistance. Reference ASTM D5264 - 98(2011)

Abrasion Test — A test designed to determine the ability to withstand the effects of rubbing and scuffing. Reference ASTM D5264 - 98(2011)

Absolute Humidity — A measure of the total amount of water vapor in the atmosphere or material such as paper; also known as moisture content. It is determined by the weight difference of a sample before and after oven drying to "bone dry" or by measuring with various hand-held moisture sensing equipment (hygrometer). Excessive atmospheric relative humidity changes may affect a paper's structural properties resulting in print misregister, wrinkles and other converting problems. See Relative Humidity.

Achromatic Color — A neutral color (white, gray or black) that has no hue.

Acid — Any chemical that undergoes dissociation in water with the formation of hydrogen ions. Acids have a pH less than 7.0; lower numbers indicate greater acidity. Among its properties is a corrosive action on many materials and sour in taste. Will turn litmus paper red.

Additive Color — Color produced by the mixing of light falling onto a surface, as compared to subtractive color. The additive primary colors are red, green and blue.

Addressable Output Resolution — Maximum number of image positions along a straight line one inch in length that can be addressed by a bar code designer. This resolution would exclude further resolution enhancing techniques performed by the imaging device or software that are beyond the control of the designer.

Aging/Fade Resistance — Ability of a paper and/or ink to resist changes in its optical, chemical or structural properties over time. Accelerated aging, yellowing, brightness loss and fading upon exposure to ultraviolet light and humidity can be determined with instruments such as a fadeometer or weatherometer. Also known as permanence, colorfastness and lightfastness.

American National Standards Institute (ANSI) — The USA member of the International Standards Organization (ISO) that develops voluntary standards for business and industry. See Appendix A for contact information.

Analog Proof — A proof typically produced using film(s).

Anilox Roll — Engraved ink metering roll used in flexographic presses to provide a controlled film of ink to the printing plates, which print onto the substrate.

ANSI — See American National Standards Institute.

Apparent Trap (Preucil)

An estimate of how well an ink overprints a previously printed ink. It is the Ratio of the difference between the density of the overprint and the density of the first down ink to the density of the second down ink; all densities are measured with the complementary color (major) filter of the second down ink. For example, in measuring red created by overprinting yellow over magenta, one would use the blue filter on the densitometer, the complement of yellow.

$$\% \text{ Apparent Trap} = 100 \times \frac{D_{OP} - D_1}{D_2 - D_O}$$

Where: D_O = Density of the substrate
D_{OP} = Density of the overprint
D_1 = Density of the first-down ink
D_2 = Density of the second-down ink

Archival

Pertaining to the long-term storage of data.

ASCII

American Standard Code for Information Interchange. A 7-bit standard code adopted to facilitate the interchange of data among various types of data processing and data communications equipment.

ASCII File

A digital file encoded in the industry-standard ASCII representation for text. An ASCII file contains only plain text and basic text-formatting characters such as spaces and carriage returns, but no graphics or special character formatting.

Attribute

Distinguishing characteristic of a sensation, perception or mode of appearance.

Baggy Web

A condition where one side of the web, as it runs through a press or converting equipment, has uneven tension from side-to-side. This can result in printing and/or converting problems.

Bar-Width Reduction (BWR)

A prepress decrease in bar-code image width to compensate for normal image growth as predetermined by press fingerprinting and production monitoring.

Base Alignment

In setting type, a mode specifying that the lower reference edge of all letters in a line of mixed sizes or styles should be horizontally even; also called baseline alignment.

Basis Weight

Paper weight in pounds per ream of a given grade, sheet size and number of sheets (usually 500) in North America. Reported in lbs./ream using TAPPI Method T410. Common ream sizes and grades include the following.

Basis Weight of Paper Grades

Paper Grade	Sheet Size/Quality	Square Ft. Area/Ream
Bond, Computer, Copier	17" x 22" / 500	1,300 (Actual 1,298.6)
Book, Offset, Text	25" x 38" / 500	3,300 (Actual 3,298.6)
Bristol, Tag	22.5" x 28.5" / 500	2,700 (Actual 2,226.6)
Cover	20" x 26" / 500	1,806 (Actual 1,805.6)
Index	25.5" x 30.5" / 500	2,700 (Actual 2,700.6)
Linerboard	1,000 Square Ft.	1,000
Bag Paper, Newsprint, Paperboard, Tissue, Wrapping	24" x 36" / 500	3,000
Printing, Writing	17" x 22" / 500	1,300 (Actual 1,298.6)

*International operations report basis weight in Grams per Square Meter.

$$\text{Conversion formula: } \frac{\text{Basis Weight} \times 1406.5}{\text{Basic Size}} = \text{Grams per Square Meter}$$

BCM — Abbreviation for billion cubic microns; a measurement of the average volume per square inch of engraved ink-carrying cells on an anilox. 1 bcm = 1 microliter.

Binary — A coding or counting system with only two symbols or conditions, such as on/off or zero/one; the format for storing data in computers.

Black and White — Original art or proof in single color (black image on a white background), as distinguished from multicolor.

Bleed — Image or color that extends beyond the trim edge of the finished printed piece.

Blend Vehicle — A clear fluid material which is mixed with dispersions to generate a finished ink. Also see Vehicle.

Blocking —
1. An undesired adhesion between touching layers of materials such as might occur under moderate pressure and/or temperature in storage or use;
2. The extent to which damage to at least one surface is visible upon their separation.

Blushing — Milky, foggy, or matte appearance in an ink or coating.

Brightness — A measure of reflectance in the blue region of the visible light spectrum, specifically at a wavelength of 457 nm, as specified by TAPPI Method T452 using directional 45°/0° geometry. This method is an industry standard for the determination of the brightness of white, near-white, and naturally colored paper and paperboard. The non-USA standard for paper brightness is measured with diffuse illumination and diffuse reading using spherical geometry. Higher numbers on a 0 –100 scale indicate brighter surfaces that increase the perception of print contrast, brilliance and paper quality, especially when viewed under blue-white illumination common with fluorescent lighting. High brightness papers can improve bar code contrast and scannability.

Bump Curve — Highlight compensation applied to avoid imaging dots in the plate that are too small to allow full dot formation on the plate during main exposure and processing. A bump curve can also be referred to as a tonal value increase of any portion of or the entire curve to calibrate the proofing process with the printing process.

BWR — See Bar Width Reduction.

C1S — Coated, one side.

Caliper — Thickness measurement of a single sheet of paper as defined by TAPPI Method T411 and reported in mils or thousandths of an inch (1 mil. = 0.001"). Multiply inches by 25.4 micrometers and round to the nearest whole number to find metric thickness expressed in microns (μ) or micrometers. Also used to identify thickness of other printing materials such as plates, mounting tape, etc. See Gauge for flexible film substrate thickness and see Point for paperboard thickness.

Camera Ready — Copy and/or artwork that is ready for the photography step to make a film negative for platemaking in the printing process.

CCNB — See Clay Coated News Back.

CEPS — See Color Electronic Prepress System.

CGATS — See Committees for Graphic Arts Technologies Standards.

Chambered Doctor Blade System — An ink chamber made up of two doctor blade assemblies and end seals. Ink is pumped into the assembly, fills the anilox roll cells, and is metered by the doctor blades – popular on wide-web applications.

Character Count The number of characters included in a block of text. In graphic arts, spaces are counted but other nonprinting characters usually are not. In information processing, both printing and nonprinting characters are usually included.

Character Set The entire set of characters that can be either shown on a monitor or used to code computer instructions. In a digital printer, the entire set of characters that the printer is capable of printing.

Chatter, or Banding A print defect where a darker line across the entire printed form is seen on a particular unit where there is slurring of halftone dots and/or solids due to a mechanical vibration. Often attributed to gears and mechanics, it can also be due to other factors including form layout, plate durometer, mounting tape attributes, tension problems, and others.

Chroma Attribute of color used to indicate the degree of departure from a gray of the same value. Correlates with the dimension of saturation. Chroma is on of three coordinates in the LCH color model. See Saturation.

CIE Commission Internationale de l'Eclairage. See Appendix A for contact information.

CIE Standard Illuminant A spectrally-based numerical definition of various light sources as defined by the CIE in terms of relative spectral power distribution. Examples include Illuminant A, C, the D-series (D50, D65, etc.). F-series (F2, F7, F11, etc.). Used in conjunction with one of the CIE Standard Observers and the spectral reflectance curve of a measured sample to calculate colorimetric values. The illuminant utilized should always be specified along with the standard observer when communicating color information.

CIE Standard Illuminants

Illuminant	Description
Illuminant A	Incandescent lighting at a color temperature of about 2,856° Kelvin.
Illuminant B	Direct sunlight at about 4,874° Kelvin.
Illuminant C	Tungsten illumination simulating daylight at about 6,774° Kelvin.
Illuminant D50	Graphic arts standard viewing condition at about 5,000° Kelvin.
Illuminant D65	Used by textile, paint and ink industries at about 6,500° Kelvin.
Illuminant F2	Cool white fluorescent lamp at about 4,200° Kelvin.
Illuminant F7	Broadband daylight fluorescent lamp at about 6,500° Kelvin.
Illuminant F11	Narrow band white fluorescent lamp at about 4,000° Kelvin.

CIE Standard Observer Color matching functions defined by the CIE characterizing the visual response of a typical human observer. Used in conjunction with a standard illuminant and the spectral reflectance curve of a measured sample to calculate colorimetric values. There are two standard observers: the 1931 2° standard observer and the 1964 10° standard observer. The standard observer utilized in calculating colorimetric values should always be specified along with the illuminant when communicating color information.

Clarity/Haze Material characteristics permitting distinct images to be observed through it; typically a visual comparison to a standard clear transparent material. Poor formation and other related properties could negatively affect clarity and apparent print quality.

Clay Coated News Back (CCNB) Paperboard made from recycled newsprint base fiber with a clay coated surface to improve printability.

CMYK Cyan, magenta, yellow, black; the four process color printing inks.

CNK See Coated Natural Kraft.

Co-Extruded (COEX) A multi-layer film or coating in which each distinct layer is formed by a simultaneous extrusion of hot polymers through a single die.

Coated Natural Kraft (CNK) Unbleached paperboard, usually clay coated on the side to be printed for folding cartons.

Coated Recycled Board (CRB) Paperboard made from recycled fiber (newspapers, office waste paper, old corrugated cartons, etc.), clay coated on one or both sides, and printed as folding cartons.

Coefficient of Friction (CoF) Measure of static and/or kinetic slip resistance of one material against another. COF has limited effect on printability, but it is critical in converting and bag filling operations as well as end use applications.

Color A visual sensation produced in the brain when the eye views various wavelengths of light. Light is transmitted, reflected and/or absorbed. For example, if a printed sheet of paper is sufficiently thick, all light will be either absorbed or diffusely reflected; there should be no significant amount of light transmitted. Color viewing is a highly subjective experience that varies from individual to individual. Lighting and viewing standards help ensure the accuracy of color reproduction in the graphic arts industry. The most widely used objective color measurement system is defined by the CIE and known as CIELAB which is a 3-dimensional coordinate system where L* represents lightness, a* represents redness/greenness, and b* represents yellowness/blueness. Color is measured with either a spectrophotometer or a colorimeter.

Color Break Designation of ink colors to be used for specific image areas.

Color Correction Any method (masking, dot-etching, screening, scanning, etc.) used to change reproduction of the color original (photograph, transparency, chrome, 35mm slide, digital photo, painting, etc.).

Color Difference A calculation intended to represent the perceived color difference between two samples. It can be expressed in multiple ways utilizing differences in L*a*b* or L*C*ho, or with an overall color difference calculation known as Delta E, or DE. Refer to section 20.1.1 for more information.

Color Electronic Prepress System (CEPS) A high-quality, proprietary computer-based system that may include equipment for page make-up, scanning color separations and making color corrections. PC-based color scanning and manipulation systems, often referred to as desktop publishing systems, usually lack the capabilities and sophistication of CEPS.

Color Key An overlay proof (analog or digital) made of layers of acetate or polyester attached in register to a backing substrate. Each overlay film carries the colored image from a film negative. Color breaks and traps can be judged, but exact color match to the final printed product cannot be made.

Color Managed Comp A prototype of a contract proof on the finished structure representing the design intent in the context of the target color space. See Contract Proof and Comprehensive Layout, Comp.

Color Monitor An RGB or composite monitor that uses separate video signals of red, green, and blue, the three primary additive colors. It uses these signals to display almost any number of hues, depending upon the computer software and calibration. This type of monitor usually produces clearer, sharper colors and images than can be reproduced by printing CMYK process inks. Composite monitors use one signal that combines the three primary colors.

Color Resolution The number of different colors or gray-scale values a system can work with or present. The value is usually given in bits; each added bit doubles the number of available colors. For example, 8-bit color displays show 256 colors (or shades of gray).

Color Saturation — The component of color that represents the purity of the color and departure from grayness for a particular hue and lightness. Saturation is one of three coordinates in the HSL and HSV color models. See Chroma.

Color Separation — The process by which original artwork is separated into individual color components for printing. The final digital file includes color to color trapping, press mandatories (marks), color modification for specific inks and substrates, as well as halftone screening to enable printing a uniform tone scale with proper gray balance from extreme highlights through midtones and shadows to maximum solid color.

Color Target — A proof or swatch that represents the customer's expectation for color. Also known as a color reference.

Color Tolerance — An acceptable color difference between a standard (reference) and a sample.

Color-Managed Proof — See Profiled Contract Proof.

Colorimeter — An optical measurement instrument that responds to color in a manner similar to the human eye by filtering reflected light into its dominant regions of red, green and blue. A color's numeric value is then determined by using the CIE XYZ color space or one of its derivatives such as CIE L*a*b* or CIE L*u*v*.

Combination Plate — Printing halftones or screen tints and solid line or text copy using the same plate; may require print quality compromises because halftone dots require minimum impression and ink film thickness whereas solids need maximum impression and ink film thickness for optimum printability; often can be avoided with advance planning.

Combination Run — A press run that contains multiple designs on the same form/plate. The use of expanded gamut printing and shorter press run volumes has increased the frequency of combination runs.

Committees for Graphic Arts Technologies Standards (CGATS) — Formed in 1987, this group reports to ANSI and is charged with the overall coordination of graphic arts standard activities and the development of graphic arts standards where no applicable standards developer is available. The IT8 Committee, developer of digital data exchange standards, was merged under CGATS in 1994. Information about existing and pending CGATS activities is available , see Appendix.

Common Print Stations — A common image remains throughout a pressrun; plate or color changes are made for different design elements such as weight marks, UPC codes, ingredients, nutritional labeling, etc.

Compensation Curve — Curves appiled in platemaking derived from data that indicates the halftone dot area needed to compensate for dot gain/TVI throughout the entire tone scale during the printing process. This data is specific to particular printing materials and process conditions.

Comprehensive Layout, Comp — A mock-up of a printed piece showing all type and pictures in rough form but in the right size and in the correct position; used for evaluating a design before final type and artwork are produced.

Concept Proof — A proof generated early in the creative process used to capture input from all partners in the supply chain during initial design development. It is typically not color-managed and, therefore, not used for matching color.

Cones — A photoreceptor cell in the retina of the eye that is responsible for color vision and color sensitivity, functioning best in bright light.

Continuous Tone — An image containing a range of color tones from light to dark. Appear as pixels on a color monitor or silver/pigment particles on a photograph. Must be converted to halftone dots in order to be printed. See Continuous Tone (CT).

Continuous Tone (CT)
A picture file format; conveying the concept that halftone screening can be performed on this file upon output, as when screening CTs at a specific size and screen ruling on an image setter. CT files are created by either scanning a picture into the system or by generating a CT image within an application.

Contract Proof
A proof representing the customer's complete content and color expectations for the final printed product and used as a contract for determining compliance to expectations.

Contrast
The difference between extreme highlight and shadow areas of a continuous tone original or halftone reproduction. Image contrast usually is compressed to bring an originals density range to what can be reproduced on a printing press.

Cross Direction (CD)
The direction across the width of a machine – 90-degrees from machine direction (MD). The direction at right angle to the paper grain or flow of material through a machine (paper machine, extruder, printing press, etc.). See Machine Direction.

CT Merge
A photo retouching term describing the function of combining two CT files in such a manner that they appear to vignette together smoothly without a noticeable break between images.

Curl
The tendency of a substrate not to lay flat.

Cut-back Curve
See Compensation Curve.

D-Max
The highest measured density on a sample – not to be confused with the maximum density achievable by the material.

D-Min
The lowest measured density on the clear/non-image area of a sample – not to be confused with the minimum density achievable by the material.

DDCP
See Direct Digital Color Proof.

DDES
See Digital Data Exchange Standards.

Delta E (ΔE)
An overall color difference calculation used typically to determine pass/fail criteria for printed products. Delta E is a single number that represents the 'distance' between two colors in a specific color space. There are multiple derivations of Delta E formulas currently in use. Therefore, the Delta E method being utilized should always be specified for communicating color difference results.

Densitometer
A photoelectric instrument that measures the optical density of images or colors. A reflection densitometer measures the amount of incident light that is reflected from the surface of a substrate, such as ink on paper or film. A transmission densitometer measures the amount of light that is transmitted through film from a measured light source.

Density, Absolute
Optical density referenced to a perfect reflecting diffuser through calibration procedures; typically referred to as "density with paper/film included."

Density, Optical (Reflection Density)
The light absorbing property of a material, expressed as the logarithm of the reciprocal of the reflectance factor (i.e., higher density indicates more light is absorbed or a darker surface). Also called print density.

$$\text{Reflection Density} = \log_{10}\left(\frac{1}{R}\right)$$

Reflectance:
100% = 0.0
10% = 1.0
1% = 2.0
0.1% = 3.0
0.01% = 4.0

Density, Optical (Transmission) The light absorbing property of a material, expressed as the logarithm of the reciprocal of transmittance (i.e., higher density indicates more light is absorbed).

$$\text{Reflection Density} = \log\left(\frac{1}{T}\right)$$

Where:
T = Transmittance

Density, Relative The absolute (optical) density of the sample minus the absolute (optical) density of the substrate; typically referred to as density minus paper.

Deuteranomaly A type of abnormal, defective color vision. Deficient in green response for certain color mixtures. See Abnormal Color Vision.

Deuteranopia A type of abnormal, defective color vision. Specific to the Red-Green color region, with most of the deficiencies in the green region.

Device Independent Color Space A color space that can be used to describe all the colors seen by the human eye, independent of the colorants used to reproduce colors for a specific device.

Device Specific Color Space A color space that is defined based on how a specific device reproduces color. RGB and CMYK are both device specific color spaces.

Digital Bar Code file A bar code symbol that is designed and stored in a digitized format.

Digital Data Exchange Standards (DDES) A body of standards developed for the graphic arts industry by the ANSI accredited Image Technology Committee (i.e., ANSI IT8) and the ISO accredited graphics technology committee (i.e., ISO TC130). DDES provides standardized exchange formats for the digital information developed and used in printing design and production.

Digital Proof A proof produced directly from digital data through a digitally controlled imaging system – without the use of film(s).

Dimensional Stability Ability of a substrate (paper, board, corrugated, film) to retain its dimensions and its shape despite changes in its moisture or mechanical stressing. Moisture changes are caused by differences in ambient relative humidity from the internal relative humidity of the substrate. Converting, printing and ink drying processes may apply mechanical stresses (roll build, baggy/slack edges, etc. Changes in surface moisture and relative humidity tend to cause curl, wavy edges, frame shrinkage, etc. These can degrade print registration.

Direct Digital Color Proof (DDCP) Prepress color proof that is imaged directly from digital data without the intermediate steps of film and contact exposure.

Dirt/Gels Apparent dirt area on a substrate affecting its aesthetic appearance and possibly resulting in print defects/voids. Size, frequency, color and location are typical criteria for measuring dirt/gels visually against a standard agreed upon between the customer and the supplier.

Dirty Print A print defect. Also identified as dot bridging, feathering and unwanted print. A general description for printed areas showing a build up or excessive ink transfer of unwanted printed dots.

Dispersion A uniform distribution of solid particles in a vehicle by mixing or milling.

Display Type In composition, type set larger than the main reading body text. Used to attract attention; for example, a headline.

Distortion The amount a design/prepress art file is reduced to compensate for cylinder wrap & the stretch of plate material during the mounting process.

Distortion Factor A multiplier that compensates for normal flexographic image shrinkage with rubber plates and image stretch when any type of flexographic plate is made flat and mounted around a cylinder for printing.

Dot Area (%), apparent (Tone Value in ISO Documentation)
The dot area of a printed halftone element that is computed from reflection densities of the printed element and area of solid, continuous coverage. The computation of apparent dot area makes use of the Murray-Davies equation. It accounts for the physical area covered by the dot pattern plus optical effects that cause the dots to appear larger in size (optical gain). This approximates the visual impression of the printed area. Also identified as Tonal Value Increase (TVI) and Dot Gain.

Dot Area (%), Film Printing (Tone Value in ISO Documentation)
The area that will print as the final dot on the substrate. For making calculations, the following applies:

- The film printing dot area for positive separations is that value measured as the opaque dot on the input film.
- The film printing dot area for negative separations is that value measured as the opaque dot in the input film subtracted from 100".

Dot Bridging
A print defect. A type of dirty print. Appearance of a dirty, grainy effect as a result of two or more process dots linking together. The gaps between dots are bridged by ink.

Dot Gain/TVI
A physical and/or optical measurement and theoretical calculation of the apparent increase in dot area from one medium to another. Normally expressed as the difference between a midtone (nominal 50%) dot area on the digital data/film and the printed dot area; for example, a 50% film dot area which prints as a 78% dot has a 28% dot gain. Dot gain (and loss) is normal and must be controlled throughout the prepress and printing process. Also identified as Tonal Value Increase (TVI) and Dot Area.

Dot Gain/TVI (Apparent, Equivalent or Total)
The difference in dot area between the digital data/film dot area and the apparent dot area measured on the printed sheet. The computed value includes both physical changes in dot size and optical effects which increase the apparent size of the printed dot (e.g., a 72% apparent printed dot area from a 50% input film dot area is reported as 22% total dot gain). Also identified as Tonal Value Increase (TVI) and Dot Area.

Dot-Gain Curve
Graphic illustration/model of dot gain/TVI data throughout the entire highlight (non-image) to extreme shadow (solid image) tone scale.

Double Bump
Application of two layers of ink in the same printed area to achieve greater opacity or more intense color.

DPI
Dots per inch. DPI is used to measure the resolution of an image both on screen and in print. DPI measures how many dots fit into a linear inch. The higher the DPI, the more detail can be shown in an image.

Dyne Level
A measurement of the surface energy, typically associated with film, but can be used to describe the surface energy of other solid surfaces (i.e. anilox, plates).

EAN/UPC Symbols
A family of bar-code symbols using UPC Version A, UPC Version E, EAN-8, and EAN-13.

EB
See Electronic Beam.

Electron Beam (EB)
EB is a type of radiation ink curing process, avoiding the need for traditional solvent usage in curing inks.

Encapsulated PostScript (EPS)
A file format that carries both a description of an image in the PostScript page-description language and an optional bitmap equivalent for screen display. Commonly used for image interchange on the Macintosh.

Enclosed Inking Chamber
See Chambered Doctor Blade System.

EPS
See Encapsulated PostScript.

Expanded Gamut The addition of more colors to the typical four process colors (CYMK) for the purpose of enlarging the available color space.

Extrusion The production of a continuous sheet or film (or other shapes) by forcing hot thermoplastic material through a die or orifice.

Fade See Vignette and Aging/Fade Resistance.

Feathering A print defect. A ragged, feathery appearance of type or image edges on the printed material due to ink drying on plates and/or anilox rolls.

FFTA See Foundation of Flexographic Technical Association.

Fill-In (Reverses) A print defect identified as ink bridging across small print and non-printing gaps in design.

Fisheyes A print defect. Tiny round discontinuities in printed image.

Flatness Departure of a substrate from a flat plane to the extent that contributes to misregistration or other printing/converting quality degradation.

FLEXO Color Guide Edition X (10) Color Specification Guide produced by the Glass Packaging Institute (See GPI:Glass Packaging Institute) to help identify brand color standards between similar/same designs that print on glass, along with accompanying Flexo printed items like corrugated containers and other structures. The FLEXO Color Guide was initiated in 1949 by the Glass Container Manufacturers Institute (GCMI), now known as the Glass Packaging Institute (GPI).

Flexographic Technical Association (FTA) Member supported nonprofit organization that promotes, develops and maintains the advancement of flexographic processing and/or printing. For contact information, see Appendix A.

Float The material which floats on top of an ink or coating.

Fluorescence The ability of a substrate and/or ink to absorb ultraviolet light waves and reflect them as visible light.

FOGRA The FOGRA Graphic Technology Research Association, located in Munich, Germany, is focused on research and development for printing technology. FOGRA's tasks are research, development, transfer of know-how to industry, development of standards, consultancy and technical reports.

Font A complete set of characters in one design, size, and style. In traditional typography usage, font may be restricted to a particular size and style or may comprise multiple sizes, or multiple sizes and styles, of a typeface design.

Formation Distribution of fibers in paper. Excessive non-uniform distribution or flocking of fibers can contribute to print mottle. Although instruments exist to measure paper formation and print mottle, these characteristics are typically measured visually against a standard agreed upon between the customer and the supplier.

Foundation of Flexographic Technical Association (FFTA) Organization exclusively for educational purposes benefiting members of the Flexographic Technical Association and the flexographic industry.

FPO (For Position Only) The temporary image used for contextual reference in a design that will be replaced in prepress production with a high resolution image.

Frequency Modulation (FM) Screening An alternative to conventional (AM) halftone screening where the frequency of same-size microdots (typically 10 to 35 microns in diameter) are varied, or modulated, to create various tones. Also known as Stochastic Screening. Classifications include First Order and Second Order.

FTA — See Flexographic Technical Association.

G7™ — A near-neutral calibration technique developed and offered by IDEAlliance. Refer to the Appendix for contact information.

Gamut — The range of colors that can be produced on various output devices such as printing presses, proofing system, displays, etc. with a given set of colorants.

Gauge — The thickness of flexible packaging film substrates (100 gauge = 0.001"). Also a measurement sometimes used to identify the thickness of printing materials such as substrates, plates, mounting tape, etc.

GCR — See Gray Component Replacement.

Gear Side — The side of the press that contains the gearing or drive mechanisms.

Gels — See Dirt/Gels.

General Requirements and Applications for Commercial Offset Lithography (GRACoL®) — Guidelines for sheetfed offset litho prepress, press and binding/finishing operations, were introduced in 1996 and as of 2006, is up to GRACoL 7 – available from IDEAlliance; for contact information, see Appendix A.

Ghosting — A print defect. Presence of a faint image of a design in areas that are not intended to receive that portion of the image. Usually a repeat pattern in the press machine direction.

Gloss — Specular reflection of light from a surface, measured by a variety of instruments (like a glossometer) and reflection angles, reported as percentage with higher values indicating higher gloss. Film, ceramics & aluminum gloss is often specified at 45°; most paper is manufactured to specifications of 75° (ISO/TAPPI standard for paper); print gloss, metals & plastics are commonly measured at 60°, matte surfaces are measured at 85°; and very high gloss is commonly measured at 20° to correlate with visual perception.

GPI: Glass Packaging Institute — Founded in 1919 as the Glass Container Association of America, GPI is the trade association representing the North American glass container industry. On behalf of glass container manufacturers, GPI promotes glass as the optimal packaging choice, advances environmental and recycling policies, advocates industry standards, and educates packaging professionals. (www.gpi.org) GPI Produces the FLEXO Color Guide Edition X (10)

GRACoL® — See General Requirements and Applications for Offset Lithography.

Grade — Paper classification based primarily upon end use and brightness.

Gradient — See Vignette

Gray Balance — The proper combination of cyan, magenta, and yellow ink dot area, hue/density, trap, transparency, and register on a specific substrate under normal printing conditions that reproduce as a neutral gray.

Gray Component Replacement (GCR) — A color separation technique that replaces the least prevalent process color (the "gray component") with black in areas where all three (CMY) are present. An example would be removing magenta from a green and replacing it with black.

Halftone — A pictorial that has been converted from a continuous tone original image, such as a photograph, into dots of appropriate size which, when printed, give the visual illusion closely resembling the original over a gradation range from highlight to shadow.

Halftone Tint — Refers to the pattern of dots of varying sizes applied to an image of varying tones. A Flat Halftone Tint is an area of approximately equal sized halftone dots producing a uniform optical density.

Halo A print defect. An undesirable outline appearing around a printed image or type/copy.

Haze A milky discoloration of a transparent film or liquid solution such as ink or overprint coating in any printing process. Reflection haze is scattering of reflected light in directions near that of specular reflection by a specimen having a glossy surface thereby masking print quality. Transmission haze is scattering of light within or at the surface of a nearly clear specimen causing a cloudy appearance when viewed by transmission typically negatively affecting the quality of reverse printed or laminated items. See Clarity/Haze.

HDPE See High Density Polyethylene.

High-Density Polyethylene (HDPE) Film that has excellent moisture barrier and stiffness so it is used in applications such as cereal and cracker packaging. It is frequently co-extruded with heat-seal layers such as Surlyn to make a finished packaging material. Blown HDPE film has better stiffness and moisture barrier than cast HDPE, but is hazier. Extrusion coated HDPE resins are generally used to improve grease resistance. The density of high-density polyethylene can range from 0.93 to 0.97 g/cm^3.

Highlight The lightest or whitest parts in an image or photograph represented in a halftone reproduction by the smallest dots or no dots. Also, the range of the tonal reproduction area between 0-15%.

Hooking A term that describes a rapid, undesirable change in hue angle of a process color or overprint color due to an increase in ink film thickness or pigment load in an attempt to maximize the color gamut.

HSL Hue, Saturation, Lightness. Recognizing that the geometry of the RGB model is poorly aligned with the color-making attributes recognized by human vision, computer graphics researchers developed two alternate representations of RGB, HSV and HSL (hue, saturation, value and hue, saturation, lightness)

HSV Hue, Saturation, Value. See HSL.

Hue The attribute of color that distinguishes one shade from another (The name of the color: i.e. red, green, yellow, blue, etc.).

Hybrid Screening A type of screening for flexo platemaking that combines AM and FM screening. Typically, FM screening will be utilized at the extreme highlights, then transition to AM screening. The approach takes advantage of FM screening's capability to render fine highlight tone levels and AM screening's smoothness at higher tone levels. The technique can also be used to smooth the transition from a 95% to 100% that utilizes a solid cell screening technology.

ICC International Color Consortium – formed in 1993 to develop a color management system that would function transparently across all operating systems and software packages.

ICC Profile See Profile.

Ink Absorbency Ability of an ink to penetrate a substrate surface to a desired level promoting adhesion, high density, high gloss and ink lay uniformity.

Ink Bleed A print defect. Color spreads into subsequently applied coating or adhesive.

Ink Trap Percent A measure of how well one ink prints over another, calculated from print densities measured using the filter for the second ink printed to form the overprint. Higher numbers are desirable indicating the ability of an ink to transfer equally to an unprinted substrate and to a previously printed ink film. "Perfect" 100% trap is rarely achieved due to the inherent measuring geometry and data additive failure. Calculated as follows from print densities taken using the complementary filter for the second ink printed. Formula (next page):

$$\% \text{ Ink Trap} = \frac{\text{Overprint Ink (D3) - First Printed Ink (D1)}}{\text{Second Printed Ink}} \times 100$$

International Organization for Standardization (ISO) — A worldwide group from 100 countries with a mission to promote the development of international standards for intellectual, scientific, technological and economic activity. The ISO Technical Committee for graphic arts is TC 130. See Appendix A for address.

ISO — See International Organization for Standardization.

Kerning — Modifying the normal space between letters during typesetting; can be plus or minus letter spacing in computerized typesetting; traditionally involved reducing space between only selected characters, such as the L and Y in ONLY, to be more readable or pleasing to the eye. See Letter Spacing.

Keyline — An outline on finished art indicating the exact shape, position, and size for elements such as halftones, line art, UPC symbols, etc.

Kick Out — Coagulated ink, with solid lumps or particles in ink

Kiss Impression — The minimum amount of pressure necessary on press to properly transfer ink between anilox and plate and/or plate and substrate/impression. In printing, a very light impression, just enough to produce an image on the paper.

Lamination — A printed or unprinted construction made by adhering two or more substrates together.

LCH — Lightness, Chroma, Hue. LCh is a variant of the CIELab color space. It is sometimes easier to visualize 3D color space in terms of LCh than Lab. In LCh, the further out from the center of the space you go, the more saturation (Chroma) is represented. Also the 'perimeter' of this 3D cylinder becomes a color wheel, which is represented by h (hue angle).

LDPE — See Low Density Polyethylene.

Letter Spacing — Adding space between characters and spaces during typesetting; also known as "tracking" in some typesetting software. See Kerning.

Line Color — Color printed as solid areas without halftone screen.

Line Copy, Line Art, Line Drawing, Line Film, Line Work — Any image or design element reproduced with solid ink and without the use a halftone screen.

Linear Low Density Polyethylene (LLDPE) — Film having the same features as LDPE but is stronger with better hot tack strength. Film resins are more expensive than LDPE, and extrusion coating grades are even more expensive. LLDPE has a density of 0.92 g/cm3.

Linear Medium Density Polyethylene (LMDPE) — Film is similar to LLDPE but provides improved stiffness, gloss, and reduced flavor adsorption. LMDPE is defined by a density range of 0.926–0.940 g/cm3

Lines Per Inch (LPI) — The number of dots per linear inch along the angle of imaging in a halftone. Dot size varies from very small highlight dots to large shadow dots. More lines per inch increases resolution detail and dot gain/TVI. Lines per centimeter are specified outside the USA. Also used to define the screen line count in anilox rolls.

LIVE — Indicates a scan or illustration in an electronic document that is ready for production of the platemaking film negative.

LLDPE — See Linear Low Density Polyethylene.

LMDPE — Linear medium density polyethylene. Film is similar to LLDPE but provides improved stiffness, gloss, and reduced flavor adsorption.

Loose Color Proof Also known as a random proof. Process color proof with no line copy or special ink colors.

Low Density Polyethylene (LDPE) LDPE is defined by a density range of 0.910–0.940 g/cm3. Low-cost resin LDPE film has good moisture barrier, heat sealability, and strength. Extrusion LDPE has an excellent bond to paper and varying bonds to other substrates.

LPI See Lines Per Inch.

Machine Direction (MD) Flow or movement of material through a machine. Cellulose paper fibers are often oriented somewhat parallel to the direction of flow through a papermaking machine. See Cross Direction.

Mask
1. Outline of an image on original art;
2. Opaque material used to protect open or selected areas of a printing plate during exposure.

Masstone The reflected color of a bulk ink. See Undertone.

MB See Megabyte.

MD See Machine Direction.

MDPE See Medium Density Polyethylene.

Medium Density Polyethylene (MDPE) A type of polyethylene defined by a density range of 0.926–0.940 g/cm3.This film provides better barrier and chemical resistance than LDPE, but is less dense than HDPE.

Megabyte (MB) A unit of measure equal to 1,048,576 bytes, or 1,024 kilobytes; commonly used to specify the capacity of computer memory.

Metamerism A phenomenon exhibited by a pair of colors that match under one or more set of conditions (real or calculated), but do not match when these conditions are changed.

Metamerism Index (MI) A formula that calculates the difference between two colors under two different light sources. An MI of greater than 2.0 usually indicates that the metameric is visible to the human eye.

Min-Dot or Minimum Dot The smallest tone value, or dot area level that can be consistently resolved, typically when referring to the printed piece.

Moiré An optical interference pattern caused when two screened images are superimposed at inappropriate angles. It is possible for an anilox roll screen pattern, or a substrate with an inherent pattern to be one of the sources of screen interference.

Moisture Content See Absolute Humidity and Relative Humidity.

Monochromatism A type of abnormal, defective color vision. No discrimination of hue or saturation. See Abnormal Color Vision.

Mottle Print Defect. Result of uneven ink lay or non-uniform ink absorption across the paper surface, especially visible in mid-tone imagery or areas of uniform color such as solids and continuous-tone screen builds. Also known as orange peel, pigment flocculation, striations, etc.

Murray-Davies (M-D) Equation Used to calculate tonal value increase/dot gain/dot area. This measurement approximates the total of physical dot size plus apparent (optical) dot gain due to insufficient light absorption of the ink and extra light absorption of the substrate, thus the term "apparent dot area." Calculations can be made from densitometric or colorimetric data input.

N-Factor — A coefficient or correction factor that is an empirical calculation specifically for each printing substrate, and is used in the Yule-Nielson equation for factoring out Optical Dot gain. Typically the 50% tint is used. It is a best practice to communicate that an n-factor was used, especially if your data will be compared with others.

Nanometer (nm) — Unit of measure equal to one millionth of a millimeter. Color wavelengths are measured in nanometers.

Near-Neutral Calibration — A technique of calibrating a printing or proofing system to a defined condition in terms of gray balance and tone reproduction. One of the most common approaches is G7(TM) from IDEAlliance, however there are multiple alternate approaches.

OCR — Acronym for Optical Character Reader; a device that allows a computer to read printed or written material.

One Color Moiré — An interference pattern that appears in a one color screened area of a print typically caused by a coarse line screen of the anilox roller.

Opacity — Comparison of the percentage of light reflected by a sheet of paper with a black backing compared to the light reflected with a white backing. Higher values indicate higher opacity (less undesirable show-through of an image printed on the opposite side of a sheet). Can be measured with an opacimeter or a spectrophotometer.

Operator Side — The side of the press where the operator typically interfaces with the press - opposite of the gear side of the press

POP — See Oriented Polypropylene.

Opponent Color Theory — Opponent Color Theory explains conceptually how the human visual system perceives color. To the human visual system, red and green are opposites and yellow and blue are opposites. This means that if something is red, it has no green in it (but it may also be blue or yellow) and if something is yellow, it has no blue in it (but it may also be red or green). This theory is the basis for most uniform color spaces, such as CIELab and CIELCh.

Optical Scanner — A device that analyzes the light reflected from or transmitted through copy, art, or film and produces an electronic signal proportional to the intensity of the light or color.

Orange Peel —
1. A variety of mottle.
2. A finish resembling the dimpled appearance of an orange peel.

Oriented Polypropylene (OPP) — A clear, stiff film with good heat resistance and good moisture barrier. Coated grades also have good oxygen barrier or good heat sealability.

Ortho Response — Specified as Type 2 in ISO 5-3:1995: Photography – Density measurements – Part 3: Spectral conditions. This is generally used for measuring densities when printing to orthochromatic (blue/green sensitive) materials with sensitivities of 350 nm to 520 nm with a peak at approximately 435 nm.

Pantone Matching System® (PMS) — The company/brand name of a system for specifying colors; a standard practice in the printing industry.

Parker Print-Surf — An instrument that uses an air leak principle to estimate substrate surface micro roughness by the average mean pore depth in microns using TAPPI method T555. Higher numbers indicate a rougher surface.

PE, Poly — See Polyethylene, HDPE, LDPE, LLDPE, LMDPE, MDPE.

PET Polyester (Polyethylene Terephthalate) — Oriented PET film has excellent stiffness, clarity, heat resistance and dimensional stability good oxygen barrier, and some moisture barrier.

pH — A measurement of acidity or alkalinity. pH is often associated with the chemistry of waterbased inks. A value of 7 is neutral in a scale ranging from 0 to 14. Solutions with values below 7 are acid, above 7 are alkaline.

Phthalocyanine — Official name for phthalic acid commonly referred to as "Phthalo Blue" or "Phthalo Green". A bright greenish blue crystalline compound C32H18N8; metal derivatives that are brilliant fast blue to green dyes or pigments.

Pica — Unit of measurement principally used in typesetting. One pica equals 12 points or approximately 1⁄6 of an inch.

Pick Resistance — A balance of substrate surface cohesive strength being higher than the force necessary to split a wet ink film.

Picking — Print Defect. Rupture of the surface being printed that occurs when the force necessary to split (transfer) an ink film is greater than the surface strength of the substrate being printed. Occurs when pulling force (tack) of ink is greater than surface strength of paper. Transfers from the substrate web to the image carriers & rollers.

Pinholing — Print Defect. Failure of a printed ink to form a completely continuous film. This condition appears in the form of small holes or voids in the printed area.

Pixel — In electronic imaging, a basic unit of digital imaging. A picture element, or the smallest unit (cell, dot, square) on a color monitor display screen grid that can be displayed, stored, or addressed. A picture is typically composed of a rectangular array of pixels.

Plate Break — Non-print area where the two ends of a flexographic plate butt together after being wrapped around the plate cylinder on the printing press.

PMS — See Pantone Matching System®.

Point — 1. A typesetting measurement indicating type size. One point equals 0.01383"; 2. Paperboard thickness measurement (20 pts. = 0.020").

Poly — See Polyethylene.

Polyethylene (PE, Poly) — A polymerized ethylene resin used for packaging films or molded for a wide variety of structures. See HDPE, LDPE, LLDPE, LMDPE, MDPE.

Polypropylene (PP, Polyprop) — Film has the highest melting point of the economical polyolefin family. It has excellent optics, high stiffness, and good moisture barrier. Copolymer polypropylenes give improved low temperature impact resistance and sealability. PP can be oriented (OPP) to make films with improved stiffness, barrier and optics.

Polyvinylidene Chloride (PVDC) — Film with excellent water, oxygen and flavor barriers. In emulsion form, it can be used as a barrier coating.

Porosity — The resistance of substrate to the passage of air, oil or water; it can affect ink absorbency, drying and adhesion. Porosity is measured quantitatively as either the length of time it takes for a quantity of air to pass through a paper sample, or the rate of the passage of air through a sample, using either a Gurley densometer (in the first case) or a Sheffield porosimeter (in the second case).

PostScript — Adobe® Systems' trademarked page description language.

PP — See Polypropylene.

Print Contrast

An indicator of a printing system's capability to hold image detail in the shadow region. Most desirable (highest) print contrast occurs with the simultaneous highest solid ink density and the lowest dot gain/TVI. Calculated using the ratio of the difference between the printed solid area density and a printed shadow tint area (traditionally 75% for offset lithography and 70% for flexography) to the density of the solid, expressed as a percentage.

$$\text{Print Contrast} = \frac{D_S - D_T}{D_S}$$

Where:

D_S = the Density of Solid Ink/Printed Solid
D_T = the Density of the Halftone Tint.

Note: Solid and tint must be the same color.
% Print Contrast = Print Contrast x 100

Printing Industries of America

A member-supported, nonprofit, scientific, technical, and educational organization serving the international graphic communications industries See Appendix A for contact information.

Process Colors (CYAN, MAGENTA, YELLOW & BLACK)

Cyan, magenta, yellow, and black; inks used in four-color process printing; hue may be modified to meet specific needs. Incorporated into past and current editions of FIRST, process colors have been colorimetrically specified and are mono-pigmented. All process inks must be a transparent. This will allow for the blending of varying amounts of each of the process colors to achieve the visual appearance of the many thousands of shades capable of being printed by flexography.

Profile

A file that serves to describe how a particular output device (press, proofer, etc.) renders and reproduces color. It is used to allow conversions from RGB to CMYK and CMYK to CMYK for transferring of images from one output device to another with the most accurate color matching.

Profiled Contract Proof

Also known as a color-managed proof, this type of proof is created using a color management system and is typically meant to represent a specific printing condition. Proofing systems often have larger color gamuts than most printing systems, so their gamuts need to be constrained, or limited, so as to not produce colors that cannot be achieved on press. It is often used as a contract proof, or agreement between a supplier and a customer.

Proof

A physical sample of a print or ink. A proof is typically intended to be a representation of the finished printed product and is used to judge the acceptability of an intermediate step in the process.

Proofer

The hardware device used to generate a proof. (Graphic target or drawdown.)

Protanomalous

A type of abnormal, defective color vision. Deficient in red response for certain color mixtures. See Abnormal Color Vision.

Protanopia

A type of abnormal, defective color vision. Specific to the Red-Green color region, with most of the deficiencies in the red region. See Abnormal Color Vision.

PVDC or PVdC

See Polyvinylidene Chloride.

Quiet Zones

Areas free of printing that precede the leftmost bar and follow the rightmost bar in a barcode symbol.

Radiation Curing

Radiation curing can be UV (ultra-violet) or EB (electron beam). These systems generate energy which will transform a liquid ink or coating into a solid.

Raster Graphics

Sometimes also referred to as "bitmapped graphics". Graphics that are made up of pixels and have a specific resolution. Subject to changes in resolution when scaled which can lead to quality concerns for sharpness and clarity of the image (both edge quality and internal quality)

Raster Image File Format (RIFF) A file format for paint-style graphics, developed by Letraset® USA. RIFF is an expanded version of the TIFF format used by many scanner makers.

Raster Image Processor (RIP) A computer device or program that translates digital information in a page description language to the pattern of dots to be delivered by the output unit of the system.

Raster Scan The generation of an image on a display screen made by refreshing the display area line by line.

Reference In evaluating color difference, the reference is the color against which all measurements are compared. Also referred to as Standard.

Reflection Densitometry The measurement technology that determines the amount of light absorption of materials by measuring reflectance and calculating and reporting optical density.

Register (Registration) Proper alignment or positioning of two or more elements with each other. This includes print-to-print and print-to-cut applications.

Relative Humidity A percentage of the amount of water vapor the air or a material such as paper can hold at a given atmospheric temperature and pressure. Most paper is made to be dimensionally stable in equilibrium with the atmosphere at 35% to 50% relative humidity. Excessive variation from this general range can result in non-flat paper, print misregister and other converting complications.

Rendering Intents Methods established by the ICC to define the objective for a color conversion. The ICC specification includes four different rendering intents: absolute colorimetric, relative colorimetric, perceptual, and saturation.

Repeat The printing length of a plate cylinder determined by one revolution of the plate cylinder gear.

Representative Contract Proof A contract proof representing a group of similar designs. For example, a customer may ask for one proof to be made to represent two or more pieces of art, typically as a cost saving effort. See Contract Proof.

Resin A sticky flammable organic substance, insoluble in water, exuded by some trees and other plants.

Resolution A measure of image sharpness, usually expressed in lines or dots per inch or millimeter. On a prepress visual display terminal, the number of pixels per unit of linear measure, e.g.,12 pixels per millimeter is a RES 12. Normally, the resolution (RES) of a file is the same vertically and horizontally, thus a square millimeter contains 12 x 12 = 144 pixels for a RES 12 file. The higher the RES, the better the image detail; but the file will be larger and will require longer processing time.

Reverse To change the tonal orientation of an image, making the darker elements lighter and the lighter darker. Note that to physically reverse the spatial orientation of an image is known as "flopping" the image.

Reverse (Knock-Out) The process of dropping a surprinted image out of the background color so type, for example, will appear white with a color surround.

Reverse Angle Doctor Blade Doctor blade used with light pressure and a reverse angle on the anilox roll.

Reverse Print 1. Printing wrong-reading on the underside of transparent film which, when laminated to another substrate the image becomes right-reading when viewed through the sheet it was printed upon. See Surface Print.
2. Design in which the "copy" is "dropped out" and the background is printed.

RGB — Red, green, and blue, the primary additive colors which are the backbone of computer color visual display monitors and prepress color separation. They also are the complementary or secondary subtractive ink colors which produce red by overprinting magenta and yellow, green by trapping cyan and yellow, and blue by overprinting cyan and magenta.

Rheology — The science of the flow of matter for liquids and soft solids. The ability of the liquid or soft solid to flow or be deformed.

RIFF — See Raster Image File Format.

RIP — See Raster Image Processor (Processing).

Rods — A photoreceptor cell in the outer edges of the retina in the eye that is responsible for peripheral vision and night vision

Rollout — Fluid ink print on a substrate using a Meyer rod applicator.

Rounding Errors (Bar Codes) — The process of allocating imaging device dots to bar or space modules in an uneven manner.

Rub Resistance — See Abrasion Resistance.

Sample — In evaluating color difference, the sample is the color to be measured and compared to the target reference/standard.

Sans Serif — Text characters without serifs, which are the fine lines that curve out from the main strokes of a letter. An example is the font Arial.

SBS — See Solid Bleached Sulfate.

Screen Ruling — See Lines Per Inch.

Screen Tint — See Halftone Tint.

Screening — Small voids in print in image areas, often has very regular shape consistent with anilox pattern.

Scuff — A mark made by scraping or grazing a surface or object. See Abrasion Resistance.

Serif — A small decorative line added as embellishment to the basic form of a character. An example is the font Times New Roman.

Set-Off — Ink transfers from image side to back side of substrate when unrolled in subsequent operations. Set-off is a less severe case of blocking.

Sharpen — To decrease in color strength, as when halftone dots become smaller; opposite of dot spread, dot gain/area or Tonal value increase.

Sheffield Smoothness — Macro smoothness of a substrate surface (typically uncoated paper, corrugated, etc.) measuring the rate of surface air flow as specified by TAPPI T538 using a Sheffield instrument. Reported as Sheffield units, values are inversely related to smoothness; the higher the value, the rougher the surface.

Shell Cup — A device for measuring viscosity.

Simultaneous Contrast — The phenomenon that occurs when the surrounding color influences how a color is perceived.

SKU — See Stock-Keeping Unit.

Slur — Print Defect. A condition caused by slippage at the moment of impression between any two of the following: substrate, plate, blanket.

Smoothness — Extremely important substrate surface uniformity requirement for high quality flexographic printability that affects ink lay and ink transfer. Measured mostly by a variety of air-leak instruments and profilometers. See Parker Print-Surf, Sheffield.

Snowflaking — Print Defect. Condition of a printed area characterized by very small dots of unprinted areas showing throughout a deposited ink.

Solid Bleached Sulfate (SBS) — Paperboard made from bleached wood pulp, usually clay coated on one or both sides to improve printability.

Solvent — A substance that is liquid at standard conditions and is used to dissolve or dilute another substance; this term includes, but is not limited to, organic materials used as dissolvers, viscosity reducers, degreasers or cleaning agents. Water is considered the universal solvent.

Solvent Release — In ink, the ability of a binder to influence the rate of evaporation of a solvent.

SPC — Acronym for statistical process control.

Specific Gravity — The ratio of the weight of a body to the weight of an equal volume of water at the same specified temperature.

Specifications for Web Offset Publications (SWOP) — A set of production specifications developed for those involved in heatset web offset litho magazine publication printing. First published in 1975, the eleventh edition was released in 2007, and is available from IDEAlliance. See Appendix A for contact information.

Spectral Reflectance Curve — The spectral reflectance curve graphically depicts the color composition of an object. The x-axis shows the wavelengths, starting with 380nm and ending with 700nm, and the y-axis shows the relative reflectance (the amount of light reflected from the object in %).

Spectral Response (Densitometer Response) — Spectral response is the product of the spectral power distribution of the lamp, attenuation of the optics and filters, and the spectral response of the detector used. The aim responses for spectrophotometers/densitometers are contained in ISO 5-3:1995, Photography – Density measurements – Part 3: Spectral conditions. The status responses of interest for densitometer response to the graphic arts are Status E, Status I, and Status T.

Spectrodensitometer — A color measurement device that is truly a spectrophotometer, but includes all of the densitometric functions along with the colorimetric functions in one device.

Spectrophotometer — A device that measures the spectral reflectance of a sample and generates a data set that is utilized in combination with a CIE Standard Illuminant and a CIE Standard Observer to generate colorimetric values to describe color quantitatively. Instrument classifications include; inline, hand-held and desktop.

Spot Color — A non-process color, typically made up of a combination of two or more pigments. Associated with Brand Colors, metallics, whites, fluorescent pastels and can be used as a line color or with a halftone tint.

Spots (Bar Codes) — Undesirable presence of ink or dirt within the space of a bar code symbol.

Stain Level — In carbon-mask-based platemaking systems, this describes the amount of carbon left on the plate after CtP imaging/ablation. It is measured with a transmission densitometer and is an important quality consideration, similar to Dmin for film, in its impact on the effectiveness of the main exposure.

Standard — In evaluating color difference, the standard is the color against which all measurements are compared. Also referred to as Reference.

Standard Illuminant — See CIE Standard Illuminant.

Standard Observer	See CIE Standard Observer.
Standard Reference Material	A physical sample with characteristics traceable to an accepted primary standard or set of standards. An example would be a T-Ref from IDEAlliance which is used to verify conformance to the Status-T densitometric response.
Standard Viewing Conditions	Conditions defined in ISO 3664 that provide specifications for viewing proofs and printed products so that a fair and consistent visual evaluation can be made.
Stochastic Screening	See Frequency Modulation, FM screening.
Stock-Keeping Unit (SKU)	An assortment or variety of wholesale items shipped in one physical case.
Streaking	Not wiping clean, leaving stripes or lines of color on web, rollers or printed product (Print Defect)
Strength	Usually refers to intensity of a color of ink.
Striation	A print defect seen as weak ink or no ink in print direction of image. A fine streaky pattern of parallel lines, usually in the direction of the web.
Stroke of Oscillation	The distance the doctor blade oscillates.
Substrate	The material that is printed upon, i.e., film, paper, paperboard.
Subtractive Primaries	Yellow, magenta and cyan, the hues used for process color printing inks.
Surface Print	Conventional flexographic printing resulting with a right-reading image on the top surface of the web. See Reverse Print.
Surface Strength	See Pick Resistance.
Surface Tension	Measurement of surface energy that affects ink transfer and adhesion to a substrate. (The tendency of a liquid surface to contract rather than flow out.) Commonly measured with a dyne indicator solution applied to a film substrate surface. Substrates typically should be 8 to 10 dynes/cm higher then the ink.
SWOP	See Specifications for Web Offset Publications.
Tack	In printing inks, the property of cohesion between particles; the separation force of ink needed for proper transfer and trapping on multicolor presses. A tacky ink has high separation forces and can cause surface picking or splitting of weak papers.
TC 130	See International Organization for Standardization.
Telescoping	Transverse slipping of successive winds of a roll of material so that the edge is conical rather than flat.
Tensile Strength	The maximum load in tension that a material can withstand without failure.
TIFF (Tagged Image File Format)	A file format for exchanging bitmapped images between applications developed by Aldus, Adobe, and Apple that is particularly suited for representing scanned images and other large bitmaps. The original TIFF saved only black and- white images in uncompressed forms.
Tinctorial Strength	The relative ability of a pigment or dye to impart color value to a printing ink.
Tint	See Halftone Tint.
TIR	See Total Indicated Runout.

Tolerances
The specification of acceptable variations in print attributes like color, register, density, dot size, plate or paper thickness, concentration of chemicals and other printing parameters. Suppliers and customers should communicate acceptability limits or process capabilities to each other.

Total Area Coverage (TAC)
Also known as Total Ink Limit – a description of the total percentage of coverage, typically in four-color process printing. This is typically quantified when profiling, or characterizing, a printing system to determine at which point additional coverage yields no additional darkness, or density. A target is printed with a matrix of CMYK at different amounts to assist in this assessment.

Total Indicated Runout (TIR)
A measure of the out-of-roundness of a printing press roller or cylinder. The difference in the lengths of a roller's radius as measured from the center to the outside surface. A perfectly round roller would have zero TIR.

Tracking
A print defect associated to an ink that appears in area where there is no print.

Transmission Densitometry
The measurement technology that characterizes the light absorption of materials by measuring transmittance, and calculating and reporting optical density.

Transparent Ink
A printing ink that does not conceal the color under it. Process & EG inks are examples of transparent inks allowing inks to blend to form other colors.

Trapping (Image)
To compensate for registration variation, two adjacent colors butting each other must be altered to allow for normal registration variances to exist without degrading the design. This is accomplished by spreading or overlapping the lighter of the two adjacent colors into the dominant, or darker, color.

Trapping (Ink)
The overprinting and adhering of one ink over another to produce desired secondary or tertiary colors. This typically refers to the overprinting of CMY to produce various shades of RGB, but is not limited to this. There can also be trapping of and with spot colors.

Tristimulus
The magnitudes of three standard stimuli needed to match a given sample of light. A method for communicating or generating a color using three stimuli (colorants such as RGB or CMY) or three attributes (such as lightness, Chroma and hue).

Tritanopia
A type of abnormal, defective color vision. Specific to the Blue-Yellow color region, with most of the deficiencies in the blue region. Confusion of blue with green and Yellow with violet. See Abnormal Color Vision.

Truncated
Shortened. Decreasing the height of the bars in an UPC bar code symbol below the normal specification decreases the symbol's ability to be read omni directionally and should be avoided.

Two Roll Metering
A method of metering ink in which a rubber roller is used to meter the ink off of an anilox.

Ultraviolet (UV)
Radiant energy below just below the visible spectrum.

Undercolor Addition (UCA)
A color separation technique that adds more cyan, magenta, and yellow to the neutral shadow areas of a process color image.

Undercolor Removal (UCR)
A color separation technique that reduces the amount of cyan, magenta, and yellow in neutral areas of a process color image and replaces them with an appropriate amount of black.

Undertone
Color of an ink printed in a thin film. See Masstone.

Uniform Code Council (UCC)
An organization responsible for overseeing and administering the Universal Product Code.

Universal Product Code (UPC) A 12- or 8-digit code number that identifies a wide range of products; printed on packages as the UPC bar code symbol which can be read electronically by a scanner at retail store checkout counters.

UV (Ultraviolet) Response Refers to that response specified as Type 1 in ISO 5/3. This is generally used for measuring densities when printing to UV/blue sensitive materials. Type 1 (UV) printing density was standardized to provide printing density values for use when exposing diazo and vesicular films normally sensitive in a narrow band of the blue and ultraviolet region of the spectrum, between 380nm and 420nm with a peak at 400nm.

UV Coating Liquid laminate bonded and cured with ultraviolet light.

UV Ink Solventless inks that are cured by UV radiation.

Varnish A thin, protective coating applied to a printed sheet for protection or appearance. Also, in ink making, it is the binder/resin component of an ink.

Vector Graphics Graphics that are defined in mathematical terms in illustration programs. As a result, the quality of the graphic is independent of any scaling that may occur – edge quality stays razor sharp regardless of any amount of enlargement. A key difference in comparison to raster graphics.

Vehicle In printing inks, the fluid/liquid component which acts as a carrier for pigment.

Vignette A halftone graphic (design element) that changes smoothly in tonal values from light to dark or vice-versa. It may or may not go all the way to a specular highlight (zero tone value). Also referred to as a gradient or blend.

Viscometer Instrument used to measure the viscosity of ink, varnish, or other solution.

Viscosity A measure of a fluid's (ink, coating) resistance to flow, which influences the amount of ink (color) printed.

Visual Spectrum Portion of the electromagnetic spectrum between 380nm and 700nm that can be seen by the human eye.

Voids (Bar Codes) The undesirable absence of ink or presence of dirt within a bar of a bar code symbol.

Wash Boarding Print defect of combined board in which the linerboard is depressed between flutes, giving the appearance of a washboard; typically measured visually against a standard agreed upon between the customer and the supplier.

Wetting Agent A substance that reduces the surface tension of a liquid, thereby causing it to spread more readily on a solid surface.

White Opaque Polyethylene Film frequently used to package frozen foods. (WhOPE, WOPE)

WhOPE, WOPE See White Opaque Polyethylene.

X-Dimension The specified width of the narrow element in a bar code symbol.

Zahn Cup: A device for measuring viscosity. Five cup specifications exist and are identified as Zahn cup #n (n=1, 2, 3, 4 or 5)

DOWNLOAD FIRST 5.1 Extras Referenced in this Section at:
http://www.flexography.org/FIRST_extras

Appendix A: Contact List

American Association of Textile Chemists & Colorists Research
Triangle Park, NC 919-549-8141
www.aatcc.org
Referenced Material:
- "Colour Index International Pigments & Solvent Dyes" -- reference book

American National Standards Institute (ANSI) -- CGATS
New York, NY 212-642-4900
www.ansi.org
Referenced Material:
- ANSI & CGATS Standards
- IT8.7/4 Characterization Target

American Society for Testing & Materials (ASTM)
West Conshohocken, PA 610-832-9500
www.astm.com
Referenced Material:
- Substrate Test Methods
- Ink Test Methods

Applied Image, Inc.
Rochester, NY 585-482-0300
www.appliedimage.com
Referenced Material:
- Calibrated Conformance Standard Test Cards for Code 39 & Code 128 Symbol Verifiers

Association of Independent Corrugated Converters (AICC)
Alexandria, VA 703-836-2422
www.aiccbox.org
Referenced Material:
- Technical Trade Association

Automatic Identification Manufacturers (AIM-USA)
Warrendale, PA 724-934-5688
www.aimusa.org
Referenced Material:
- ANSI / AIM Bar Code Specifications
- "Layman's Guide to ANSI Print Quality" – reference book

Commission Internationale del'Eclairage (CIE)
Salem, MA 978-745-6870
www.cie-usnc.org
Referenced Material:
- CIE 1976, 1994, 2000 Color Tolerancing Methods
- CIE Illuminant D50 & CIE 1931 Standard Colorimetric Observer
- "Colorimetry, 3rd Edition" -- Publication CIE 015:2004

Composite Can & Tube Institute (CCTI)
Alexandria, VA 703-823-7234
www.cctiwdc.org
Referenced Material:
- Technical Trade Association

DFTA
Stuttgart, Germany 0711-678960
www.dfta-tz.de
Referenced Material:
- Technical Trade Association
- DFTA CtP Strip V1.3 - Control Target for Digitally Imaged Photopolymer Plates

Envelope Manufacturers Association (EMA)
Alexandria, VA 703-739-2200
www.envelope.org
Referenced Material:
- Technical Trade Association

European Flexographic Technical Association
Somerset, UK +44(0)1458-241-455
www.efta.co.uk
Referenced Material:
- Technical Trade Association

Fibre Box Association (FBA)
Elk Grove, IL 847-364-9600
www.fibrebox.org
Referenced Material:
- Technical Trade Association

Film & Bag Federation, Unit of the Society of the Plastics Industry
Washington, DC 202-974-5200
www.plasticbag.com
Referenced Material:
- Technical Trade Association

Flexible Packaging Association (FPA)
Linthicum, MD 410-694-0800
www.flexpack.org
Referenced Material:
- Technical Trade Association

Flexographic Pre-Press Platemakers Association (FPPA)
Bel Air, MD 443-640-1045
www.fppa.net
Referenced Material:
- Technical Trade Association

Flexographic Technical Association (FTA)
Bohemia, NY 631-737-6020
www.flexography.org
Referenced Material:
- Technical Trade Association
- *FIRST*
- *FIRST* Certification Programs
- Flexography Principles & Practices 6.0

Ghent PDF Workgroup (GWG)
Ghent, Belgium
www.gwg.org
Referenced Material:
- GWG Packaging Specification for ISO 15930-1:2001

GS1 US
Lawrenceville, NJ 609-620-0200
www.gs1us.org
Referenced Material:
- Bar Code Standards
- Assigns Company Prefixes for Bar Codes
- "Guidelines for Producing Quality Symbols"
- Spreadsheet for Calculating Check Digits
- "UPC Printed Symbol & Quality Specifications"
- UPC / EAN Calibrated Conformance Standard Test Card for Symbol Verifiers

IDEAlliance (International Digital Enterprise Alliance)
Alexandria, VA 703-837-1070
www.idealliance.org
Referenced Material:
- IDEAlliance G7™ P2P Near Neutral Calibration Target
- IDEAlliance 12647-7 Control Strip 2009 – Digital Proof Control Target
- "Introduction to Densitometry -- Users Guide to Print Production Measurement Using Densitometry"
- T-Ref™ Calibration Target for Densitometers

International Corrugated Packaging Foundation (ICPF)
Alexandria, VA 703-549-8580
www.icpfbox.org
Referenced Material:
- Technical Trade Association
- Fibre Box Handbook

International Standards Organization (ISO) – Technical Committee for Graphics Technology (TC130)
Geneva, Switzerland +44-22-749-0111
www.iso.org
Referenced Material:
- International Standards Development
- Substrate Test Methods

National Association of Printing Ink Manufacturers (NAPIM)
Woodbridge, NJ 732-855-1525
www.napim.org
Referenced Material:
- Technical Trade Association
- "NPIRI Raw Materials Data Handbook - Volume 4 Pigments"

Newspaper & Publication Flexo Users Group (NPFUG)
Northampton, MA 413-584-5000
www.flexonews.org
Referenced Material:
- Technical Trade Association

NPES - The Association for Suppliers of Printing, Publishing & Converting Technologies
Reston, VA 703-264-7200
www.npes.org
Referenced Material:
- ANSI Standards
- CGATS Standards
- ISO Standards

Paper Shipping Sack Manufacturers Association (PSSMA)
Coopersburg, PA 610-282-6845
www.pssma.com
Referenced Material:
- Technical Trade Association

Paperboard Packaging Council (PPC)
Springfield, MA 413-686-9191
www.ppcnet.org
Referenced Material:
- Technical Trade Association

Printing Industries of America (PIA)
Sewickley, PA 412-741-6860 or 800-910-4283
www.printing.org
Referenced Material:
- Technical Trade Association
- Star/Flower Impression Target
- Test Form & Targets

Radtech International North America
Bethesda, MD 240-497-1242
www.radtech.org
Referenced Material:
- Technical Trade Association

UK Society of Dyers & Colourists –
Color Measurement Committee
Bradford, UK +44(0) 1274-725138
www.sdc.org.uk
Referenced Material:
- CMC Color Tolerancing Method
- "Colour Index International Pigments & Solvent Dyes"

Tag & Label Manufacturers Institute (TLMI)
Naperville, IL 630-357-9222
www.tlmi.com
Referenced Material:
- Technical Trade Association

Technical Association of the Pulp & Paper Industry (TAPPI)
Norcross, GA 770-446-1400 or 800-332-8686
www.tappi.org
Referenced Material:
- Technical Trade Association
- Paper & Paperboard Test Methods

Technical Association of the Graphic Arts (TAGA)
Sewickley, PA 412-259-1706
www.taga.org
Referenced Material:
- Technical Trade Association

United Nations Standard Products & Services Code (UNSPSC)
www.unspsc.org
Referenced Material:
- Commodity Codes -- Classification of Products & services

United States Postal Service (USPS)
Washington, DC 800-275-8777
www.usps.com
Referenced Material:
- Intelligent Mail Bar Code (CB4)
- USPS-B-3200C Intelligent Mail Bar Code Specification
- Bulk Mailing Guidelines

Appendix B: Referenced Standards, Specifications and Publications

ANSI 2.30 – 1989	Color Viewing Conditions
ANSI CGATS.4 - 1993	(Graphic Technology - Graphic Arts Reflection Densitometry Measurements - Terminology, Equations, Image Elements and Procedures)
ANSI CGATS.5 - 2003	(Graphic Technology - Spectral Measurement and Colorimetric Computation for Graphic Arts Images)
ANSI CGATS.9 - 1994	(Graphic Technology - Graphic Arts Transmission Densitometry Measurements - Terminology, Equations, Image Elements and Procedures)
ANSI IT8.7/4	Characterization Target
ANSI X3.182	(1990 Bar Code Print Quality Guideline)
ANSI CGATS 17	Exchange format for colour and process control data using XML or ASCII text
ANSI CGATS 21-1	Printing from digital data across multiple technologies - Part 1: Principles
ANSI CGATS 21-2	Printing from digital data across multiple technologies Part 2: Reference characterization data-2013
ANSI CGATS TR015	Technical Report that defines a methodology for establishing individual printing tone reproduction and near-neutral gray-scale aims, and families thereof, based on a shared near-neutral gray-scale definition
ANSI CGATS TR016	A Technical Report that defines a process that can be used in evaluating the conformance of printed material to a set of reference color characterization data, which are used as the intended printing aim
ASTM D1475	Ink - Specific Gravity - Test Method for Density of Paint, Varnish, Lacquer & Related Products
ASTM 2196	Ink - Test Method for Rheological Properties of Non-Newtonian Materials by Rotational (Brookfield) Viscometer
ASTM D1003	Substrate - Clarity & Haze
ASTM D1200	Ink - Test Method for Viscosity of Paints, Varnishes & Lacquers
ASTM D1319	Ink - Fineness of Grind of Printing Inks by the NPIRI Grindometer
ASTM D1640	Ink - Standard Test Method for Drying, Curing or Film Formation
ASTM D1647	Ink - 1996 Standard Test Method for Resistance of Dried Films of Varnishes to Water & Alkali
ASTM D1729: 2003	Ink - Standard Practice for Visual Appraisal of Colors and Color Differences of Diffusely Illuminated Material
ASTM D1894-95	Ink & Substrate - COF - Horizontal Plane Method
ASTM D2244	Ink - Standard Practice for Calibration of Color Tolerances and Color Differences from Instrumentally Measured Color Coordinates
ASTM D2248-01	Ink - Standard Practice for Detergent Resistance of Organic Finishes
ASTM D2457-08	Substrate - Gloss (20 degree)
ASTM D2793	Ink - Standard Test Method for Block Testing
ASTM D3359	Ink - Test Method for Measuring Adhesion by Tape
ASTM D3424-01	Ink - Standard Test Method for Evaluating the Relative Lightfastness and Weatherability of Printed Matter
ASTM D3519	Ink - 88: Standard Test Method for Foam in Aqueous Media
ASTM D4212	Ink - Test Method for Viscosity by Dip Type Viscosity Cup
ASTM D4518	Ink - Standard Test Method for Measuring Static Friction of Coating Surfaces
ASTM D4713	Ink - Test Method for Non-Volatile Content in Printing Inks, Resin Solutions & Vehicles
ASTM D523	Ink - Standard Test Method for Specular Gloss

ASTM D5264/92	Ink - Standard Test Method for Sutherland Rub Test
ASTM D589-97	Ink - Standard Test Method for Opacity of Paper - 15-degree Diffuse, Illuminant A, 89% Reflectance Backing & Paper Backing
ASTM D645	Substrate Caliper & Gauge
ASTM E-462	Ink - 1989 Test Method for Odor & Taste Transfer from Packaging Film
ASTM E70	Ink - 97 Standard Test Method for pH of Aqueous Solutions with the Glass Electrode
ASTM	E97 Ink - Standard Test Method for 0/45-degree - Directional Reflectance of Opaque Specimens
CGATS TR 011-2002	(Graphic Technology - Package Development Workflow – Design Concept Through Approved Production File)
CGATS TR 012-2003	(Graphic Technology - Color Reproduction and Process Control for Package Printing)
CGATS	"Developing a Color Characterization Data Set - Analyzing & Reporting"
CIE 015:2004	Colorimetry, 3rd Edition
DFTA	Digitally-Imaged Photopolymer Platemaking Control Strip CtP Strip V1.3 - Vector & Pixel Versions
FTA	"Flexography Principles and Practices 6.0"
ICPF	"Fibre Box Handbook"
Ghent PDF Workgroup	Packaging Specification 2012
GS1 US	"Calibrated Conformance Standard Test Card"
GS1 US	"Guidelines for Producing Quality Symbols"
GS1 US	"UPC Printed Symbol & Quality Specifications"
IDEAlliance	"Introduction to Densitometry - Users Guide to Print Production Measurement Using Densitometry"
IDEAlliance	T-Ref™ Densitometer Calibration Standard
ISO 534	Substrate - Caliper & Gauge
ISO 3034	Substrate - Caliper & Gauge
ISO 5631	Substrate - Color
ISO 11475:2004	Substrate - Color
ISO 11556:2005	Substrate - Flatness/Curl/Warp
ISO 12639:2004	(Graphic Technology - Prepress Digital Data Exchange - Tag Image File Format for Image Technology - TIFF/IT))
ISO 12647-2	(Graphic Technology - Process Control for the Manufacture of Halftone Colour Separations, Proofs and Production Prints – Part 2: Offset Lithographic Printing)
ISO 12647-6:2012	(Graphic Technology - Process Control for the Production of Halftone Colour Separations, Proofs and Production Prints – Part 6: Flexographic Printing)
ISO 12647-7:2007	(Graphic Technology - Process Control for the Manufacture of Halftone Colour Separations, Proofs and Production Prints – Part 7: Proofing Processes Working Directly from Digital Data)
ISO 12647-8:2007	(Graphic Technology - Process Control for the Manufacture of Halftone Colour Separations, Proofs and Production Prints - Part 8: Validation Prints (Design Proofs)
ISO 3664:2009	Graphic technology and photography -- Viewing conditions
ISO 18620	Graphic technology -- Prepress data exchange -- Tone adjustment curves exchange
ISO 15930-1:2001	(Graphic Technology - Prepress Digital Data Exchange - Use of PDF Part 1: Complete Exchange Using CMYK Data PDF/X-1 and PDF/X-1a)

ISO 15930-7:2010	Graphic technology -- Prepress digital data exchange using PDF -- Part 7: Complete exchange of printing data (PDF/X-4) and partial exchange of printing data with external profile reference (PDF/X-4p) using PDF 1.6)
ISO 17972-1	Graphic technology — Colour data exchange format (CxF/X) — Part 1: Relationship to CxF3 (CxF/X-1)
ISO 17972-2	Graphic technology —Colour data exchange format (CxF/X) — Part 2: Scanner target data (CxF/X-2)
ISO 17972-3	Graphic technology — Colour data exchange format (CxF/X) — Part 3: Output target data (CxF/X-3)
ISO 17972-4	Graphic technology — Colour data exchange format (CxF/X) — Part 4: Spot colour characterization data (CxF/X-4))
ISO 15937	Graphic technology — Communication of Graphic Paper Properties
ISO 13655:2009	Graphic technology -- Spectral measurement and colorimetric computation for graphic arts images
ISO 15415:2000	(Information Technology - Automatic Identification and Data Capture Technique - Bar Code Print Quality Test Specification - Two-Dimensional Symbols)
ISO 15416:2000	(Information Technology - Automatic Identification and Data Capture Technique - Bar Code Print Quality Test Specification - Linear Symbols)
ISO 15420:2000	(Information Technology - Automatic Identification and Data Capture Techniques - Symbol Specification - EAN/UPC)
ISO 15426-1:2006	(Information Technology - Automatic Identification and Data Capture Technique - Bar Code Verifier Conformance Specification - Part 1: Linear Symbols)
ISO 15426-2:2005	(Information Technology - Automatic Identification and Data Capture Technique - Bar Code Verifier Conformance Specification - Part 2: Two-Dimensional Symbols)
ISO 2470:1999	Substrate - Brightness
ISO 2471:1998	Substrate - Opacity
ISO 3200-1	(Document Management - Portable Document Format - Part 1: PDF 1.7)
ISO 3783:2006	Substrate - Surface Strength - Wax Pick Test
ISO 8254-1:1999	Substrate - Gloss (75 degree)
ISO 8254-3:2004	Substrate - Gloss (20 degree)
ISO 8791/3	Substrate - Smoothness - Sheffield Method
ISO 8791/4	Substrate - Smoothness - Print-Surf Method
NPIRI	"Raw Materials Data Handbook, Volume 4 Pigments"
TAPPI T411 om-89	Substrate - Caliper & Gauge
TAPPI T412 om-02	Substrate - Moisture Content
TAPPI T425 om-96	Ink & Substrate - Opacity/Transparency - 15/d Geometry, Illuminant A/2*, 89% Reflectance Backing and Paper Backing
TAPPI T433	Substrate - Sizing - Dry Indicator Method
TAPPI T437 om-96	Substrate - Dirt & Gels
TAPPI T441 om-98	Substrate - Sizing - Cobb Test
TAPPI T452 om-98	Substrate - Brightness
TAPPI T453 sp-97	Substrate - Aging / Fade Resistance - Dry Heat
TAPPI T459 om-93	Substrate - Surface Strength - Wax Pick Test
TAPPI T460 om-96	Substrate - Porosity - Gurley Method

TAPPI T480 om-92	Substrate - Gloss (75 degree)
TAPPI T499 um-591	Substrate - Surface Strength/Pick Resistance
TAPPI T514 cm-92	Substrate - Surface Strength - Wax Pick Test
TAPPI T524 om-94	Substrate - Color
TAPPI T538 om-96	Substrate - Smoothness - Sheffield Method
TAPPI T544 sp-97	Substrate - Aging / Fade Resistance - Moist Heat
TAPPI T547	Substrate - Porosity - Sheffield Method
TAPPI T549	Substrate - COF - Horizontal Plane Method
TAPPI T551	Substrate - Caliper & Gauge
TAPPI T552 pm-92	Substrate - Surface Tension / Treat Level - Mayer Rod Technique
TAPPI T555 om-99	Substrate - Smoothness - Print-Surf Method
TAPPI T562	Substrate - Color
TAPPI T575 om-07	Substrate - Smoothness - Emveco Method
TAPPI T653 om-98	Substrate - Gloss (20 degree)
TAPPI T815	Substrate - COF - Inclined Plane Method
USPS B-3200C	United States Postal Service Intelligent Mail Bar Code Specification

Appendix C: Quick Reference Control and Test Targets

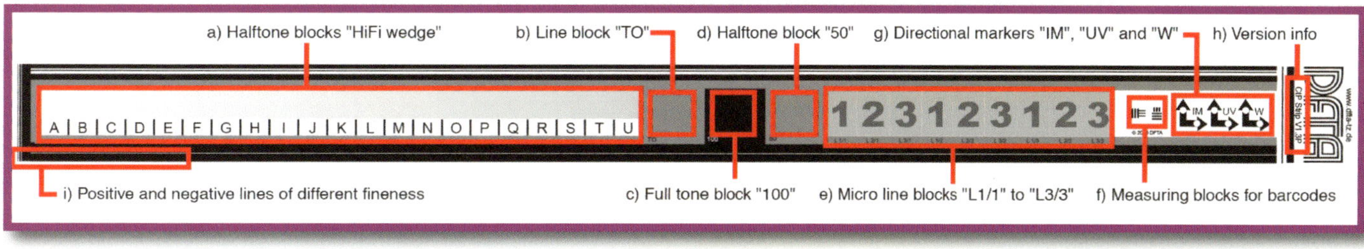

DFTA Control Strip for Digitally-Imaged Photopolymer Plates

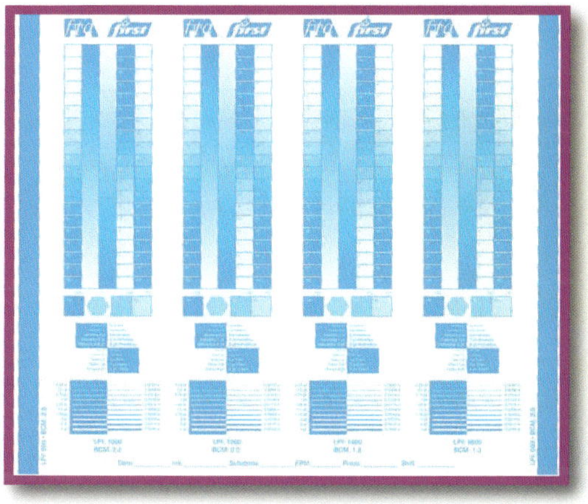

Banded Anilox Optimization Test Form

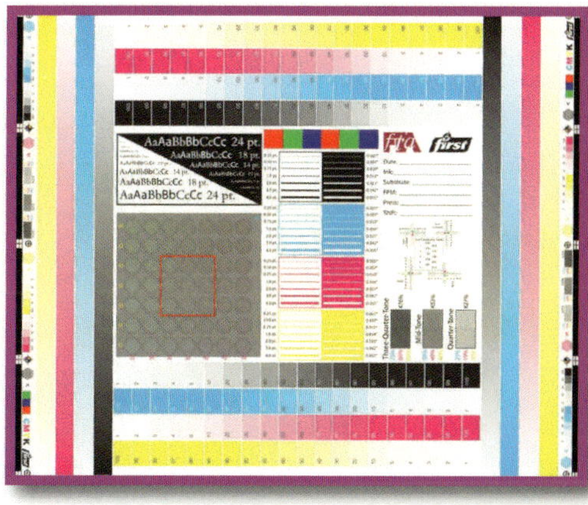

Densitometric Fingerprint Test Form

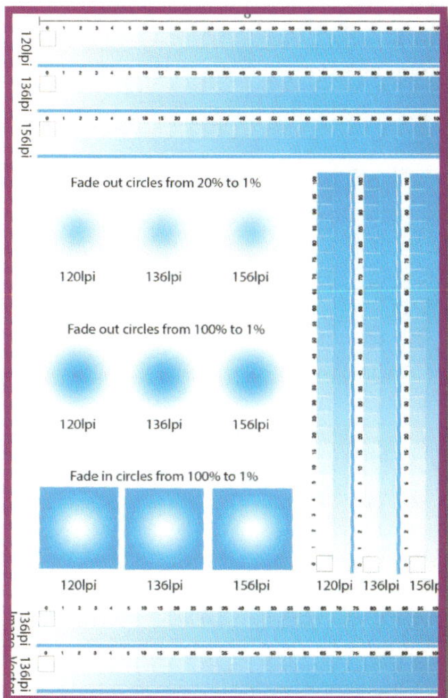

Vignette Fingerprint Test Form

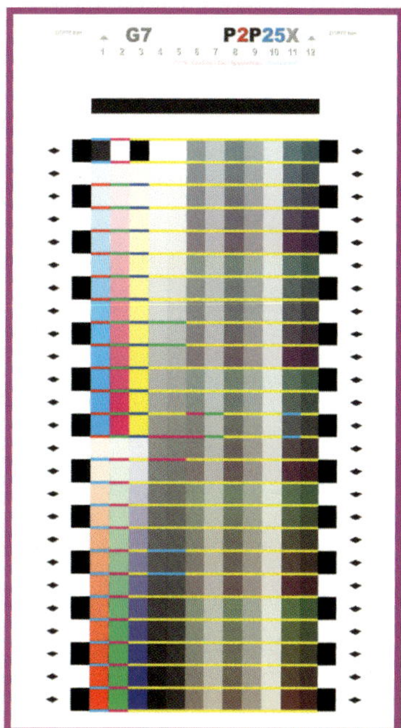

IDEAlliance Near Neutral Calibration P2P Target

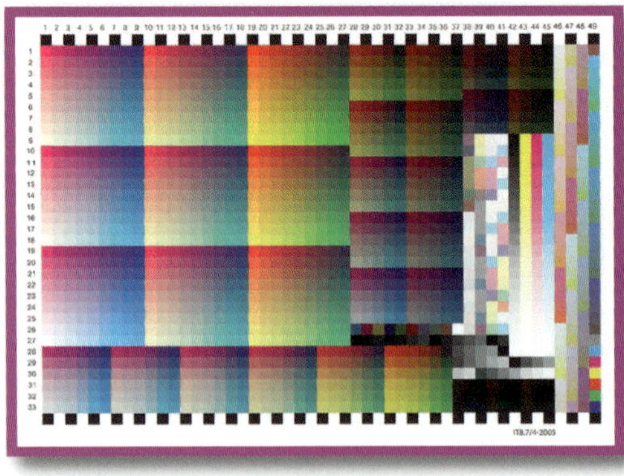

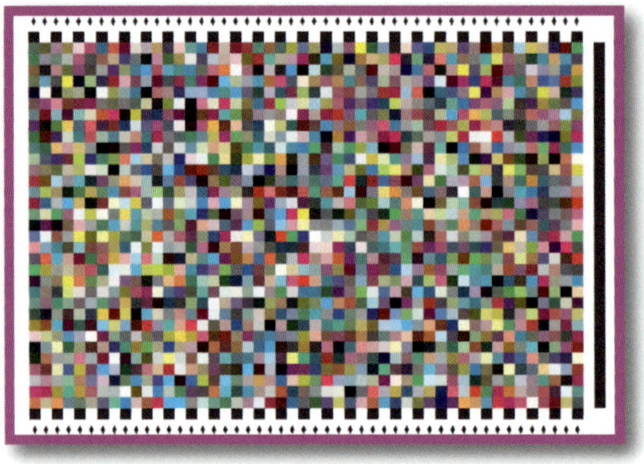

ANSI IT8.7/4 Color Characterization Targets (Ordered and Random Layouts)

ISO IDEAlliance 12647-7 Digital Control Strip 2009

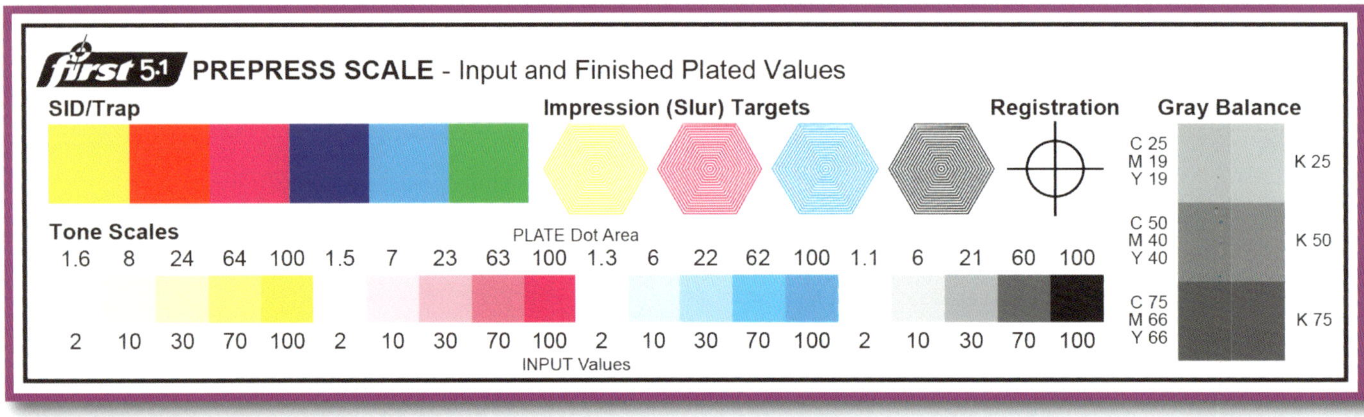

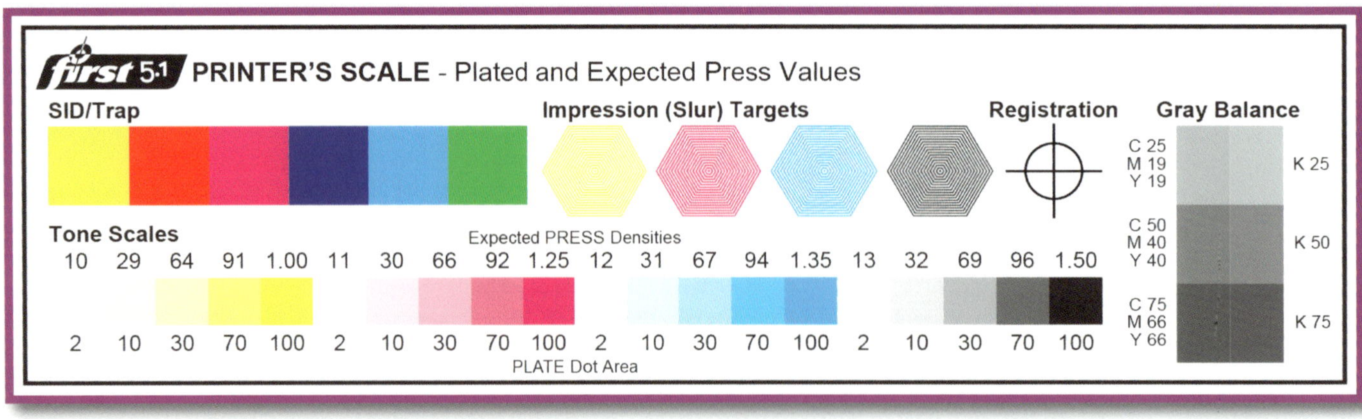

FIRST Prepress & Printer Control Targets

Appendix D: Proofing and Measurement Methods Used to Create L*a*b* Values for the *FIRST* Recommended Process Pigments

In a global effort to identify flexographic process ink sets, multiple samples in triplicate were proofed for each recommended process pigment. Each was proofed at several concentration levels and measured by leading flexographic ink suppliers in compliance with ISO 2846-5 (Graphic Technology – Colour & Transparency of Printing Ink Sets for 4-Colour Printing – Flexographic Printing). They were proofed with a mechanical proofer equipped with a #3 bar. All inks were proofed on Leneta card stock as the standard substrate.

All of the readings for the *FIRST* recommended process ink pigment data were taken using a calibrated X-Rite 938 spectrophotometer (0/45 geometry) with a 4mm aperture. Measuring procedures were in compliance with CGATS.5-1993 (Graphic Technology Spectral Measurement and Colorimetric Computation for Graphic Arts Images). The data was graphed in L*a*b* color space using a 5000 Kelvin (D50) light source. Three measurements were taken across each sample and then averaged to create the final pigment color values. All samples were plotted using the absolute values against a neutral gray standard and plotted using X-Rite QA Master software.

The actual printable color gamut will vary depending on:
- Ink Chemistry: Water, Solvent or UV/EB
- Ink Formulation: Lamination or surface print
- Pigment Concentration: Actual pigment load of finished ink
- Substrate: Paper or film
- Instrumentation: Geometry and measurement procedure used

Note: The above equipment manufacturer was reported as a basis of the information outlined. FIRST is not recommending X-Rite or any other manufacturer as a preferred instrument manufacturer and encourages the reader to explore all available technologies and brands.

Appendix E: How to Create the *FIRST* "Printer" Tone Scale with Integral Mask Flexo Plates

Goal: To provide the printer with a standard tone scale (with predictable plated dot values) to control the printing process independent of the prepress provider, platemaking system, bump curves, etc.

Objectives:
1. To identify the dot percentage (%) required in the 1-bit TIFF file in order to plate a specified dot percentage (ie. 2%, 10%, 30%, 50%, 70%) for a given set of platemaking conditions (platemaking equipment, processing line, type of plate material, bump curve, line screen, screen angle, etc.).
2. To identify the printed dot percentage for each plated tint under a given set of print conditions (anilox, stickyback, substrate, ink system, viscosity, pH, press, print deck, etc.).

Test Procedure:
1. Prepress Provider: Create a 1-bit TIFF file containing 99 tint patches (1%-99%). Use the bump curve that will be used on the live job. Do not apply any cutback curves, near neutral calibration curves, ICC profiles, etc.
2. Platemaker: Make a plate from the 1-bit TIFF file. Use the platemaking equipment, type of plate material, etc. that will be used on the live job.
3. Platemaker: Once the test plate has been processed, measure each tint patch and record the plated dot percentage. Identify the TIFF values that most closely result in 2% (or minimum dot), 10%, 30%, 50% and 70% in the plate.
4. Prepress Provider: Create a modified PRINTER control target using the 1-bit TIFF file values that yield finished plate values of 2%, 10%, 30%, 50% and 70%. Label this modified tone scale PRINTER and indicate the type of plate material.
5. Prepress Provider: Include the modified PRINTER tone scale on the press characterization test form. The modified PRINTER tone scale must RIP with settings that will yield the 1-bit TIFF values selected in above step 3. Include the 1-bit TIFF file dot percentages that correspond to the desired plated dot percentages (2, 10, 30, 50 and 70%) on the CoA sent to the printer along with the press characterization test plates.
6. Prepress Provider: Include a second tone scale on the press characterization test form that is unmodified. Label the unmodified tone scale PREPRESS or PROFILE. The unmodified tone scale should contain tint patches with 1-bit TIFF file values of 2, 10, 30, 50 and 70%.

7. Prepress Provider: The unmodified PREPRESS scale must RIP with the same settings as the rest of the job and yield the TIFF values that will provide the anticipated cutback curves results.
8. Platemaker: When the plates are made for the press characterization, verify the scale values in the plate measure 2, 10, 30, 50 and 70% in the PRINTER tone scale and that the PREPRESS tone scale reflects the expected cutback results. Record both and include on CoA to the printer.
9. Printer: Print the characterization test form under standard press conditions (process deck set-up or combo deck set-up) with the desired inks, substrate, anilox rolls, type of plate material and mounting tape.
10. Printer: For each color in each tone scale, measure the printed dot value or tint density of each tint and record as the target print values. Provide the prepress provider with all of the target print values.
11. Prepress Provider: Update both the PRINTER tone scale and the PREPRESS tone scale with the target print values captured from the press characterization test.

Result:
1. Upon completion of this test, the prepress provider will know what dot percentage to place in the file for each tint patch on the printer tone scale in order to achieve the desired plated dot percentage.
2. The prepress provider will be able to label each tint on the printer tone scale with the plated dot percentage and the printed dot percentage for clarity (if desired by printer). Optional: The printer could elect to use print density values instead of dot percentages.
3. The prepress provider can create a look-up table relating the file dot percentage to the plated dot percentage. This table can be used on all future jobs using the same prepress/plating conditions to determine the appropriate dot percentage to place in the file.

Live Job Procedure:
1. Prepress Provider: Include the PRINTER and the PREPRESS tone scales on all live jobs.
2. Prepress Provider: Verify in the production platemaking 1-bit TIFF files that the PRINTER tone scale contains the values identified in step 5 above.
3. Prepress Provider: Verify in the platemaking 1-bit TIFF files that the PREPRESS tone scale values match the values from step 7 above.
4. Platemaker: Confirm both tone scales in the final plate match the values recorded for each in step 8 above.
5. Printer: Set up and run the production jobs to the "target print values" labeled on the PRINTER tone scale.

Refer to Sections 12.9.2 and 19.4.3 for additional information on how to use the PRINTER and PREPRESS tone scales.

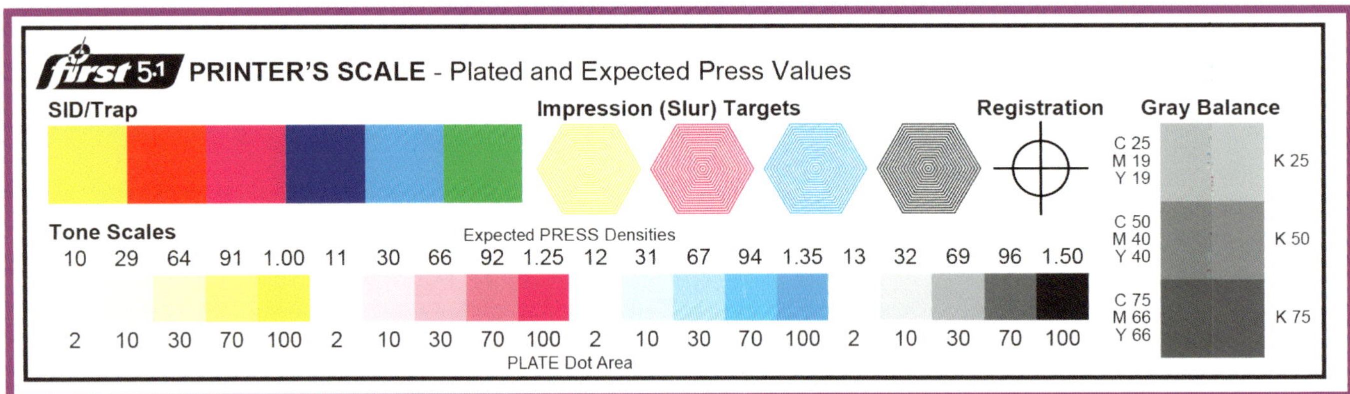

Values in the example tone scales above are for demonstration purposes only. Actual values will vary with each application.

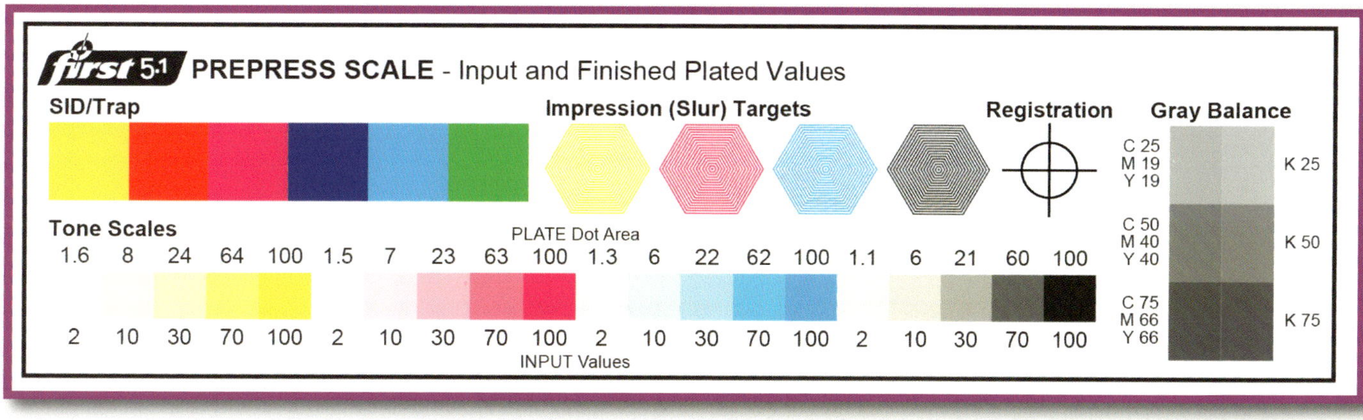

Printer's Tone Scale: Test Plate - EXAMPLE

Prepress Provider: ABC Company	Platemaking Equipment: Flexosync F
Line Screen & Screen Angle: 120 lpi; 22.5 degrees	Plate Material & Bump Curve Applied (if any): 0.067 DPR

File Dot %	Plated Dot %	Printed Dot %	Tint Patch	File Dot %	Plated Dot %	Printed Dot %	Tint Patch	File Dot %	Plated Dot %	Printed Dot %	Tint Patch	File Dot %	Plated Dot %	Printed Dot %	Tint Patch	File Dot %	Plated Dot %	Printed Dot %	Tint Patch
1%	0.9			11%	6.0			21%	15.0			31%	22.4			41%	29.1		
2%	1.3			12%	6.6			22%	15.5			32%	23.2			42%	29.1		
3%	1.6			13%	7.8			23%	16.2			33%	23.6			43%	30.0	67%	
4%	1.7			14%	8.6			24%	17.4			34%	24.8			44%	30.2		
5%	2.3	12%		15%	9.7	31%		25%	18.2			35%	25.4			45%	31.3		
6%	2.4			16%	10.5			26%	18.8			36%	25.6			46%	32.2		
7%	2.9			17%	11.0			27%	19.7			37%	26.6			47%	33.8		
8%	3.8			18%	12.0			28%	21.0			38%	26.9			48%	34.0		
9%	3.9			19%	12.9			29%	21.4			39%	26.9			49%	34.8		
10%	5.0			20%	13.7			30%	21.6			40%	27.9			50%	35.7		
File Dot %	Plated Dot %	Printed Dot %	Tint Patch	File Dot %	Plated Dot %	Printed Dot %	Tint Patch	File Dot %	Plated Dot %	Printed Dot %	Tint Patch	File Dot %	Plated Dot %	Printed Dot %	Tint Patch	File Dot %	Plated Dot %	Printed Dot %	Tint Patch
51%	35.9			61%	49.6	74%		71%	63.5			81%	71.7			91%	83.1		
52%	36.4			62%	51.4			72%	64.4			82%	73.0			92%	84.7		
53%	38.2			63%	52.9			73%	64.9			83%	73.9			93%	84.9		
54%	39.0			64%	54.2			74%	66.8			84%	75.0			94%	87.5		
55%	40.3			65%	59.5			75%	67.2			85%	75.9			95%	88.6		
56%	41.2			66%	59.5			76%	67.3			86%	76.8			96%	90.2		
57%	41.4			67%	60.7			77%	68.8			87%	78.3			97%	91.4		
58%	41.9			68%	61.2			78%	69.6			88%	79.1			98%	91.7		
59%	43.4			69%	62.4			79%	70.2	94%		89%	80.3			99%	93.1		
60%	46.2			70%	62.5			80%	71.0			90%	81.1			100%	1.00D		

Printer Variables To Document

Anilox (cell count, volume & angle): 650 / 1.9”	Substrate: 1.25mil Polyethylene
Sleeve / Stickyback: 1020	Ink (system, viscosity, pH): XYZ-81024; 30sec. @ #2 zahn
Press & Print Deck: P11 - D5	Other: Yellow

Appendix F: Expanded Gamut: Reasonable Measurement For Process Control

Introduction: Expanded gamut printing is often referred to as "fixed ink set" or "multi-color process" printing. It's really process color printing with 5, 6, 7 or more colors instead of 4 colors. As such, the same procedures for process control of 4c printing described in this edition of *FIRST* apply to 5, 6 and 7 color process printing as well (see Sections 1.3, 12.7, and 12.9). The missing part of process control beyond 4 color CMYK printing is control of the other three colors – Orange, Green, and Violet (OGV). This section will concentrate on process control of Orange, Green and Violet inks.

Process Control vs. Printing to a Specification

The concept of process control is similar in many ways to the concept of printing to a specification, the same measurement devices and methods are used and the same data is recorded. The difference is one of emphasis. To match a specification (ie. ISO 12647-6) the numbers you measure must be within an agreed upon tolerance of the specification. For example, ISO 12647-6 calls for Magenta with CIELAB values of (L*,a*,b*) = (52, 69, 1) on the printed sheets. Given a tolerance of 3 ΔE00 2000, a measurement of L*=52 a*=69 b*=1 would have a ΔE 2000 of 2.4 and would therefore be in specification. However, a measurement of L*=53 a*=69 b*=1 would have a ΔE00 of 3.3 and would therefore be out of specification (Image A.G.1).

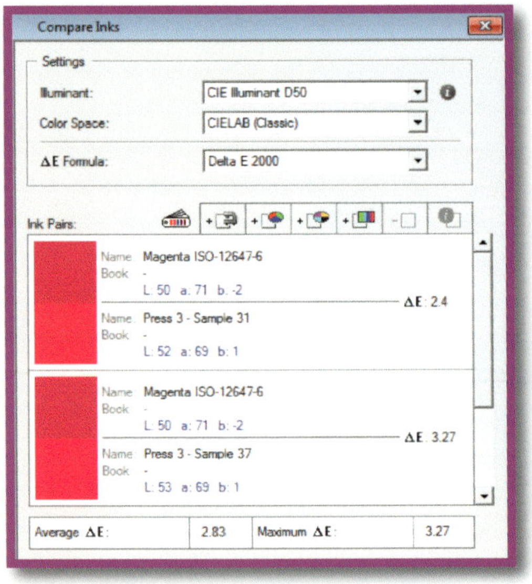

A.F.1

The concept of process control is similar to the concept of printing to a specification. Two numbers are required, a color at which to aim and a tolerance.

The concept of process control is the same as matching a specification. The difference is that the user has a choice of aiming at an industry specification or defining their own numbers at which to aim. The full debate as to when a user should aim for an industry standard specification (ie. ISO 12647-2 or ISO 12647-6) or make their own specification is beyond the scope of this Section of *FIRST*. The short answer is to aim for an industry standard specification if you can hit it. Develop a "custom" specification only if you, for legitimate reasons, can not hit the industry specification. An extreme example of when it is logical to define your own specification is printing on dark brown substrates such as kraft paper or corrugated board. In such a case, there's no way to match an industry specification that was made on white paper or white poly. It is therefore not possible to apply to the concept of printing to a specification.

However, is still possible to apply process control. All that is required is to agree on a realistic number that can be achieved on the corrugated board and can be seen in Image A.G.2.

In the example Image A.G.2, the corrugated board white point is 31.83 ΔE00 away from the ISO 12647-6 white point. Aiming for the Magenta L*a*b* values from a specification based on a very white substrate is not realistic and will lead to a reduction in quality. Note that the closest a printer can match the ISO 12647-6 Magenta is 11.23 ΔE00. This new "Magenta" color can then be used as the aim point for process control. In the example in Image A.G.2, the measurement from the production sheet are compared to the new aim point. The goal of the pressman is to print as closely as possible to the new aim point.

Process control can be applied even in cases where it is not possible to print to an industry specification. A new set of numbers can be developed for a custom set of printing conditions.

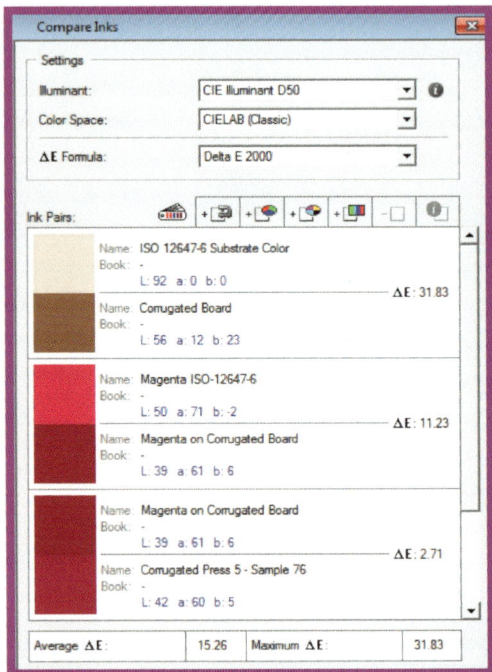

A.F.2

The example in Image A.G.2 illustrates process control in its most basic form. The basic concept is to:
1) Agree on a number at which to aim (called the "desired result").
2) Agree on a tolerance around that number.
3) Measure the production sample to see if it is within the tolerance of the desired result.

Images A.G.1 and A.G.2 show this only for a solid and in units of CIELAB. For complete process control of the Orange, Green, and Violet of expanded gamut, it's best to measure tints as well as solids. It's also helpful to use metrics other than/or in addition to CIELAB. These will be described further below.

Two Areas of Measurement: Ink Specification and Press Specification
A study of CMYK process control specifications and practices around the world would reveal two types of color specifications: ink specifications and print specifications. There is a need to apply process control to the incoming ink before it gets to press and there is a need to control the color as printed on the press.

Process Control of Incoming Ink
For process control of inks, only the solid is relevant, tints are not relevant as they are a function of press and prepress settings. The techniques of process control of incoming ink that are used for both CMYK as well as for spot colors (Section 20.2) are equally applicable to the Orange, Green and Violet of expanded gamut. A sample of incoming ink "proofed" directly on a standard substrate is known as an ink drawdown. To be of use for process control, an ink drawdown must be: similar in ink film thickness to how it will be printed on press and be consistent. Once you establish your CIELAB numbers, even if they are not the same as you will later achieve on press, you have the ability to apply process control to incoming ink.

There are a few devices on the market that offer the ability to apply a consistent ink film to the substrate that is similar in ink film thickness to the press. Devices with automated motion and pressure control offer greater consistency than devices that use manual motion and pressure control and are therefore the recommended.

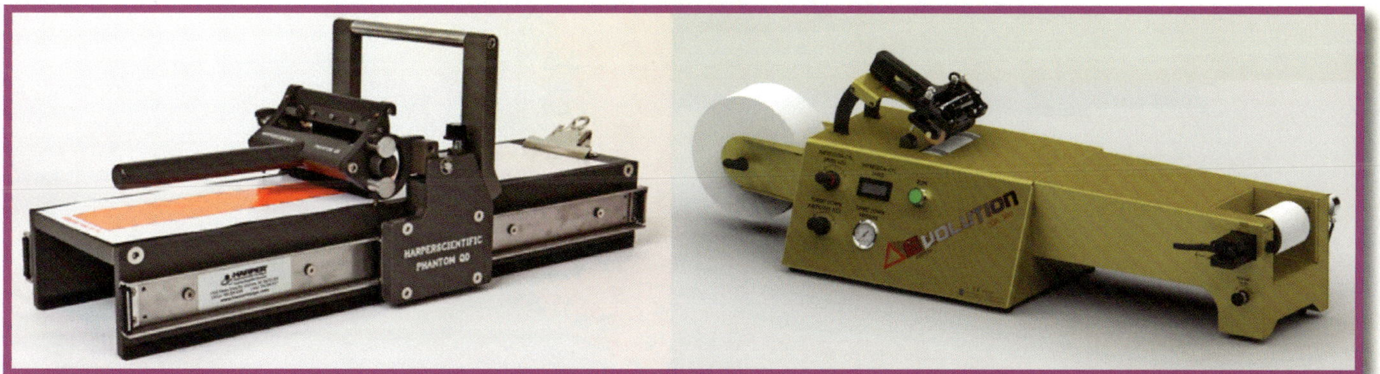

A.F.3: *Ink drawdown devices that offer automated motion and pressure control offer the consistency required for process control.*

The Ink Specification/Print Specification Debate
It should be noted that there is some debate in the industry as to the usefulness of using ink drawdowns to assure that incoming ink matches a specification. Part of this debate centers around the fact that the ISO ink specifications (ie. 2846-5) specify a different substrate than ISO print specifications (ie. 12647-6). As such, the CIELAB values of the ISO inks specifications are different than the ISO print specifications. Theoretically, ink that matches the ISO ink specification CIELAB values when proofed on the "ink spec" substrate will also match the corresponding ISO print specification when printed on press on the "print spec" substrate. The debate is to the extent to which this is true.

It should also be noted that there is no debate as to the usefulness of using ink drawdowns for process control (ie. to assure incoming ink is consistent). It is extremely useful and far more economical to catch defective ink in the ink room than on press.

Substrate for Drawdowns
The substrate for making drawdowns of flexographic inks that was part of the ISO 2846-5 inks spec is no longer manufactured. The industry has yet to agree on a suitable replacement. However, an industry standard substrate is not required to apply process

control to incoming inks. The goal is to find a substrate that is consistent and similar to the substrate used in production. A logical choice is to use the most common substrate for a printing operation, as long as it is consistent. If there are multiple substrates, choose only one. If there are known consistency issues with the most common substrate, choose a substrate that is less common and more consistent. Remember, consistency is the goal of process control. The key is to keep parameters the same. For example, if a substrate on the press requires corona treatment or primer, then the proofing substrate (if drawn from this same source) must have similar treatment. Similarly, if the print substrate is uncoated and highly absorbing, it will be difficult to predict the color of a print if the ink proof is on gloss coated label stock.

The Concept of a Wet Sample

One way to develop an internal color specification for ink is to base it on your color specification for press. With your press setup to production conditions and your ink matching your print specification as measured on the final printed sheet (a "dry" sample), take a liquid sample (a "wet" sample) from your ink fountain back to your ink lab, being careful not to loose too much solvent, and make a test print on the substrate you have chosen as your drawdown sample. This becomes your new ink standard, the target at which to aim for ink process control.

Checking Opacity

When printing with the Orange, Green and Violet pigments described in Section 20.2.5, adequate transparency is almost assured. Still, it is good practice to drawdown the ink on white over a black printed bar on an occasional basis. There are not specifications for opacity for expanded gamut inks at this point, but a good "acid test" is that the ink drawn down over Black should be darker than the Black bar itself. This means darker to the eye when observed by a human and a lower L* reading when measuring CIELAB values.

A.F.4: *Orange 54°, Green 181°, Violet 307°*

Orange, Green, and Violet

Record the L*a*b* and C*h° of a "known" set of Orange, Green and Violet inks used in production. On an on-going basis, your goal will be to measure incoming inks to be sure they match the known set of inks to reasonable tolerance. A typical tolerance would be about 3 ΔE00. Note that the specification to which you are aiming is based on your own ink drawdown. It should noted, however, that as of 2013, the FTA Flexo Quality Consortium (FQC) committee has recommended hue angles for the three inks as follows: Orange 54°, Green 181°, Violet 307° (Section 20.2.5) but has stopped short of recommending L and C values as there are just too many sets of conditions to recommend a single standard. Recall also that ink drawdowns do not necessarily match printed reproductions (see The "Inks Specification/Print Specification" Debate). However, hue values of these three inks are known to remain fairly constant through a wide range of ink film thicknesses. As such, the hue angle numbers in Image A.G.4 can serve as a guide for ink drawdowns.

Process Control on Press

Taking samples through the duration of the press run, measuring the samples and adjusting press settings to keep the print sample as close to the aim as possible achieve process control on press. The biggest difference between process control on press and process control of incoming ink is the measurement of tint values. While incoming ink can be controlled through solid measurement alone, control on press requires measurement of solids and tints.

Control Elements to Include

A minimum of three points are required for process control of the Orange, Green, and Violet on press: a solid, a midtone patch, and a minimum dot patch. More points provide even more information and are recommended if space permits. Each of these patches provides information on a specific characteristic of printing. The solid patch is an indicator of ink and inking, the tint patch is an indicator of prepress curve accuracy and impression on press and the minimum dot patch is an indicator of numerous attributes related to dirty printing.

A.F.5

The value of the solid and minimum dot patch are fairly straightforward. The solid should be a 100%, but should include ink transfer cells if ink transfer cells are used in the work. The minimum dot should be a value close to the minimum value that can be held on plate. This value should be no larger than 2.0%, but a smaller value can be used if it can be held on plate. For example, values of 0.1% to 0.4% or 0.5% can be used if the stakeholders in the prepress and press areas agree that 0.4% or 0.5% is a good indication of the highlights in the production work. The preferred value of the midtone patch is a 50% value in the original digital file cutback using the same curve that is used for the production run. In all cases above, the values specified are the values in the original digital file (ie. Adobe Illustrator or standard PDF). The values in the screened file and on plate will likely be different.

Consider first, the minimum dot value. A value in the original digital file of 0.1% may seem too small, but recall that this value will be modified by the plate curve when the job is ripped and screened. This plate curve should be made in a way in which the 0.1% value in the original digital file produces a dot that can plate and print cleanly. Hence, an 0.1% dot in the digital file will produced an acceptable dot on plate. If you are uncertain about the plate curves used and their exact effect on the 0.1% dot, a larger values should be used.

Consider next, the midtone dot value. As mentioned above, the best practice is a 50% dot in the original digital file. This value will almost always be significantly smaller on the finished plate (typically down around 35%). Such a reduction in value may be achieved by the natural dot sharpening on the digital plate in the case of "round top dot" digital plates or the application of large press compensation curves in the case of "flat top dot" digital plates. In either case, the midtone patch of the printed result should appear approximately half way between substrate and solid to the human eye when printing to a "reasonable" Tonal Value Increase (TVI) specification. To apply the principle of printing to a TVI specification, the midtone value should read within an agreed upon tolerance to the TVI of the desired specification. Hence, both the prepress curve and the press settings are "embedded" in this measurement. If the value does not match the desired specification within the desired tolerance, this reading alone will not reveal whether the problem in prepress (ie. curves) or press (ie. impression settings) and further troubleshooting will be required. However, if the value does match the specification, it confirms that both prepress curves and press setting are correct and all can be assured that the printed reproduction will match a proof made to that same specification.

Process Control on Press

Process control in the CMYK world has been slowly converting from wide band density based metrics (ie. Murray-Davies Tone Value) to CIELAB based metrics. One example is G7™ where solid colors and CMY neutral tints are specified in CIELAB coordinates instead of density.

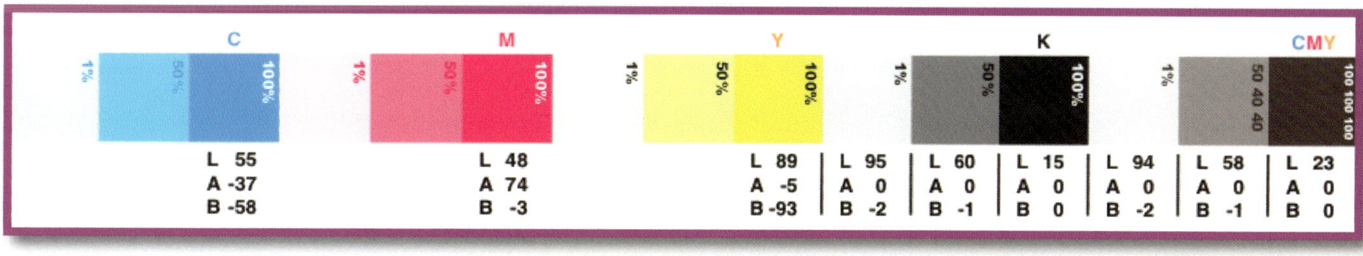

A.F.6: *G7™ Specification Based on L*a*b**

CIELAB based approaches have the potential to eliminate the need for quantifying TVI of tints. While there are industry-accepted methods of measuring TVI of tints for CMYK, there are not widely accepted methods of measuring TVI of tints for non-CMYK inks, including the Orange, Green and Violet of expanded gamut.

However, it's also recognized that even if the "ultimate" process control systems of the future are based solely on CIELAB units, these systems are not "drop-in" systems that can be used by all printers today. As such, this Appendix will present an overview of the direct CIELAB approach to process control. It will then describe three different methods for computing TVI for Orange, Green and Violet inks so that the methods used today for process control of CMYK can be used in a similar way for process control of Orange, Green and Violet.

Direct CIELAB Matching

Process control based on direct CIELAB measurement has three requirements: obtaining desired CIELAB values for each of the elements (solids and tints) to be controlled on press, measuring the elements on the production sheets and providing guidance

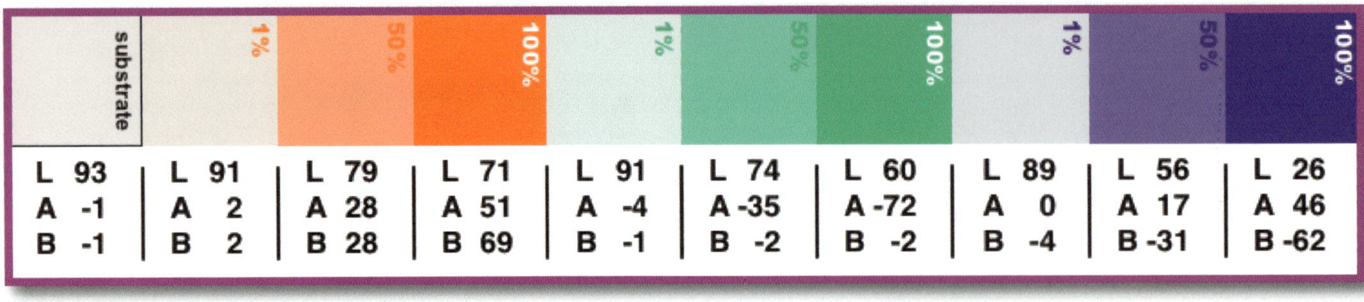

A.F.7

to press personnel or automated QC systems for adjusting press or prepress settings to match the measured values to the desired values.

Process control in expanded gamut printing will ultimately be based on direct L*a*b* matching. One of the requirements is obtaining L*a*b* values for each of the elements in the chart.

An example user interface from a direct CIELAB matching system is shown in Image A.G.8. Before the press run, software must obtain the desired CIELAB values (usually from the expanded gamut press profile) for each point to be measured and controlled. During the press run, the software accesses the measurement data (usually through direct connection to an on-line spectrophotometer). The software then displays the relationship between the measured value and the desired value and ideally provides helpful information as to how to make the measured color closer to the desired color.

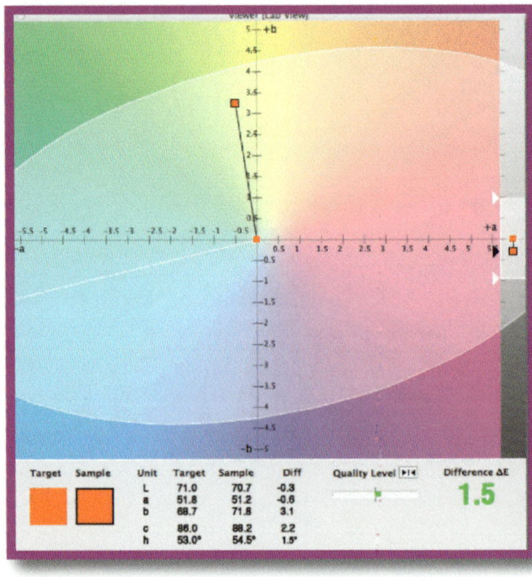

A.F.8

Of the three requirements cited above, the industry has had the requirement of measuring the elements on the production sheets for a number of years. It has been the lack of the ability to meet the requirement of obtaining desired L*a*b* values for each of the elements and the requirement of providing guidance to pressman, that has limited the acceptance of direct L*a*b* matching for process control. The industry is now on the brink of solving these requirements.

Obtaining Desired L*a*b* Values for Control Elements

CIELAB values for all relevant expanded gamut dot combinations can be extracted from color profiles (ie. ICC, Esko, GMG etc.). Most control patches contain only single color builds of Orange, Green and Violet. For example, the L*a*b* value of a 50% Orange from a given 7c profile may be L*=79, a*=28, b*=28. This value needs to be entered into your direct L*a*b* matching software as am aim point (or desired value) to which measured values will be compared.

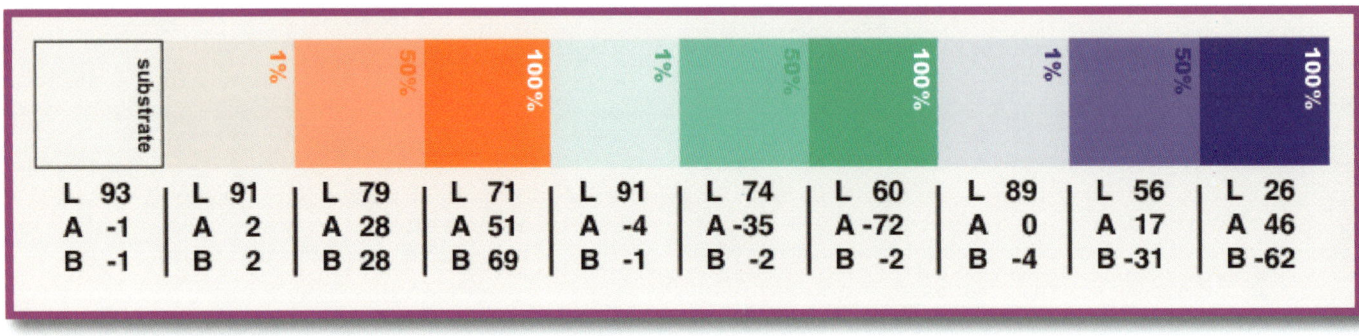

A.F.9: *Desired L*a*b* Values (or values at which to aim) can extracted from 7c profiles.*

A Warning on Substrate Cast

An issue that has limited the acceptance of direct CIELAB matching is the effect of substrate color. If the production substrate is different than the substrate from the profile, it will effect tint measurement at which to aim. With density or tone value approaches, the substrate can easily be set to null (commonly called "zeroed out"). No similar adjustment exists for CIELAB measurement.

A method known as "tristimulus linear correction" is gaining increasing acceptance as a way to adjust for substrate color for L*a*b* matching (see "Substrate Correction in ISO 15339-1, Robert Chung and Quanhui Tian, TAGA Conference, March 2011). A key part of the recommended procedure is that the substrate correction is applied to the "desired value", not to the measured value. As this method becomes built into software, a major obstacle to direct L*a*b* matching will be removed.

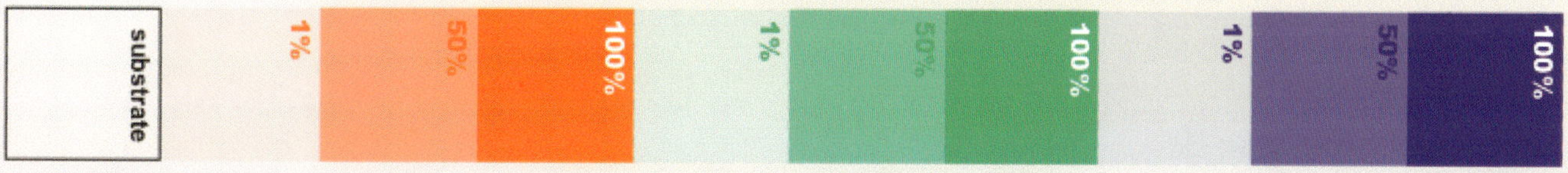

substrate	1%	50%	100%	1%	50%	100%	1%	50%	100%
L 93	L 91	L 79	L 71	L 91	L 74	L 60	L 89	L 56	L 26
A -1	A 2	A 28	A 51	A -4	A -35	A -72	A 0	A 17	A 46
B -1	B 2	B 28	B 69	B -1	B -2	B -2	B -4	B -31	B -62

A.F.10: *Desired L*a*b* Values can be altered for paper color based on "tristimulus linear correction". The values in the bottom have been calculated based on substrate correction from the values at the top.*

Feedback for Press Adjustment or Prepress Curve Adjustment

A second issue, which has limited the acceptance of direct CIELAB matching, is the "non-intuitive" nature of measured CIELAB readings. With density or tone value approaches, a single number provides an intuitive reference for press operators. If the number is too high, the pressman knows that the adjustment must lower the number. With direct CIELAB matching, differences in 3 numbers make for a less intuitive adjustment process. Logical methods to train pressman to make adjustments based on L*a*b* exist, but most experts believe dedicated direct CIELAB matching software is the ultimate answer to the acceptance of this superior technology.

Direct L*a*b* Matching: A Present or Future Technology?

When deciding on the content of *FIRST* recommendations and specifications, FTA associates must occasionally balance recommending technologies which are widely available at the time of writing with recommending technologies which are superior and promise to be widely available within a few years. Density/Tone Value Increase (TVI) matching is less powerful then direct L*a*b* matching but is more widely available and more similar to the way CMYK colors are controlled today. As such, it may be simpler to implement for many companies just starting into expanded gamut printing. As such, Density/TVI methods will be described further in this Appendix and considered equally acceptable as direct L*a*b* matching.

Density/TVI Matching and Color Filters

The basic premise of Density/TVI matching is simple: establish "desired" density values for all solids and "desired" tone values for all tints, then measure and control press and prepress to hit these values. This approach has been used around the world for CMYK for the last 40 years. It's logical to use the same approach for Orange, Green and Violet. Even though industry standard filters for Cyan, Magenta and Yellow can produce strange results when used for colors other than CMY, experience has shown they work acceptably for process control of OGV inks.

Density/TVI Matching VIA Status T Wide Band Filters

Common filters used in the United States are ISO Status T. ISO Status E, I, T and G are also used in other parts of the world or for specific applications. These filters are all slightly different variations of Red, Green and Blue, the "compliments" to Cyan, Magenta and Yellow. Errors associated with using these filters for other colors can be significant. However, for process control of the Orange, Green and Violet inks used for expanded gamut, such filters have been shown to work acceptably.

	substrate	1%	50%	100%	1%	50%	100%	1%	50%	100%
Cyan Density	0.08	0.08	0.09	0.10	0.10	0.64	1.49	0.10	0.65	1.49
Magenta Density	0.08	0.09	0.53	1.00	0.08	0.21	0.32	0.10	0.74	1.83
Yellow Density	0.07	0.09	0.69	1.65	0.08	0.28	0.47	0.09	0.39	0.64
Black Denisty	0.08	0.09	0.25	0.35	0.09	0.36	0.61	0.10	0.69	1.48

A.F.11: *Status T Density has been shown to be an acceptable measure for process control of Orange, Green and Violet inks on press.*

A simple test can be run to determine the "fit for use" of a given filter set for process control of a given ink. First, that filter should produce a significantly high reading when measuring the solid of that ink. By "significant", it is meant readings as high or higher then C, M, or Y process inks. Second, the highest filter reading for the solid should also be the highest filter reading for all tint values of that same color.

The Status T density readings from Image A.G.11 were taken from a live press run at Clemson University using inks that were a close (but not perfect) match to the designated hue angles for Orange (53°), Green (182°) and Violet (307°). Status T filters meet both criteria described.

A related discussion is the question of the "desired values", the values at which to aim. Such values cannot be derived from ICC Profiles, because the spectral data from which the density calculations are derived is not available (only CIELAB and CIEXYZ are available in an ICC Profile). Such values can be derived from measured characterization data BEFORE the ICC Profile has been made. Or, a user can designate "abstract" values and alter prepress and press to match these "abstract" values. However, setting "abstract" values for tint density is very non-intuitive. One way to arrive at reasonable "abstract" density values is to designate abstract tone valuc values (such as a 50% digital values should print to be 70%) then derive the density values from the tone values and solid values using the Murray-Davies formula in reverse.

Tone Value Matching VIA Status T Density

A more intuitive way to work with Status T density for process control of Orange, Green and Violet inks of 7c process is to use "Tone Value". Tone Value (TV) or Tone Value Increase (TVI) is one of the most common metrics for CMYK printing in the United States and other parts of the world. The most common formula is the Murray-Davies equation which is used to convert density into Tone Value.

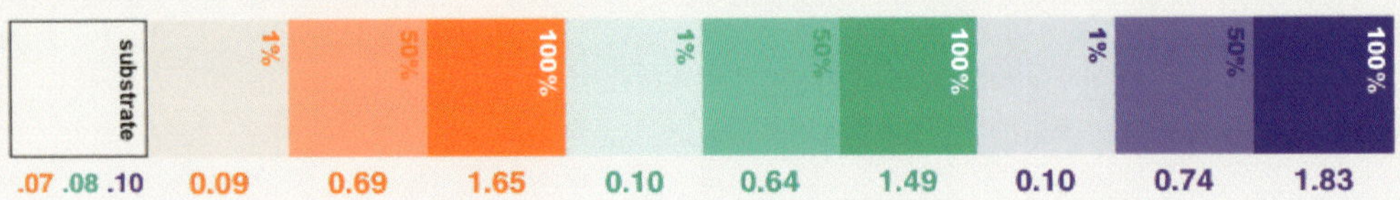

A.F.12: *Status T Density – Maximum Filter Reading*

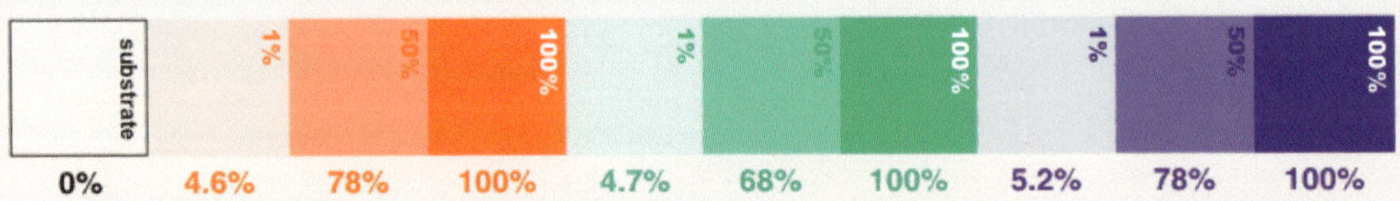

A.F.13: *Tone Value Based on Status T Density*

A related discussion is the question of desired values, the values at which to aim. Such values cannot be derived from ICC Profiles for the reasons described earlier. The most common practice is to pick a set of "abstract" values. ISO has designated 6 curves (ISO-A through ISO-F) as reasonable values at which to aim for CMYK inks. They range from ISO-A at 50% = 64.3% to ISO-F at 50%=77.5% (a designation of 50%=77.8% means that a 50% value in the digital file be reproduced as a 77.8% on press). There also exists Annex-B in ISO 12647-6. This curve specifies that a 50% value in the digital file be reproduced as a 68.8% on press.

The values in Images A.G.12 and A.G.13 that were achieved in a live press run at Clemson University using inks that were a close (but not perfect) match to the *FIRST* designated hue angles and printed with similar TVI to the CMYK matching G7™ are shown above. Note that for Green, similar values are measured (50%=68%) but that for Orange and Violet, much higher numbers are measured (50%=78%). Hence, Status T density appears to over-exaggerate TVI for Orange and Violet inks. There are two approaches for working with this anomaly. One is to just aim for a higher value for Orange and Violet (ie. aim for 50%=78%). The other is to aim for a similar value as for CMYK as well as Green (something around 50%=68%). Either way is acceptable as long as prepress and proofing are handled accordingly.

Density/TVI Matching VIA Spectral Density

The accepted practice for using density to measure ink colors that do not have matching filters (all colors other than CMYK) is to use "spectral density". Spectral density is simply the measure of density at the band of lowest reflectance (highest density).

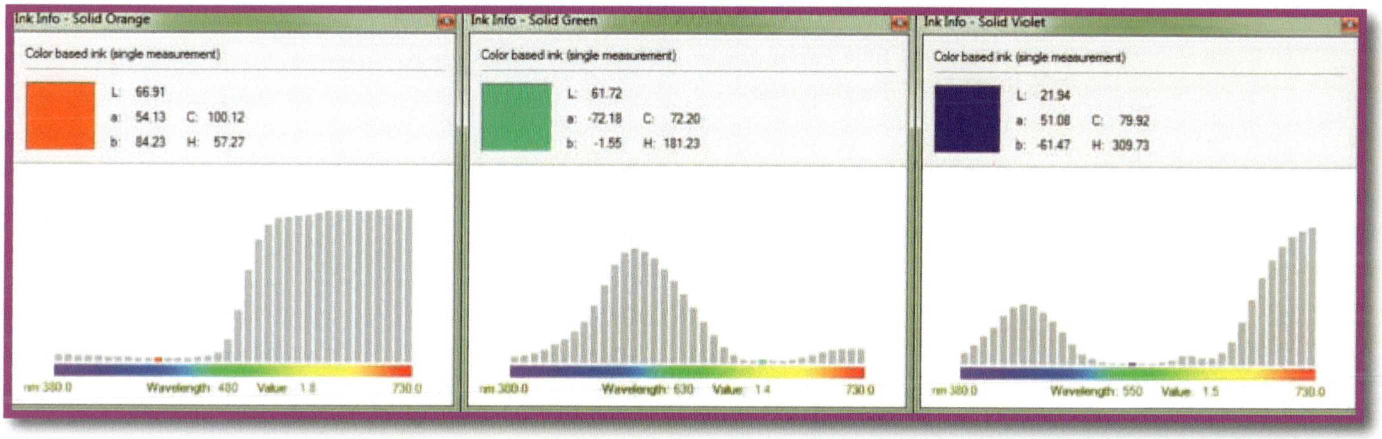

A.F.14: *Spectral density is common way to measure non-CMYK inks.*

Modern spectrophotometers and color measurement software have the ability to measure spectral density. Many report the wavelength of lowest reflectance as well as the reflectance and density at that wavelength.

Spectral density can be used for process control in the same way as wide band density is used for process control. The issues of selecting desired values (values at which to aim) apply to spectral density in the same way as they apply to Status T density. Given desired values for solids and tints in units of spectral density, process control can be applied.

One item worth noting is that with spectral density, different tint values often have a different band of minimum reflectance than the solid value. The recommended procedure is to use the minimum reflectance band for the solids for all tint values as well.

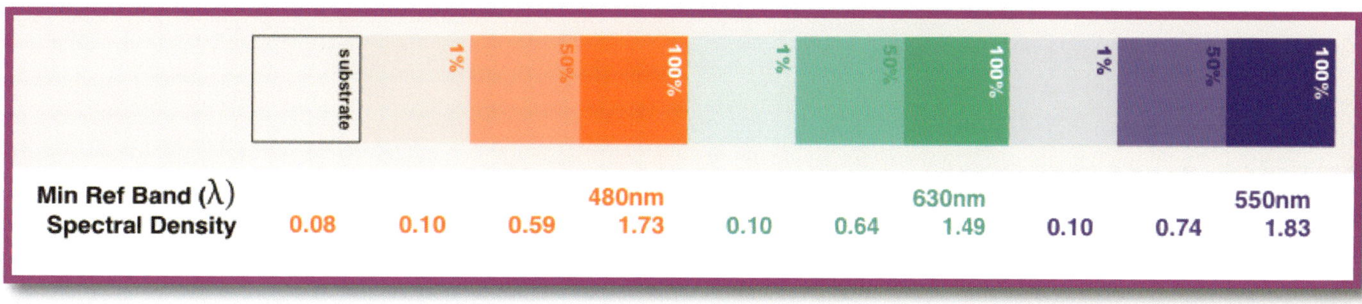

A.F.15: *Spectral Density Values for typical O, G, V inks run under typical flexo conditions.*

	substrate	1%	50%	100%	1%	50%	100%	1%	50%	100%
Min Ref Band (λ)				480nm			630nm			550nm
Spectral Density	0%	4.6%	71%	100%	3.7%	69%	100%	5.2%	78%	100%

A.F.16: *Murray-Davies TVI – Calculated from Spectral Density.*

Just as Status T density can be converted into tone value, spectral density can also be converted into tone value. The spectral density numbers in Image A.G.15 taken from a live press run at Clemson University, using inks that were a close (but not perfect) match to the *FIRST* recommended designated hue angles and printed with similar TVI to the CMYK matching G7™ are shown in Image A.G.16.

Note that in this case, a tone value of around 70% would be a reasonable desired value for Orange and Green, but would appear a bit light for Violet where a value of 78% might be more reasonable. There are two approaches for working with this anomaly. One is to just aim for a higher value for Violet (ie. aim for 50%=78%). The other is to aim for a similar value as for CMYK as well as Orange and Green (aim for something around 50%=68%). Either way is acceptable as long as prepress and proofing are handled accordingly.

ΔE-P Matching

As the industry moves towards direct CIELAB matching and toward using industry standard specifications as well as custom profiles as "desired results" at which to aim, a metric called ΔE-P (pronounced "Delta E to P") is gaining increasing acceptance. ΔE-P is simply a comparison of the color (solid or tint) to the paper in units of ΔE 76. Note that unweighed ΔE* (ΔE*76) is used because the numbers can be very large (often over 100 for Orange and Violet solids) and the weighed formulas (ie. ΔE00) are designed for use cases where the colors are fairly close (under 10 ΔE*).

The CIELAB data and ΔE-P metrics in Image A.G.17 were taken from a live press run at Clemson University using inks that were a close (but not perfect) match to the *FIRST* designated hue angles and printed with similar TVI to the CMYK matching G7™ are shown in Image A.G.17.

	substrate	1%	50%	100%	1%	50%	100%	1%	50%	100%
L of LAB	92.7	92.3	76.3	66.7	92.3	75.3	61.6	91.1	52.6	20.7
A of LAB	0.3	1.7	33.4	54.9	-1.2	-35.9	-73.9	0.88	19.6	52.0
B of LAB	-1.4	-0.7	36.0	85.6	-1.9	-2.3	-0.67	2.86	-31.9	-62.7
ΔE - P	0.0	1.6	52.6	105.9	1.6	40.4	80.5	2.1	54.1	107.6
% ΔE - P	0%	1.6%	50%	100%	2.0%	50%	100%	1.9%	50%	100%

A.F.17: *ΔE-P and %ΔE-P calculated from CIELAB.*

ΔE-P can be converted to percentage and be used in similar ways as tone value simply by dividing the ΔE-P of the given tint patch by the ΔE-P of the solid of the same color.

Like all metrics, ΔE-P has strengths and weaknesses. One strength is that it can be obtained from CIELAB data, widely available in both measurement devices as well as ICC Profiles and other types of color profiles. Another is that when converted to the percentage scale, "linear" reproduction becomes a reasonable desired value. For example, aiming to achieve a reproduction where a 50% value in the digital file reproduces as a 50% ΔE-P value on print produced results very similar in dot gain as ISO-C (50% = 70% in Murray-Davies TVI) or GRACol 7.

A disadvantage of ΔE-P is that at high ink film thicknesses, some Violet inks behave "monotonically" in the solids. When increasing ink film thickness on press of a Violet ink, at some point, the ΔE-P reading will begin to reverse and go back down. Even when it is clear that the color is getting darker and even when Status T or spectral density values continue to increase. This occurs at levels above the ink film thickness at which Violet should be printed. By monitoring the LCH values (particularly the C), care can be taken to assure that the ink film thickness is well below this "mass tone" point. Never the less, it needs to be mentioned as a weakness of ΔE-P.

Another disadvantage is that some spectrophotometers won't report ΔE numbers above 100. Without a solid ΔE-P number, it's not possible to convert tint ΔE-P into the percent scale. Also, spectrophotometers don't have the %ΔE-P formula built into them (as they have, for example, Murray-Davies % tone value). Still, many printers find ΔE-P to be the best metric for process control or Orange, Green and Violet inks on press.

Other Spot Color TVI Metrics
At the time of this writing, there are a few more promising "tone value" metrics that work for non-CMYK inks and there is an industry committee designated to selecting one as a standard. In terms of *FIRST* Specifications, all of these metrics can be used under the condition that: a reasonable "desired result" can be designated for solids and tint values, measurements can be made in the proposed metric and the information can be used to control the press in such a way as to guide it back within tolerance when it moves out of tolerance.

	substrate	1%	50%	100%	1%	50%	100%	1%	50%	100%
L of LAB	92.7	92.3	76.3	66.7	92.3	75.3	61.6	91.1	52.6	20.7
A of LAB	0.3	1.7	33.4	54.9	-1.2	-35.9	-73.9	0.88	19.6	52.0
B of LAB	-1.4	-0.7	36.0	85.6	-1.9	-2.3	-0.67	2.86	-31.9	-62.7
ΔE - P	0.0	1.6	52.6	105.9	1.6	40.4	80.5	2.1	54.1	107.6
% ΔE - P	0%	1.6%	50%	100%	2.0%	50%	100%	1.9%	50%	100%
C of LAB	1.6	1.8	49.1	102	2.3	35.9	74.9	2.9	37.5	81.4
H of LAB	281°	336°	47°	57°	239°	184°	281°	287°	302°	310°
Status T Cyan Density	0.08	0.08	0.09	0.10	0.10	0.64	1.49	0.10	0.65	1.49
Status T Magenta Density	0.08	0.09	0.53	1.00	0.08	0.21	0.32	0.10	0.74	1.83
Status T Yellow Density	0.07	0.09	0.69	1.65	0.08	0.28	0.47	0.09	0.39	0.64
Status T Black Denisty	0.08	0.09	0.25	0.35	0.09	0.36	0.61	0.10	0.69	1.48
Tone Value (from Density)	0%	4.6%	78%	100%	4.7%	68%	100%	5.2%	78%	100%
Minimum Reflectance Band (λ)				480nm			630nm			550nm
Density at λ	0.08	0.10	0.59	1.73	0.10	0.57	1.84	0.10	0.74	1.83
Tone Value (from Density at λ)	0%	4.6%	71%	100%	3.7%	69%	100%	5.2%	78%	100%

A.F.18: *A summary of 6 metrics than can be used for process control or O, G, and V. Note: "The above numbers are examples only based on real press sheet measurement at Clemson University and are not to be construed as industry specifications or standards in any way."*

Appendix G: 2D Codes (QR Codes, DataMaritx Codes and Snap Tags)

2D codes:
The Quick Response (QR) code was developed in the early 90's by a division of Toyota called Denso Wave to track vehicles in manufacturing. GS1 defines them as, "Two-dimensional barcodes encode data in two dimensions, in the patterns of dark and pale squares across the width and height of the symbol. The complete symbol must be seen by the reader before it can be decoded." QR codes have become popular recently because they have the ability to store lots of data and can be scanned very quickly with mobile devices such as smartphones. QR codes can now provide the consumer with fast access to CPC's websites, entertainment and transportation ticketing (you can use your mobile device as a boarding pass), product loyalty opportunities (mobile coupons that can be scanned at register from the mobile device), etc.

Anyone can generate a QR code from a variety of websites and put in specific web site or text information that the consumer can link to. The consumer can download one of many applications (typically for free) onto their mobile device to scan, or read the QR code to instantly take them to the destination chosen by the CPC. The other benefit of QR codes being open to the public are that they are now an ISO Standard that has been adopted globally.

QR codes can hold much more information than a bar code. Bar codes typically can hold 20 digits worth, while QR codes can hold over 7,000 character bits of information. In addition, QR codes have up to 30% error correction. This is very convenient for CPC's if the QR code is damaged (for example, a label is wrinkled, or torn).

QR code placement is not as strict as traditional bar codes, the finder patterns allow for a variety of placement positions. QR codes can be reproduced in print and digital media as well. Billboards, movie screens, packaging and magazines are some of the vehicles for the QR code to engage the consumer. From a print reproduction standpoint, however, there are certain parameters that must be met to ensure a viable QR code scan.

The QR code needs good contrast to the background or substrate on which it is being printed. The QR code does not need to be black and white necessarily, but a good contrast of "dark" vs. "light" non-image area. The reason for this is that many of the QR code readers and applications for mobile devices need to universally read the code. Therefore, having a stark contrast to image vs. no image is important. At the end of the day, the code should be tested on a variety of applications and readers to ensure positive scans. Avoid knockouts with a QR code. In addition, there needs to be sufficient white space surrounding the QR code (quiet zone). The purpose of this is to ensure that no surrounding copy interferes with the ability of the code to be scanned.

Other parameters that are important for QR code reproduction: traditionally, it was recommended that a QR code have a minimal size of 1.5" x 1.5" to ensure all mobile devices could be used to scan the codes. Smaller codes are obviously used on a daily basis, but the recommendation is to test to ensure positive reads. For flexographic reproduction, a vector file is preferred in order to scale it appropriately for the ink/press/substrate being printed (ie. .eps, .ai, or .pdf file). The printing parameters for Bar Codes as they exist in *FIRST* apply to QR code reproduction as well.

DataMatrix Codes:
A DataMatrix code is a 2D code not used in point of sale, but for identifying parts and inventory. According to GS1.org, GS1 DataMatrix is a two-dimensional (2D) bar code that holds large amounts of data in a relatively small space. These bar codes are used primarily in aerospace, pharmaceuticals, medical device manufacturing and by the U.S. Department of Defense.

Key advantages include the ability to encode a variety of information, such as date or lot number. Furthermore, the bar code is readable in a 360-degree orientation. The GS1 DataMatrix bar codes also have a sophisticated error correction algorithm, which compensates for lost or missing data, extraneous marks, or code damage. This means that print quality and contrast are much less critical than with traditional linear or stacked bar codes. With error correction, GS1 DataMatrix bar codes can reconstruct up to 20% of damaged characters. Items marked with GS1 DataMatrix symbols are not intended to pass through retail point-of-sale.

Snap Tags:
Snap Tags are a mobile code produced by SpyderLink. The codes differ from QR codes in that they can use a company logo or design, which is surrounded by a ring with three notches in it. The consumer takes a picture (doesn't require a smart phone), which can be linked to web sites, coupons and video trailers. Like QR codes, Snap Tags engage consumers with a mobile device, but they are limited to substrates that are flat, and typically consumers need to be in a 2 foot distance. File preparation is similar to any flexo prepress; vector artwork is preferred for appropriate scaling. Snap Tags also require a minimum size of ~ 1.5" x 1.45". That frame includes a white space requirement of 1.25" around the code ring of 0.86".

Appendix H: Process Control of Spot Colors

The Murray-Davies equation has been used to convert densitometric data into halftone values for almost 80 years. While this equation has been widely accepted for measuring and controlling press gain on process colors (CMYK) it has proven to be less than adequate for spot colors. With the ever increasing need to control brand colors tints and the growing popularity of expanded gamut (EG) ink sets (CMYKOGV) it became clear a new methodology was required.

The new equation for measuring the tone value of a spot color is below:

$$t = \sqrt{\frac{(fx_t - fx_p)^2 + (fy_t - fy_p)^2 + (fz_t - fz_p)^2}{(fx_s - fx_p)^2 + (fy_s - fy_p)^2 + (fz_s - fz_p)^2}}$$

Where: t = tone value of the patch measured
s = solid
p = paper (substrate)

The particular functions (f) can be calculated either from Lab values

$$fx = (L + 16)/116 + \frac{a}{500}$$

$$fy = (L + 16)/116$$

$$fz = (L + 16)/116 - \frac{b}{200}$$

or tristimulus values

$$fx = f\left(\frac{X}{X_n}\right)$$

$$fy = f\left(\frac{Y}{Y_n}\right)$$

$$fz = f\left(\frac{Z}{Z_n}\right)$$

While these equations may appear daunting, most are already part of the programing in today's spectrophotometers. Progress is underway to make this methodology an ISO standard (ISO 20654 Calculation of Spot Color Tone Values) and will eventually find its way into future spectrophotometers. In the meantime, a spreadsheet created to assist in performing these calculations is included in the extras download at http://www.flexography.org/FIRST_extras.

#

A

B

C

O

P

Q

R